INTRODUCTION TO MACROMOLECULAR CRYSTALLOGRAPHY

INTRODUCTION TO MACROMOLECULAR CRYSTALLOGRAPHY

Second Edition

ALEXANDER McPHERSON
University of California

WILEY-BLACKWELL
A JOHN WILEY & SONS, INC., PUBLICATION

Published by John Wiley & Sons, Inc., Hoboken, New Jersey
Published simultaneously in Canada

Wiley-Blackwell is an important of John Wiley & Sons, Inc., formed by the merger of Wiley's global Scientific, Technical, and Medical business with Blackwell Publishing.

For general information on our other products and services or for technical support, please contact our Customer Care Department within the United States at 877-762-2974, outside the United States at 317-572-3993 or fax 317-572-4002.

Wiley also publishes its books in a variety of electronic formats. Some content that appears in print may not be available in electronic formats. For more information about Wiley products, visit our web site at www.wiley.com.

Library of Congress Cataloging-in-Publication Data:

McPherson, Alexander, 1944-
Introduction to macromolecular crystallography / Alexander McPherson. –
2e.
p. ; cm.
Includes bibliographical references and index.
ISBN 978-0-470-18590-2 (pbk.)
1. Macromolecules–Structure. 2. X-ray crystallography. I. Title.
[DNLM: 1. Macromolecular Substances. 2. Crystallography, X-Ray. QD
381.9.S87 M478i 2009]
QD381.9.S87M36 2009
548'.83–dc22

2008040417

Printed in the United States of America

10 9 8 7 6 5 4 3 2 1

CONTENTS

PREFACE

In 1987 Jim Pflugrath and I initiated a course in practical protein crystallography at Cold Spring Harbor Laboratories. A few years later we were joined by Gary Gilliland and Bill Furey. The four of us have, with unmatched enthusiasm and enjoyment, organized and taught the class until now. It has been an unqualified success by any standard, a source of great pride to each of us, but it began on some uncertain footing.

The first year, Jim and I believed we could simply crack ahead with the practical aspects of the subject, assuming that the students knew the fundamentals of diffraction theory, crystal properties, and the basic concepts of solving the structures of macromolecular crystals. It quickly became apparent to both of us that we were sadly naive.

In an attempt to recover our pedagogical bearings, the author quickly devised a series of lectures, given in spare moments, on the underlying principles of protein crystallography, the bare essential ideas that you have certainly to understand if you intend to apply the crystallographic method in an intelligent manner. That first year, the lectures were delivered in front of a chalkboard by a stone fireplace in the ancient and revered Jones Lab at Cold Spring Harbor. As the course progressed, the lectures moved from Jones Laboratory on the waterfront to Plimpton, and their content evolved over 20 years into the material contained in this book.

The contents of this book are, I believe, the minimum you need to know if you want to practice protein crystallography, and understand why you are doing what you do. It is by no means intended as a comprehensive treatment of the subject. This book is not for professionals or experienced diffractionists. It is meant strictly for students, for scientists outside the field, and particularly for those who, like the author, struggle with mathematics and physics. Only a modest attempt is made to describe the practical aspects of data collection, the intricacies of phasing, nor is crystallographic refinement addressed in the detail it deserves. You will, however, find the principles of diffraction of X-rays by a crystal, how X-ray diffraction can be used to determine macromolecular structures, and the underlying theory by which X-ray crystallography has created a revolution in molecular and structural biology.

This second edition is vastly improved over the first. Numerous small errors present in the original were rooted out and banished, both in the text and the figures. The clarity of numerous figures was improved, and new tables were added and mathematical nomenclature made more uniform. This second edition includes two new chapters, one on macromolecular crystallization, and a second on X-ray diffraction data collection. Refinement and anomalous dispersion phasing are treated somewhat more extensively. More than 35 new figures have been incorporated.

The author was aided immeasurably by the fact that he was himself a miserable physics student, so he deeply sympathizes with those who share his failings in mathematics. This book is written specifically for them, from the perspective of a fellow student who claims no greater intelligence than their own, only more hard-earned experience.

The author wishes to acknowledge and thank three scientists who contributed in more ways than I can describe to this work. It could never have been written without them. They are, of course, Jim Pflugrath, Gary Gilliland, and Bill Furey, my fellow instructors, mentors, colleagues, and friends from the Cold Spring Harbor course. I also wish to thank Debora Felix who helped me organize and collate this material, and Aaron Greenwood who is responsible for many of the illustrations found in the book.

ALEXANDER MCPHERSON

CHAPTER 1

AN OVERVIEW OF MACROMOLECULAR CRYSTALLOGRAPHY

The only technique that allows direct visualization of protein structure at the atomic, or near atomic level is X-ray diffraction analysis as applied to single crystals of pure proteins. The technique has been applied to conventional small molecules now for over 80 years with extraordinary success, and very few chemical structures of less than a hundred atoms, obtainable in the crystalline state, have proven refractory. The successful application of X-ray diffraction to protein structure is relatively new, the first protein structure, that of myoglobin, having only been solved in 1960 (Dickerson, Kendrew, and Strandberg, 1961; Dickerson, 1991). Since that time nearly 50,000 additional protein structures have been added to our data base (Berman et. al., 2000), but even this collection represents only a very small fraction of the hundreds of thousands of different protein molecules that play some role in living processes. Thus the determination of protein structure by X-ray crystallography occupies the energy of several hundred laboratories in the world, and this number is ever growing as the need for more and increasingly precise structural information expands in step with the molecular biological revolution.

WHAT DO WE MEAN BY THE STRUCTURE OF SOMETHING?

In common language when we ask, "what is its structure?" we mean by that, how are the various components or elements that make up the object disposed, or placed, with respect to one another in three-dimensional space. More simply, "what does the arrangement look like?"

While seeming a straightforward question, it is one that has perplexed scientists, philosophers, and poets for centuries. Answers have been formulated, for example, by homology—

Introduction to Macromolecular Crystallography, Second Edition By Alexander McPherson

for example, "a rock, a craig, nay a peninsula[1]"—by describing the physical qualities of the object—for example, "it was a one eyed, one horned, flying, purple people eater[2]"—by analytical expressions[3]—for example, $\boldsymbol{r = a\theta^2}$—by visual illustration[4]—for example, and undoubtedly by other means as well.

However, in proper scientific terms there is only one way to precisely describe the structure of an object, be it simple, or intricate and complex. That is by specifying, as in Figure 1.1, the coordinates in three-dimensional space of each point within the object, each with respect to some defined and agreed-upon system of axes in space, namely a coordinate system. Generally, the system is chosen to be an orthogonal, Cartesian coordinate system, but it need not be. It may be nonorthogonal, cylindrical, spherical, or any number of other systems.

The object's inherent structure, being fixed, remains the same no matter how the spatial coordinate system is chosen, or where its origin is taken to be. Because the structure is invariant, even if its constituent points are transformed from one coordinate system to another, the relative positions of the points within the object remain the same. The structure is not dependent on the coordinate system we choose. Thus, if the structure of a molecule is defined by specifying the coordinates in space $\boldsymbol{x_j, y_j, z_j}$ of each atom $\boldsymbol{j}$ in the molecule, atom 5 (or 7 or 18, or whatever) maintains the same relationship in space to atom 14 (or 3, or whatever) no mater what the coordinate system.

To define the structure of a molecule in a precise manner then, we must create a list (the order is not important) of atomic coordinates $\boldsymbol{x_j, y_j, z_j}$ (and here the order is important). A molecular structure becomes a set of ordered triples $\boldsymbol{x_j, y_j, z_j}$, one for every atom. This is imminently suitable not only for translation into a visual representation, but for manipulation and analysis in a computer, and presentation on the screen of a computer graphics workstation in any number of manifestations.

When we solve the structure of a molecule, any kind of molecule, including proteins, nucleic acids, or even large assemblies such as viruses or ribosomes, in the end we seek to identify and specify the $\boldsymbol{x_j, y_j, z_j}$ coordinates of every atom in the molecule. The form, the shape, the image, must always first be defined in these simple numerical terms, as ordered triplets. Only from these can we faithfully reproduce the precise structure of the molecule in more familiar visual terms, as pictures or images.

AN ANALOGY

Let us assume that we want to determine the structure, as defined above, of some object that is invisible. It has the supernatural property that it is nonresponsive to any electromagnetic radiation such as light. But, to make the example more concrete, let's assume it is an invisible, yellow, 1963 Volkswagen beetle, like that shown in Figure 1.2. If we have never seen such a glorious object before, how can we learn of its structure? How can we visualize it?

One way we might approach this problem is to take advantage of the fact that although the Volkswagen does not reflect light, it retains all of its other physical properties. We might, for example, take a basketball and throw it at the invisible object from some direction $\boldsymbol{\bar{k}_0}$

---

[1] Cyrano de Bergerac describing his own nose.

[2] From the song *The Purple People Eater* by Sheb Wooley.

[3] The equation for the logarithmic spiral of a seashell.

[4] University of California, Irvine Anteater (official mascot).

(a vector, which has direction, is defined by some character or symbol with a line over the top), and note which direction $\bar{k}$ it bounces off the object. More informatively, we might throw a hundred, or a thousand balls at the invisible Volkswagen and note how many balls bounce in all directions $\bar{k}$. Some will hit the fender, others the hood, others the windshield, and so forth, and, depending on the orientation of the car with respect to the balls thrown along $\bar{k}_0$, some directions $\bar{k}$ for the reflected balls will be much favored over others. If the

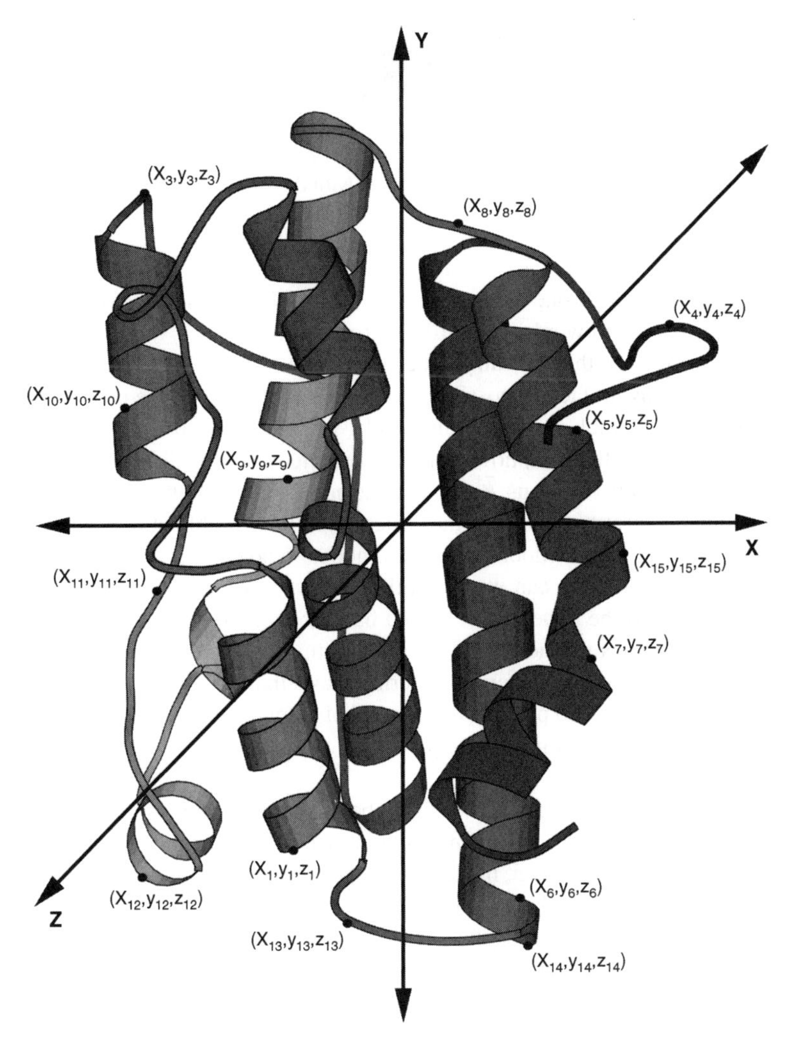

FIGURE 1.1 The structure of an object, such as the protein molecule shown here in caricature form, can be described in precise, quantitative terms by specifying every point in the object by a set of three coordinates x, y, z. The origin and orientation of the particular coordinate system is arbitrary, and is usually chosen for convenience or economy. Coordinates specified according to one linear coordinate system can always be transformed into another without alteration of the relative positions of the points making up the object. The entire structure of the object is a list of ordered triples x, y, z. For a molecule, the x, y, z coordinates are assigned to the atoms making up the structure. In X-ray crystallography, when we determine the structure of a molecule, we mean that we are determining the numerical values of the individual x, y, z coordinates that comprise the list of atoms.

FIGURE 1.2 A "continuous object," here an imagined 1963 Volkswagen Beetle regally aligned in a defined parking location in three-dimensional space, with the supple lines amenable to description in terms of bouncing balls.

direction $\bar{k}_0$ corresponds to one aiming directly at the front of the car, for example, balls bouncing off the hood and windshield will be strongly favored.

Lets assume, however, that we can walk around the invisible Volkswagen and throw the basketballs from many, in fact all possible directions $\bar{k}_0$. Each time we note carefully how many balls bounce in which direction. Ultimately, we will know for every direction of our incoming beam of basketballs $\bar{k}_0$ how many are reflected in every direction $\bar{k}$.

This ensemble of observations $\bar{k}_0, \bar{k}$ and the number, or intensity, of balls I in the direction $\bar{k}$ contains information about the structure of the invisible object, the orientations of its various external planes (doors, windows, hoods, fenders, etc.) from which the balls bounce. Now the question is, can we, from the observations, synthesize the shape of the object that gave rise to the pattern of reflected basketballs? The answer is, of course, yes. Mathematical procedures do indeed exist for extracting the shape of a 1963 Volkswagen beetle (see Figure 1.2) from a scattering pattern of basketballs. We might even invent some analogue device that we could place in a manner that it could accumulate automatically the reflected balls and somehow translate the pattern into an image of the object. We would call such a device a lens.

Now basketballs are rather large objects (probes), and when they bounce from a surface plane, they are rather insensitive to its finer details such as windshield wipers, door handles, and bolt heads. We could, however, make our investigation more sensitive by using, instead of basketballs, tennis balls, and even more sensitive still by using ping-pong balls, or even marbles (no, lets not use marbles as that would damage the paint job). Then the directions in which our probes bounce would more closely reflect the undulations of the hood, and the presence of door handles. That is, we would obtain a more refined, higher resolution image.

The approach illustrated here is not exactly what is done in X-ray diffraction, but it is similar. For example, we don't learn anything about the shape of the engine because our various balls cannot penetrate the interior of the car, whereas X radiation can penetrate and reflect from the internal atoms of molecules. But in many other ways the approach is the same.

Let us alter our analogy a bit and now assert that the reason our 1963 Volkswagen beetle is invisible, is because it is too small to see. It is smaller than the wavelength of visible light. We could, in principle, carry out the same experiment of walking around the nanoscale Volkswagen and directing a probe at it from all directions $\bar{k}_0$, then noting in every case what intensity of reflected probes $\boldsymbol{I}$ was observed for all directions $\bar{\boldsymbol{k}}$. If the probe we used sometimes penetrated into the interior parts of the object (e.g., and struck the transmission) so much the better. Although our pattern of diffracted probes would be considerably more complex, we would then learn about the structures of things inside the car as well. As long as the size of the probe is comparable to the sizes of the molecular features we wish to see (a very important point), we could do as well as we did with basketballs and ping-pong balls.

In organic molecules, the distances between bonded atoms are usually **1** to **2 Å**; hence the size of our probe must be comparable. The wavelength, λ, of X rays used in diffraction experiments are usually between **1** and **2 Å**. $\mathbf{CuK}_\alpha$ radiation produced by most conventional laboratory sources, for example, is **1.54 Å** wavelength. λ is exactly analogous to the probe size. The shorter λ is, the smaller the diameter of the ball we are using, and the greater is the detail we can resolve.

Now a single Volkswagen of molecular size, impressive though it might be, would be so very small that, in practice, it would be impossible to hit it with enough balls (or probes, or waves) from a particular direction $\bar{\boldsymbol{k}}_0$ and to measure the intensity of reflected waves, $\boldsymbol{I}$, in a particular direction, $\bar{\boldsymbol{k}}$. How could we amplify the effect so that we could measure it?

Consider an enormous parking lot full of identical 1963 Volkswagen beetles, all perfectly parked by their drivers so that every car is identically oriented and placed in exact order. That is, they form a vast periodic array of cars. If we now direct millions of basketballs at this Volkswagen array from the same direction $\bar{\boldsymbol{k}}_0$, then every car, having identically the same disposition, would reflect the balls exactly the same way. The signal, or reflected pattern of probes, would be amplified by the number of cars in the parking lot, and the end result, which we call the signal, would be far more easily detected and measured because of its strength. In our diffraction experiment, which is what we are really doing here, the Volkswagen beetles are the molecules, the basketballs analogous to X radiation, and the numbers of basketballs scattered in each direction are the intensities of the diffracted waves. Instead of an automobile parking lot, we have a molecular parking lot, a crystal.

A single voice, in a coliseum, though shouting, cannot be heard at a distance. Even a stadium full of voices cannot be heard far away if each individual is shouting a different cheer at random times. But if every voice in the stadium (or at least those favoring one particular team) are united in time in a single mighty cheer, the sound echoes far and wide.

Then from five thousand throats and more there rose a lusty yell;
it rumbled through the valley, it rattled in the dell;
it pounded through on the mountain and recoiled upon the flat;
for Casey, mighty Casey, was advancing to the bat.[5]

It is this cooperative effort of many individuals united in space and time, as occurs in a crystal scattering X rays, that makes a molecular diffraction experiment possible.

Clearly, by this analogy, we have simplified things to get at the essentials, but the details will come later. It is important to keep in mind that in carrying out our experiments on vast, ordered arrays of structurally unknown objects, we sacrifice information that might have been obtained from a single individual (i.e., there is no free lunch). In addition because our probes in X-ray crystallography are not particles, or balls, but are waves, an additional complication is introduced. This is because waves add together, or interfere with one another, in a manner unlike that of single particle probes. Thus we ultimately must consider

---

[5] From "Casey at the Bat," by Earnest Lawrence Thayer.

the diffraction pattern from our molecular array, or crystal, as sums of waves. Later this sacrifice of information will emerge as what is known as "the phase problem" in X-ray crystallography. That problem we will address in due course.

A LENS AND OPTICAL DIFFRACTION PATTERNS

In our common experience, we rarely even think of waves and how they add together, even though we depend on light (wavelength **λ = 3500 Å** to **6000 Å**) for visualizing nearly everything. We use our eyes, microscopes, telescopes, cameras, and other optical devices that depend on waves of light, yet we never, it seems, have to deal directly with waves or how they interact with one another. The reason is that we have lenses that gather together the light waves scattered by objects, and focus them into an image of the original object. The lenses of our eyes focuses light waves scattered by an object at a distance into an image of that object on our retinas. The lens of a microscope focuses the light scattered by a minute object in the path of a light beam into an image of the object, and magnifies it for us at the same time.

Figure 1.3 illustrates the essential features of image formation by a lens using a simple ray diagram. There are two unique planes where the rays emitted by the light-scattering object intersect after passage through the lens. One plane is twice the focal length of the lens (**$2f$**). There, an inverted image of the object is formed by the summation of rays from discrete points on the object converging at corresponding points on the plane. The rays converge in a different manner, however, on a second plane at a distance **f** between the lens and the image plane. In that plane, rays intersect that do not originate at the same point on the scattering object, but that have the same direction (defined by the angle **γ**) in leaving the object. The convergence of the various sets of rays, each having a different direction parameter **γ**, forms in this plane a second kind of image, which is called the diffraction pattern of the object. The diffraction pattern is known mathematically as the Fourier transform of the object.

What does the diffraction pattern of an object look like? We can visualize the diffraction pattern, the Fourier transform, of an object by making a mask about the object and then passing a collimated beam of light through the mask and onto a lens. The lens, as in

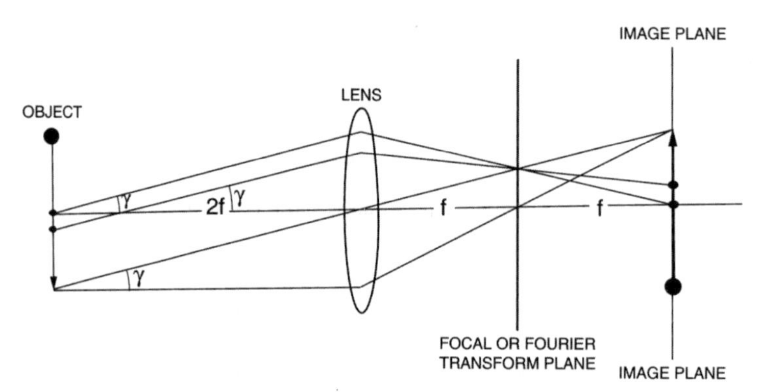

FIGURE 1.3 Formation of the Fourier transform (or diffraction pattern) of an object by a lens having focal length **f**. The rays leaving the object are caused, by the refractive properties of the lens, to converge at both the image plane (**$2f$**) and at a second "focal" plane **f**. The rays converging at each point on this transform plane at **f** are those that form a common angle with the plane of the scattering object, denoted here by **γ**; that is, they have a common scattering direction.

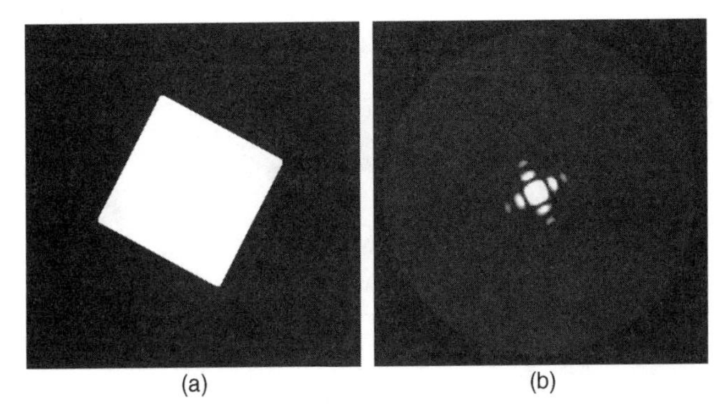

FIGURE 1.4 If a mask containing the square in (*a*) has a parallel beam of light directed through it and onto a lens, as in Figure 1.3, then if a screen were placed at one focal length *f* behind the lens, the diffraction pattern in (*b*) would be observed. At twice the focal length behind the lens, an image of the original square would appear. The pattern of light and dark seen in (*b*) is both the optical diffraction pattern of the square, and, in mathematical terms, it is the Fourier transform of the square.

Figure 1.3, then creates the diffraction pattern at a distance *f*, which we can view on a screen, or record on a film.

Figure 1.4 is a simple example. The object is the square shown on the mask in (*a*). If we look at a distance *f* behind the lens, then we see the diffraction pattern of the square in (*b*). A second example is shown in Figure 1.5. With the possible exception of the diffraction pattern of DNA, this is probably the most reproduced diffraction pattern in the history of X-ray crystallography (see Taylor and Lipson, 1964, for its origin). Here, the object in (*a*) is a duck (probably rubber), and in (*b*) we see the duck's Fourier transform, its diffraction pattern. Significantly, if we were to place the diffraction pattern in (*b*) in the place of the object in (*a*), then at distance *f* behind the lens we would now see the duck. In other words, (*b*) is the Fourier transform of (*a*), but (*a*) is also the Fourier transform of (*b*). The transform is symmetrical, and it tells us that either side of the Fourier transform contains all the information necessary to recreate the other side.

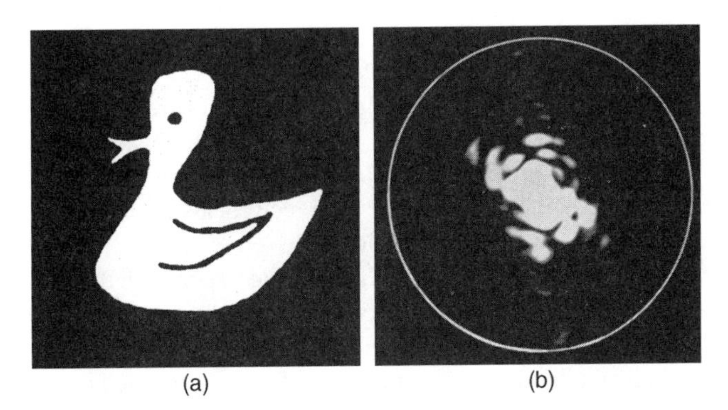

FIGURE 1.5 If the object mask contains the image of a duck, as in (*a*), its optical diffraction pattern, or Fourier transform, seen at *f*, is the seemingly meaningless pattern of light and dark in (*b*). It is important to note that the object, the duck, is a continuous object of basically arbitrary placed points (in a mathematical, not a biological sense), so the diffraction pattern is likewise a continuous function of intensity consisting of patches and islands of light. Any continuous object, such as the duck seen here, might be expected to yield a diffraction pattern having these characteristics.

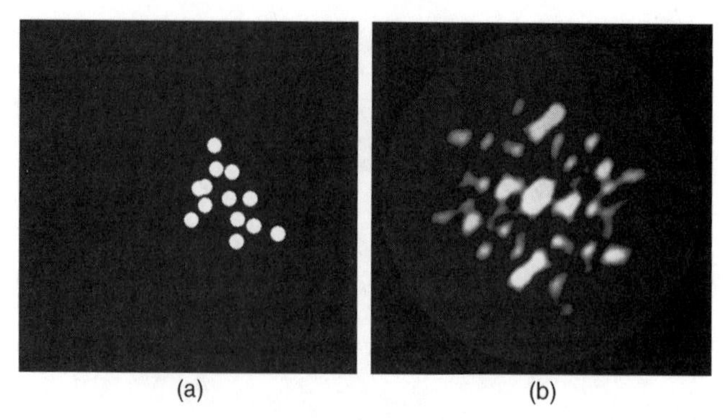

FIGURE 1.6 In (*a*) the object for the optical transform is not a continuous object, but is a set of points arbitrarily distributed in space, as we might expect to find in a molecule made up of discrete atoms. That is, they bear no fixed mathematical relationship to one another. In (*b*) the optical diffraction pattern of the set of points is again a continuum consisting of islands of light and dark. Such a transform is typical of one we might expect from any conventional organic molecule. The locations of the light and dark areas in the transform are dependent only on the x, y positions of the individual points in the object. If a point in the object were moved, the transform would change. If the entire set of points were rotated in the plane, the transform would undergo a corresponding rotation.

In the examples above, (*a*) was what we call a continuous object in that it was composed of a continuum of points covering a defined area, namely a square or the surface of a duck. The diffraction patterns were similarly continuous. Molecules, however, are not really continuous; they are composed of atoms, which serve as discrete scattering points. In Figure 1.6, for example, we have an arbitrary distribution of scattering points, like atoms in a molecule, and in (*b*) we see the diffraction pattern of the atom set. Note that even though the object is composed of unique scattering points, the diffraction pattern is still more or less continuous. Thus we should expect the diffraction pattern of a single molecule to be continuous, even if the molecule itself is not.

A final example, but of a different kind of object, is shown in Figure 1.7. This object is a discrete set of points distributed over the surface of a mask in a periodic (uniformly repetitive) array. We call such a periodic point array in space a lattice. In (*b*) is the diffraction pattern of the lattice in (*a*), and vice versa. The diffraction pattern in (*b*) is also a lattice composed of discrete points (it is what we call a discrete transform), but the spacings between the points are quite different than for the lattice in (*a*). We will see later that the distances between lattice points in (*a*) and (*b*) are reciprocals of one another. The Fourier transform of a lattice then is a reciprocal lattice.

We will further see in later chapters that it is possible to combine the two kinds of transforms illustrated here, the continuous transform of a molecule with the periodic, discrete transform of a lattice. In so doing, we will create the Fourier transform, the diffraction pattern of a crystal composed of individual molecules (sets of atoms) repeated in three-dimensional space according to a precise and periodic point lattice.

We can get some idea as to what to expect by again using optical diffraction. In Figure 1.8*a* is a pattern of scattering points having no internal symmetry or periodicity. It might well represent the set of atoms in a molecule. Its diffraction pattern is seen in Figure 1.8*b*. If the molecular motif in Figure 1.8*a* is repeated in a periodic manner in two dimensions, that is, according to a point lattice, then we can generate the array seen in Figure 1.8*c*. And what

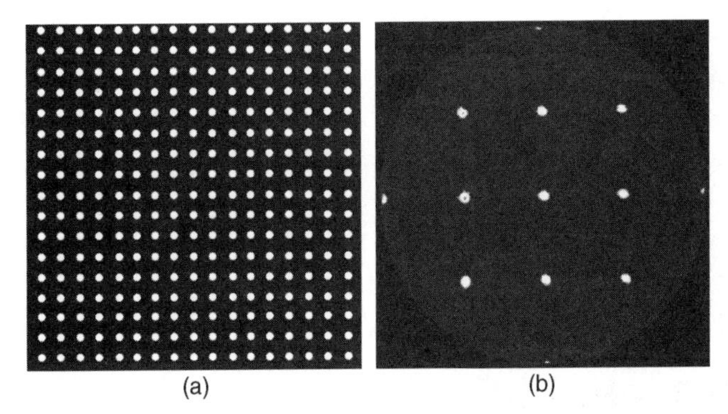

FIGURE 1.7 In (*a*) the object, again exposed to a parallel beam of light, is not a continuous object or an arbitrary set of points in space, but is a two-dimensional periodic array of points. That is, the relative x, y positions of the points are not arbitrary; they bear the same fixed, repetitive relationship to all others. One need only define a starting point and two translation vectors along the horizontal and vertical directions to generate the entire array. We call such an array a lattice. The periodicity of the points in the lattice is its crucial property, and as a consequence of the periodicity, its transform, or diffraction pattern in (*b*) is also a periodic array of discrete points (i.e., a lattice). Notice, however, that the spacings between the spots, or intensities, in the diffraction pattern are different than in the object. We will see that there is a reciprocal relationship between distances in object space (which we also call real space), and in diffraction space (which we also call Fourier space, or sometimes, reciprocal space).

is the diffraction pattern, the Fourier transform of the periodic distribution in Figure 1.8*c*? It is shown in Figure 1.8*d*.

The diffraction pattern of the molecular array in Figure 1.8*d* is also periodic, but the spacings between the diffraction intensities are reciprocals of the molecular point lattice. The intensities in Figure 1.8*d* vary from point to point as well, unlike the example shown in Figure 1.7. If the diffraction pattern in Figure 1.8*d* were superimposed on that in Figure 1.8*b*, we would find that the intensities at the discrete points in Figure 1.8*d* are identical to the intensities of the corresponding points in the continuous transform in Figure 1.8*b*, which they overlay. The discrete lattice according to which the molecules are periodically arrayed in Figure 1.8*a* has the effect of allowing us to see, or sample, the transform (diffraction pattern) of an individual molecule at periodic, specific points. The lattice appearing in diffraction space, having reciprocal spacings between points, is the corresponding reciprocal lattice.

To discriminate, or resolve, individual points in an object, as we saw in the Volkswagen parable, one must utilize a radiation of wavelength comparable to the distances between the scattering points. Thus we can use microscopy with light to resolve detail within an object that is on the order of a few thousand angstroms. We can use radio waves, as in radar, to resolve details measured in meters. If the objective is to produce an image of a macromolecule composed of atoms separated by an average bond length of about **1.5 Å**, then one is obligated to use a radiation of comparable wavelength. Conveniently, the characteristic X radiation produced by the collision of high-energy electrons with a number of different metal targets is of the range **1** to **3 Å**, precisely what is required. Less convenient is the unfortunate reality that nature has not provided us with any known lens or mechanism for the focusing of scattered X rays.

Unlike light, which, because of its refractive properties, can be focused by a properly ground glass lens, and unlike electrons, which, because of their charge, can be focused by

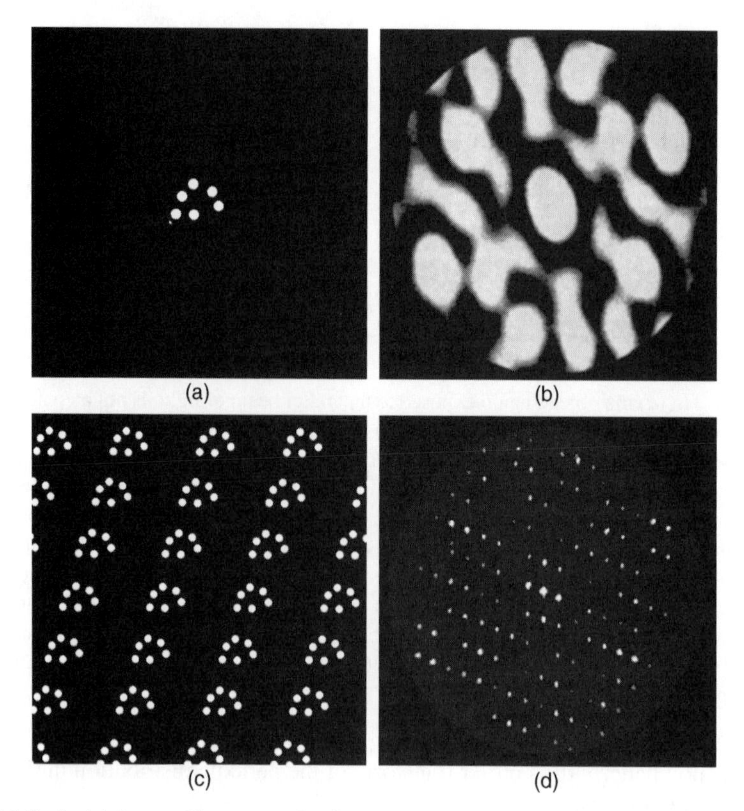

FIGURE 1.8 In (*a*) is an arbitrary set of points that might represent the atoms in a molecule, and in (*b*) is the optical diffraction pattern of that set of points. It is a continuum of light and dark over the whole surface of the screen. The mask (object) in the optical diffraction experiment in (*c*) is the periodic arrangement of the fundamental set of points in (*a*) in two dimensions (i.e., the repetition of the object according to the instruction of a lattice). The diffraction pattern of (*c*) is shown in (*d*). We would find that if we superimpose the point array in (*d*) upon the continuous transform in (*b*), the intensity at each lattice point in (*d*) corresponds to the value of the continuous transform beneath. That is, the diffraction pattern in (*d*) samples the continuous transform in (*b*) at specific points determined by the periodic lattice of (*c*).

electromagnetic fields, X rays have no properties that permit an analogous process. Thus X radiation can be scattered from the electrons of an object, just as light or electrons are scattered by a specimen as they pass through it. But contrary to the situation we enjoy with a microscope, no lens can be interposed between specimen and observer to gather the scattered radiation and focus it into a meaningful image.

The crystal lattice, however, plays a second role. It not only amplifies the diffraction signal from individual molecules, it also serves as half a lens. The X rays scattered by the atoms in a crystal combine together, by virtue of the periodic distribution of their atomic sources, so that their final form is precisely the Fourier transform, that is, the diffraction pattern that we would ordinarily observe at f if we did in fact have an X-ray lens. Thus the situation is not intractable, only difficult. We find in X-ray crystallography that while we cannot record the image plane, we can record what appears at the diffraction plane. It is then up to us to figure out what is on the image plane from what we see on the diffraction plane.

HOW X-RAY DIFFRACTION WORKS

If a collimated beam of monochromatic X rays is directed through an object, such as a macromolecule, the rays are scattered in all directions by the electrons of every atom in the object with a magnitude proportional to the size of its electron complement. This is the fundamental experiment that we perform in a diffraction experiment, and it is illustrated in Figure 1.9. If the object were composed of more or less arbitrarily placed atoms, as they are in a single macromolecule, then at any point in space about the isolated object a measurable amount of scattered radiation would be expected to be recorded by an observer; that is, the distribution of scattered rays would be a continuum of varying intensity. This was shown in Figures 1.4 through 1.6 and in Figure 1.8. The variability of intensity throughout this continuous scattering distribution, which is again the Fourier transform, or diffraction pattern of the macromolecule, would depend on the relative positional coordinates and atomic numbers of the atoms in the object, and would ostensibly be independent of any other property of the object. We will show this in Chapter 5.

If a number of identical objects were arranged in three-dimensional space in such a way that they form a periodically repeating array, the scattering distribution, or diffraction pattern from the collection of objects would tend to be less continuous, taking on observable values at some points and approaching zero elsewhere. This was illustrated in Figure 1.8. When the number of objects in the array becomes very large, as it does in a crystal, the scattering distribution, the diffraction pattern, becomes absolutely discrete.

Now, as discussed already, the scattering of X rays from a single protein or nucleic acid molecule would be immeasurably small. Due to its size alone, such an object could not

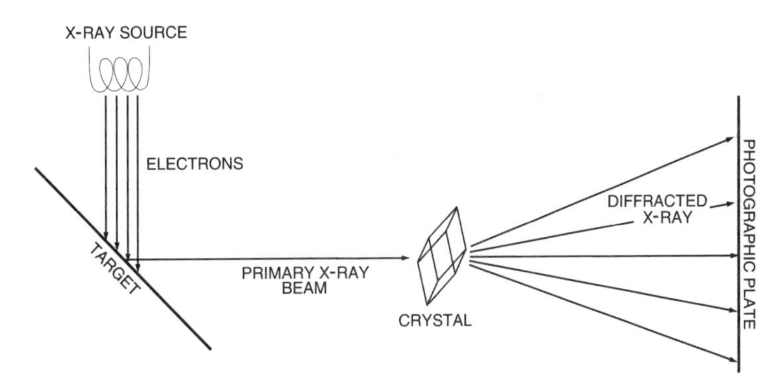

FIGURE 1.9 The basic X-ray diffraction experiment is shown here schematically. X rays, produced by the impact of high-velocity electrons on a target of some pure metal, such as copper, are collimated so that a parallel beam is directed on a crystal. The electrons surrounding the nuclei of the atoms in the crystal scatter the X rays, which subsequently combine (interfere) with one another to produce the diffraction pattern on the film, or electronic detector face. Each atom in the crystal serves as a center for scattering of the waves, which then form the diffraction pattern. The magnitudes and phases of the waves contributed by each atom to the interference pattern (the diffraction pattern) is strictly a function of each atom's atomic number and its position x, y, z relative to all other atoms. Because atomic positions x, y, z determine the properties of the diffraction pattern, or Fourier transform, the diffraction pattern, conversely, must contain information specific to the relative atomic positions. The objective of an X-ray diffraction analysis is to extract that information and determine the relative atomic positions.

be directly imaged. If, however, a vast number of such molecules are organized into an array so that their scattering contributions are cooperative, then the resultant radiation can be observed and quantitated as a function of direction in space. This is precisely what is provided by macromolecular crystals, or in fact any crystals.

Some examples of typical macromolecular crystals (see McPherson, 1999, and Chapter 2) are shown in Figure 1.10. While objects of beauty, their regular features only begin to suggest the degree of their internal order. In Figure 1.11 we see with electron microscopy, and in Figure 1.12, with atomic force microscopy evidence of their exquisite, periodic nature. In Figures 1.11 and 1.12 the individual molecules that comprise the crystals are aligned in rows and columns, indeed in all three dimensions, in perfect register, every molecule disposed, every molecule in precisely the same environment as any other. It is the molecular equivalent of our parking lot for Volkswagen beetles, but in three dimensions.

The resultant radiation scattered, or diffracted, in specific directions create the intensities we see, precisely arranged, on an X-ray diffraction photograph. Because of the uniformity

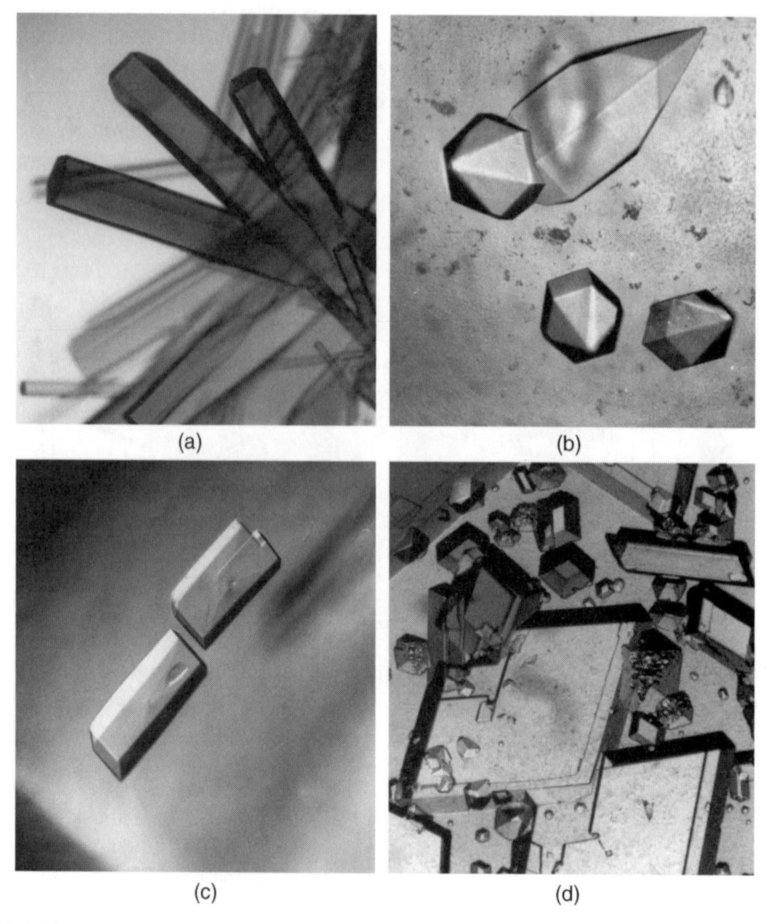

FIGURE 1.10 Crystals of a variety of proteins. In (*a*) hexagonal prisms of beef liver catalase. In (*b*) crystals of α_1 acid glycoprotein, in (*c*) Fab fragments of a murine immunoglobulin, and in (*d*) rhombohedral crystals of the seed storage protein canavalin.

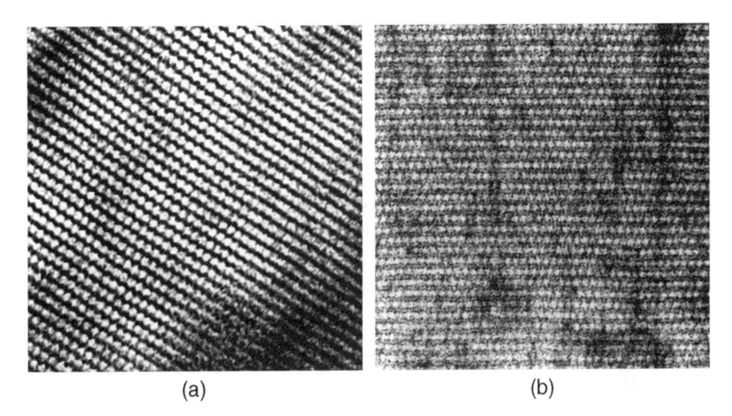

FIGURE 1.11 Electron micrographs of negatively stained crystals of (*a*) pig pancreatic **α** amylase, and (*b*) beef liver catalase. The dark areas represent solvent filled areas in the crystal, which are replaced by dense heavy metal stain; the light areas correspond to protein molecules where the stain is excluded. The underlying periodicity of the crystals is evident here, even after dehydration and staining.

of orientation and periodic position imposed on the molecules by the crystal lattice, the scattered X radiation, being waves, constructively interferes in unique directions dictated by the parameters that define the periodicity of the crystal lattice. It destructively interferes and sums to zero in all other directions. Hence we observe that the diffraction patterns from the ordered arrays that exist in crystals are absolutely discrete and that the observable diffraction pattern is an array of intensities that falls on a regular net or lattice (a reciprocal lattice). The spacings between the intensities, or reflections, and the symmetry properties that govern their distribution are manifestations of the periodic and symmetric disposition of the molecules in the crystal. Because the physical relationship between the diffracted rays and the crystal lattice is well understood, mathematical expressions, such as Bragg's law, can be written that describe the correspondence.

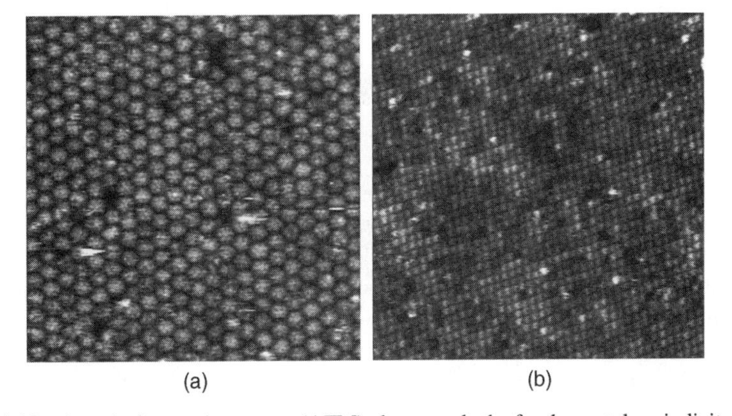

FIGURE 1.12 Atomic force microscopy (AFM) also reveals the fundamental periodicity of macromolecular crystals. In (*a*) is the surface layer of a crystal of brome mosaic virus, a particle having a diameter of about **280 Å**. In (*b*) is an AFM image of a monoclinic crystal of duodecahedral complexes of intact immunoglobulins, which have a diameter of about **230 Å**.

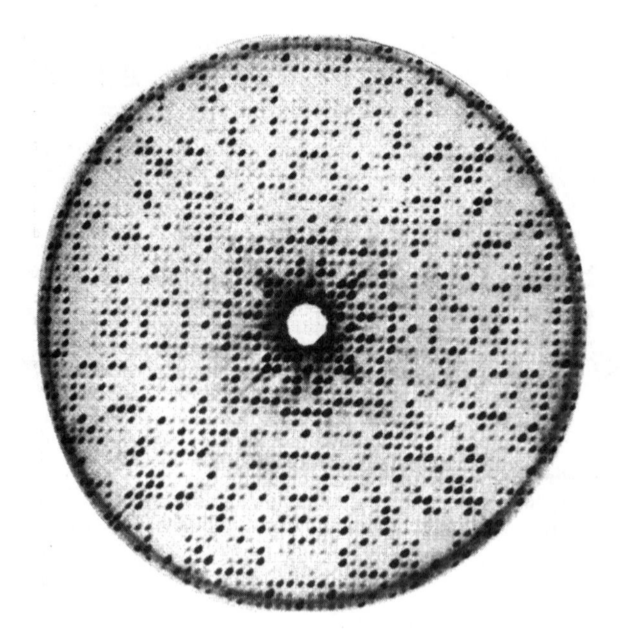

FIGURE 1.13 The ***hk*0** zone diffraction image from a tetragonal crystal of lysozyme, an enzyme from hen egg white. Here, the fourfold symmetry of the pattern is striking, and it reflects the fourfold symmetry of the arrangement of the protein molecules in the unit cells of the crystal. Again, the intensities fall on a very regular, periodic net, or reciprocal lattice. The net is based on a tetragonal axis system (see Chapter 3).

It is clear from looking at diffraction patterns obtained from real crystals, such as those in Figures 1.13 and 1.14, that all of the reflections are not equal. They span a broad range of intensity values from very strong to completely absent. It will be shown in Chapter 5 (and was already demonstrated by Figure 1.8) that the variation in intensity from reflection to reflection is a direct function of the atomic structure of the macromolecules that comprise the crystal and occupy its lattice points. That is, the relative intensities of the reflections that make up the three-dimensional diffraction pattern, or Fourier transform, of a crystal are directly related to the relative x_j, y_j, z_j coordinates of all of the atoms j that define an individual molecule, and to the relative strength, Z_j, with which the different atoms scatter X rays. Z_j is the electron complement of each atom and is, therefore, its atomic number.

The complete diffraction pattern from a protein crystal is not limited to a single planar array of intensities like those seen in Figures 1.13 and 1.14. These images represent, in each case only a small part of the complete diffraction pattern. Each photo corresponds to only a limited set of orientations of the crystal with respect to the X-ray beam. In order to record the entire three-dimensional X-ray diffraction pattern, a crystal must be aligned with respect to the X-ray beam in all orientations, and the resultant patterns recorded for each. From many two-dimensional arrays of reflections, corresponding to cross sections through diffraction space, the entire three-dimensional diffraction pattern composed of ten to hundreds of thousands of reflections is compiled.

Because the diffraction pattern from a macromolecular crystal is the Fourier transform of the crystal, a precise mathematical expression can be set down that relates the diffracted

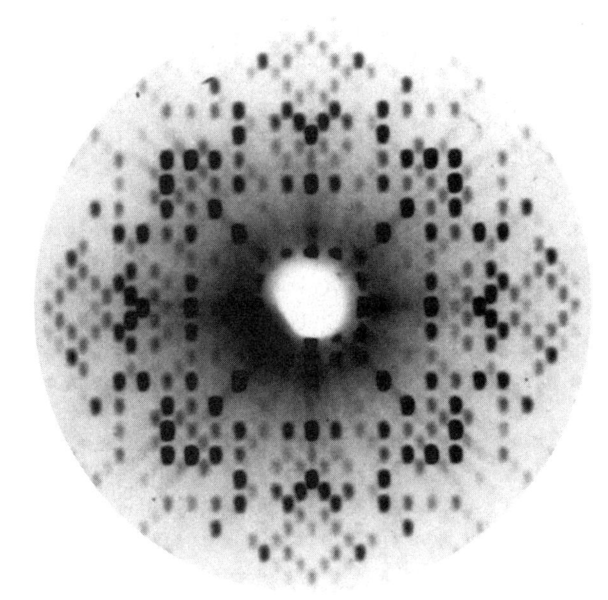

FIGURE 1.14 Seen here is the ***hk*0** zone diffraction pattern from a crystal of M4 dogfish lactate dehydrogenase obtained using a precession camera. It is based on a tetragonal crystal system and, therefore, exhibits a fourfold axis of symmetry. The hole at center represents the point where the primary X-ray beam would strike the film (but is blocked by a circular beamstop). Note the very predictable positions of the diffraction intensities. All the intensities, or reflections, fall at regular intervals on an orthogonal net, or lattice. This lattice in diffraction space is called the reciprocal lattice.

waves to the distribution of atoms in the crystal. This expression, called the electron density equation, is a three-dimensional spatial transform that we refer to as a Fourier synthesis. It is sufficient here to simply understand that it is a summation of terms, one for each reflection observed in the diffraction pattern, and that the relative intensity of each reflection is the absolute magnitude of one of the terms in the series.

THE PHASE PROBLEM

The situation, in truth, is somewhat more involved than this explanation would suggest. The individual reflections of the diffraction pattern are the interference sum of the waves scattered by all of the atoms in the crystal in a particular direction and, therefore, are themselves waves. Being waves they have not only an amplitude, but also a unique phase angle associated with each of them. This too depends on the distribution of the atoms, their x_j, y_j, z_j. The phase angle is independent of the amplitude of the reflection, but most important, it is an essential part of the individual terms that contribute to the Fourier synthesis, the electron density equation. Unfortunately, the phase angle of a reflection cannot be recorded, as we record the intensity. In fact we have no practical way (and rather few impractical ways either) to directly measure it at all. But, without the phase information, no Fourier summation can be computed. In the 1950s, however, it became possible, with persistence, skill, and patience (and luck), to recover this elusive phase information for protein crystals, thus permitting the calculation of Fourier summations and hence images of macromolecules. The technique, which is known as multiple isomorphous replacement, is based on the chemical derivatization of protein crystals with heavy metal atoms such as mercury. Its development (Boyes-Watson et al., 1947; Bragg and Perutz, 1954; Blow and Crick, 1959) was the major breakthrough in modern crystallography that ultimately made possible the determination of macromolecular structures.

In this technique, described in more detail in Chapter 8, the heavy atom, whose position in the crystal can be determined by what are known as Patterson techniques (see Chapter 9), provides a reference wave. In a derivatized crystal the resultant diffraction intensities represent the sum of this heavy atom-produced reference wave interfering with the wave arising from all of the atoms in the protein. Just as the relative phase of a specific sound wave can be deduced by "beating" it against a reference sound wave of known phase, or for light waves using interferometry, the same is done for the native diffracted wave. The mathematical construct for obtaining the phase information requires measurement of the native diffraction intensities and the corresponding intensities from crystals independently derivatized at least at two unique sites in the crystal. It is known as a Harker diagram (Harker, 1956). From Harker diagrams for each of the reflections that comprise a complete diffraction pattern, the necessary phases, or at least reasonable approximations, can be obtained. This is discussed in more detail in Chapter 8.

THE ELECTRON DENSITY

To produce an electron density image of the molecules comprising the unit cell contents of a crystal from the measured native structure amplitudes, and the approximate phase angles derived by isomorphous replacement (or by other methods, see Chapter 8), the value of the Fourier synthesis $\rho(x, y, z)$ must be computed at every point (x, y, z) in the unit cell. The unit cell, as we will see in Chapter 3, is the actual repeating unit of the crystal. The electron density $\rho(x, y, z)$ is a three-dimensional function and is continuous throughout the unit cell. A good approximation to this density continuum can be obtained by computing the value of $\rho(x, y, z)$ on a grid of points whose separations are sufficiently small. The value of $\rho(x, y, z)$ is calculated, for example, on grid points separated by distances Δx and Δy over a particular plane of constant z of the unit cell. The z coordinate is incremented by Δz, and $\rho(x, y, z)$ is computed on all grid points in the plane $(x, y, z + \Delta z)$.

Figure 1.15 contains a composite of several planes from an electron density map of a protein. By continually increasing the final coordinate by Δz, the electron density map is built up from the series of two-dimensional planes. The individual sections are plotted on some transparent material after contour lines have been drawn around areas within certain density limits. The result is a topological map of the electron density presented on sequential planes of the unit cell as a series of contour levels. When the individual planes are stacked in consecutive order, a three-dimensional electron density image is created. This is discussed in more detail in Chapter 10. Currently, however, the presentation of the electron density is considerably more sophisticated. We use automated computer graphics systems to present detailed density images in three-dimensional space as in Figure 1.16.

Although measurement of, in many cases, the hundreds of thousands of X-ray reflections that are necessary to compute the Fourier synthesis of a macromolecule has in the past been an extremely time-consuming endeavor, this has, in recent years, become a far less arduous task. This is due to the advent of very rapid data collection devices based on area detectors,

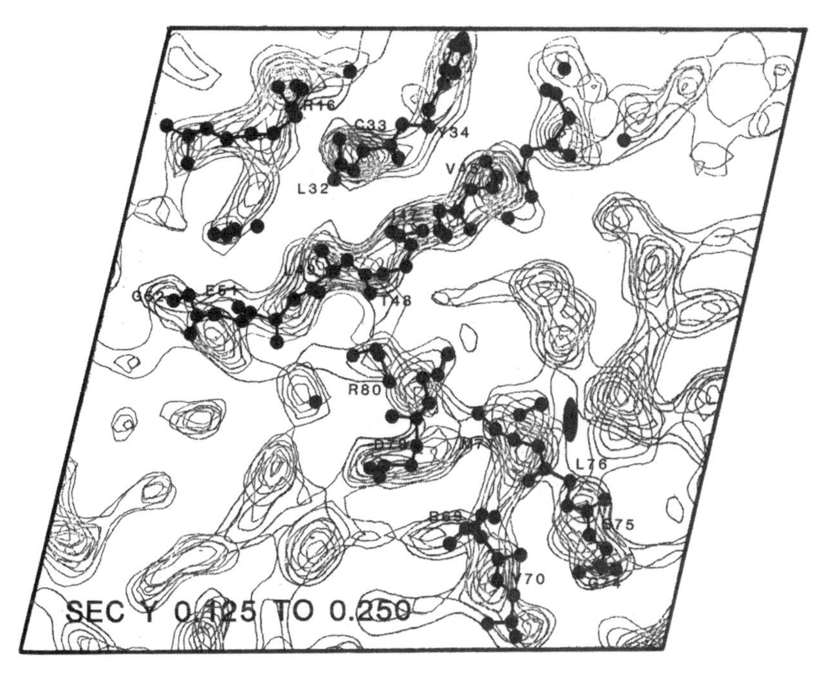

FIGURE 1.15 Electron density from a monoclinic unit cell of the Gene 5 DNA Unwinding Protein crystal, lying between $y = 0.125$ and $y = 0.250$, is projected onto a single plane. Superimposed upon this electron density is a portion of the atomic model of the Gene 5 Protein. Electron density planes, like those shown here, are the images obtained directly by X-ray diffraction from computed Fourier syntheses.

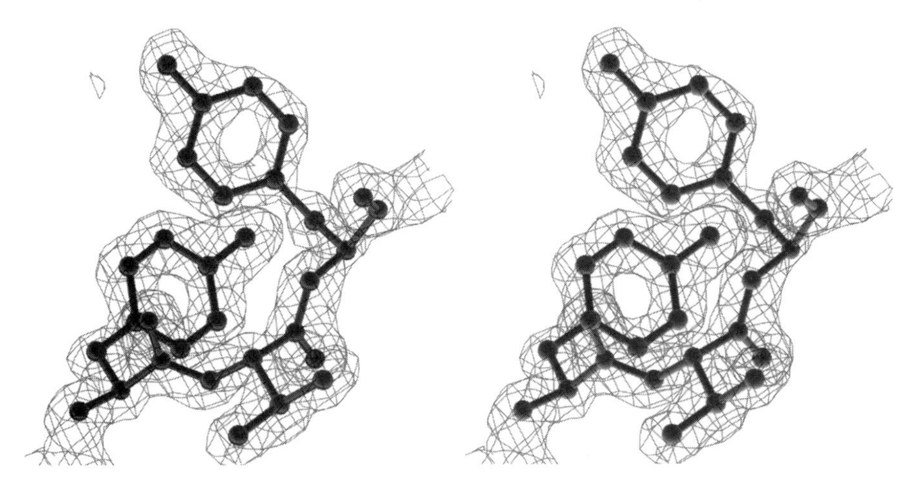

FIGURE 1.16 A more sophisticated presentation of electron density, in virtual three dimensions, is possible using computer graphics. In this stereo diagram, two tyrosines separated by a valine residue are superimposed upon their density in a **1.8 Å** resolution electron density map of the serine protease from penicillium cyclopium.

and high-flux density synchrotron X-ray sources. These, combined with enormous advances in computer technology, have now made it possible to solve the structure of virtually any protein that can be crystallized with, in the best of cases, no more than a few days to a week of concentrated effort. This time frame is continuing to shrink even now.

When the structure of a protein has been determined so that its constituent atomic positions are known to less than an angstrom, it can be precisely refined (see Chapter 10) by applying difference Fourier and various nonlinear least squares procedures (Hendrickson and Konnert, 1981; Sussman et al., 1997; Brunger et al., 1987). With these procedures, atomic coordinates are adjusted in increments of Δx, Δy, Δz so that chemical groups are made to approach ideal geometry, but in a manner that simultaneously minimizes the difference between the diffraction pattern calculated from the model structure, which we will see is possible, and the intensities actually observed experimentally. This can be done because the Fourier transform is symmetrical and allows calculation of either side of the transform from the other. By computing many refinement cycles, each accompanied by small optimizing shifts in the atomic coordinates, and using the difference between the calculated and observed diffraction pattern as a guide, convergence can ultimately be attained at a highly precise structural model having estimated errors in the atomic coordinates of less than a tenth of an angstrom. Structures refined in this way may then be utilized to carry out crystallographic experiments designed to elucidate the biochemical properties of the protein (see McPherson, 1987).

CHAPTER 2

CRYSTALLIZATION OF MACROMOLECULES

For application of X-ray diffraction to a macromolecule, the protein or nucleic acid must first be crystallized. Not only must crystals be grown, but they must be reasonably large, high-quality crystals that are suitable for a high-resolution X-ray diffraction study. The crystallization step has emerged as the primary obstacle in macromolecular crystallography. This is principally due to the empirical nature of the methods employed to overcome it, and the complexity of the molecules involved (McPherson, 1976, 1982, 1999).

Macromolecules are intricate physical-chemical systems whose properties vary as a function of environmental influences such as temperature, pH, ionic strength, contaminants, and solvent composition, to name only a few. They are structurally dynamic, often microheterogeneous, aggregating systems, and they change conformation in the presence of ligands. Superimposed on this is the limited nature of our current understanding of macromolecular crystallization phenomena and the forces that promote and maintain protein and nucleic acid crystals.

As a substitute for the precise and reasoned approaches that we commonly apply to scientific problems, we are forced, for the time being at least, to employ an empirical methodology. Macromolecular crystallization is a matter of searching, as systematically as possible, the ranges of the individual parameters that impinge upon crystal formation; finding a set, or multiple sets of these factors that yield some kind of crystals; and then optimizing the variables to obtain the best possible crystals for X-ray analysis. This is done, most simply, by conducting an extensive series, or establishing a vast array, of crystallization trials, evaluating the results, and using information obtained to improve matters in successive rounds of trials. Because the number of variables is large, and its range broad, intelligence and intuition in designing and evaluating the individual and collective trials becomes essential.

Introduction to Macromolecular Crystallography, Second Edition By Alexander McPherson

CRYSTALS GROW FROM SUPERSATURATED SOLUTIONS

In a saturated solution, including one saturated with respect to protein, two states exist in equilibrium: the solid phase and one consisting of molecules free in solution. At saturation, no net increase in the proportion of solid phase can accrue, since it would be counterbalanced by an equivalent dissolution. Thus crystals do not grow from a saturated solution. The system must be in a nonequilibrium, or supersaturated state to provide the thermodynamic driving force for crystallization.

When the objective is to grow crystals of any compound, a solution of the molecule must be transformed or brought into a supersaturated state, whereby its return to equilibrium forces exclusion of solute molecules into the solid, the crystal. If, from a saturated solution, for example, solvent is gradually withdrawn by evaporation, temperature is lowered or raised appropriately, or some other property of the system is altered, then the solubility limit may be exceeded and the solution will become supersaturated. If a solid phase is present, or introduced, then strict saturation will be reestablished as molecules leave the solvent, join the solid phase, and equilibrium is regained.

If no solid is present as conditions are changed, then solute will not immediately partition into two phases, and the solution will remain in the supersaturated state. The solid state does not necessarily develop spontaneously as the saturation limit is exceeded because energy, analogous to the activation energy of a chemical reaction, is required to create the second phase, the stable nucleus of a crystal or a precipitate. Thus a kinetic, or energy barrier allows conditions to proceed further and further from equilibrium, into the zone of supersaturation. On a phase diagram, like that seen in Figure 2.1, the line indicative of saturation is also a boundary that marks the requirement for energy-requiring events to occur in order for a

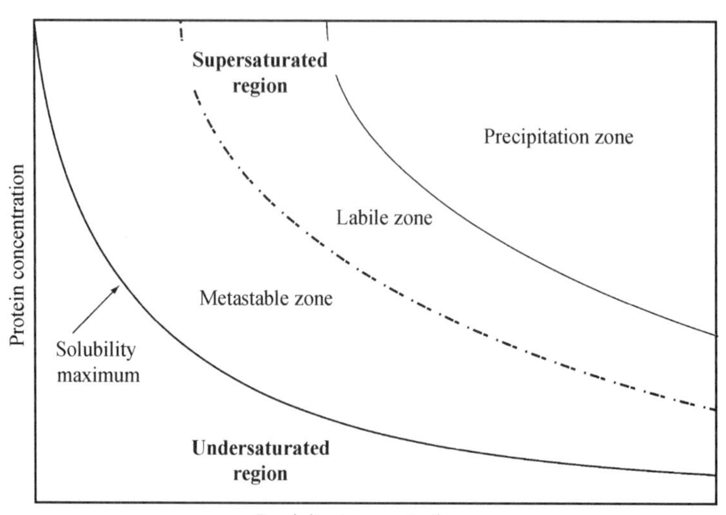

FIGURE 2.1 The phase diagram for crystallization. It consists of three regions, the undersaturated, the supersaturated, and the equilibrium line that separates them. This line denotes the maximum solubility, and the concentration of solute at which the solid state is in equilibrium with solute molecules in solution. The supersaturated region is divided into a labile region where crystals may nucleate and grow, and a metastable region where crystals are unlikely to nucleate but, if present, can grow.

second phase to be established, the formation of the ordered nucleus of a crystal, or the nonspecific aggregate that characterizes a precipitate.

Once a stable nucleus has formed in a supersaturated solution, it will continue to grow until the system regains equilibrium. So long as nonequilibrium forces prevail and some degree of supersaturation exists to drive events, a crystal will grow or precipitate will continue to form. It is important to understand the significance of the term "stable nucleus." Many aggregates or nuclei spontaneously form once supersaturation is achieved, but most are, in general, not "stable." Instead of continuing to develop, they redissolve as rapidly as they form, and their constituent molecules return to solution. A stable nucleus is an ordered molecular aggregate of such size and physical coherence that it will enlist new molecules into its growing surfaces faster than others are lost to solution; that is, it will continue to grow so long as the system is supersaturated.

In classical theories describing crystal growth of conventional molecules, the region of supersaturation is further divided into what are termed the metastable region and the labile region, as shown in Figure 2.1. By definition, stable nuclei cannot form in the metastable region just beyond saturation. If, however, a stable nucleus or solid is already present in the metastable region, then it can and will continue to grow. The labile region of greater supersaturation is discriminated from the metastable in that stable nuclei can spontaneously form. Further, because they are stable, they will accumulate molecules and thus deplete the liquid phase of solute until the system reenters the metastable and, ultimately, the saturated state.

An important point, shown graphically in Figure 2.1, is that there are two regions above saturation, one that can support crystal growth but not formation of stable nuclei and another that can yield nuclei as well as support growth. Now the rates of nucleation and crystal growth are both a function of the distance of the solution from the equilibrium position, saturation. Thus a nucleus that forms far from equilibrium and well into the labile region will initially grow very rapidly and, as the solution is depleted of nutrient, move back toward the metastable state. Then it will grow slower and slower. The nearer the system is to the metastable state when a stable nucleus first forms, the slower it will proceed to mature.

It might appear that the best approach for obtaining crystals is to press the system as far into the labile region, supersaturation, as possible. There, the probability of nuclei formation is greatest, the rate of growth is greatest, and the likelihood of crystals is maximized. As the labile region is penetrated further, however, the probability of spontaneous and uncontrolled nucleation is also enhanced. Thus crystallization from solutions in the labile region, far from the metastable state, frequently results in extensive and uncontrolled "showers" of crystals. By virtue of their number, none is favored, and in general, none will grow to a size suitable for X-ray diffraction studies. In addition, when crystallization is initiated at high supersaturation, then initial growth is extremely rapid. Rapid growth is frequently associated with the occurrence of defects, dislocations, and the incorporation of impurities. Hence crystals produced from highly saturated solutions tend to be numerous, small, and afflicted with growth defects.

In terms of the phase diagram, ideal crystal growth would begin with nuclei formed in the labile region, but just beyond the metastable. There, growth would occur slowly; the solution, by depletion, would return to the metastable state where no more stable nuclei could form; and the few nuclei that had established themselves would continue to grow to maturity at a pace free of defect formation. Thus in growing crystals for X-ray diffraction analysis, one attempts, by either dehydration or alteration of physical conditions, to transport the solution into a labile, supersaturated state, but one as close as possible to the metastable phase.

WHY CRYSTALS GROW

The natural inclination of any system proceeding toward equilibrium is to maximize the extent of disorder, or entropy, by freeing individual constituents from physical and chemical constraints. At the same time there is a thermodynamic requirement to minimize the free energy (or Gibbs free energy) of the system. This is achieved by the formation of chemical bonds and interactions that generally provide negative free energy. Clearly, the assembly of molecules into a fixed lattice severely reduces their mobility and freedom, yet crystals do form and grow.

It follows then that crystal nucleation and growth must be dominated by noncovalent chemical and physical bonds arising in the crystalline state that either cannot be formed in solution or are stronger than those that can. These bonds are in fact what hold crystals together. They are the energetically favorable intermolecular interactions that drive crystal growth despite resistance to molecular constraint. From this it is clear that if one wishes to enhance the likelihood of crystal nuclei formation and growth, then one must do whatever is possible to ensure the greatest number of most stable interactions among the molecules in the solid state.

One could ask why molecules should arrange themselves into perfectly ordered and periodic crystal lattices, exemplified by those in Figure 2.2, when they could equally well

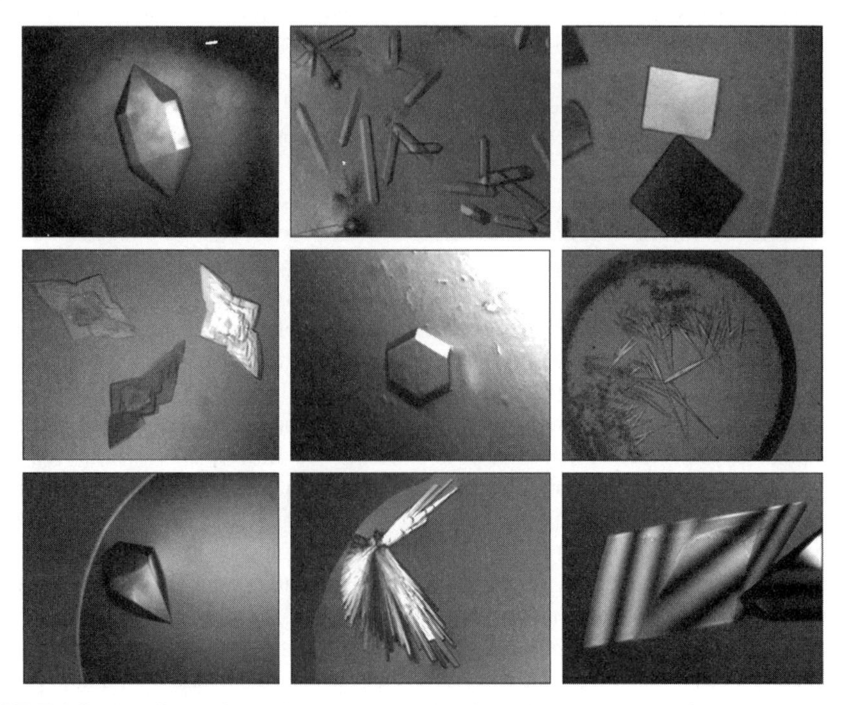

FIGURE 2.2 A collage of protein crystals, ranging from the large and perfect to the small and flawed. Courtesy of Hampton Research.

form random and disordered aggregates, which we commonly refer to as precipitate. The answer is the same as for why solute molecules leave the solution phase at all: to form the greatest number of most stable bonds in order to minimize the free energy, or enthalpy, of the system. While precipitates represent, in general, a low-energy solid state in equilibrium with a solution phase, crystals, not precipitates, are the states of lowest free energy. A frequently noted phenomenon has been the formation of precipitate followed by its slow dissolution concomitant with the formation and growth of crystals. The converse is not observed. This is one empirical demonstration that crystals represent more favorable energy states.

PROTEINS PRESENT SPECIAL PROBLEMS FOR CRYSTALLOGRAPHERS

In principle, the crystallization of a protein, nucleic acid, or virus (as exemplified in Figure 2.2) is little different than the crystallization of conventional small molecules. Crystallization requires the gradual creation of a supersaturated solution of the macromolecule followed by spontaneous formation of crystal growth centers or nuclei. Once growth has commenced, emphasis shifts to maintenance of virtually invariant conditions so as to sustain continued ordered addition of single molecules, or perhaps ordered aggregates, to surfaces of the developing crystal.

The perplexing difficulties that arise in the crystallization of macromolecules, in comparison with conventional small molecules, stem from the greater complexity, lability, and dynamic properties of proteins and nucleic acids. The description offered above of labile and metastable regions of supersaturation are still applicable to macromolecules, but it must now be borne in mind that as conditions are adjusted to transport the solution away from equilibrium by alteration of its physical and chemical properties, the very nature of the solute molecules is changing as well. As temperature, pH, pressure, or solvation are changed, so may be the conformation, charge state, or size of the solute macromolecules.

In addition proteins and nucleic acids are very sensitive to their environment, and if exposed to sufficiently severe conditions, they may denature, degrade, or randomize in a manner that ultimately precludes any hope of their forming crystals. They must be constantly maintained in a thoroughly hydrated state at or near physiological pH and temperature. Thus common methods for the crystallization of conventional molecules such as evaporation of solvent, dramatic temperature variation, or addition of strong organic solvents are unsuitable and destructive. They must be supplanted with more gentle and restricted techniques.

PROPERTIES OF MACROMOLECULAR CRYSTALS

Macromolecular crystals are composed of approximately **50%** solvent on average, though this may vary over **25%** to **90%**, depending on the particular macromolecule (Matthews, 1968; McPherson, 1999). The protein or nucleic acid occupies the remaining volume. The entire crystal is permeated with a network of interstitial spaces through which solvent and other small molecules may freely diffuse. This is seen quite dramatically in electron and atomic force micrographs of protein crystals such as those in Figures 1.11 and 1.12 of Chapter 1.

In proportion to molecular mass, the number of bonds (salt bridges, hydrogen bonds, hydrophobic interactions) that a conventional molecule forms in a crystal with its neighbors far exceeds the few exhibited by crystalline macromolecules. Since these contacts provide the lattice interactions that maintain the integrity of the crystal, this largely explains the difference in properties between crystals of salts or small molecules, and macromolecules, as well as why it is so difficult to grow protein and nucleic acid crystals.

Although morphologically indistinguishable, there are noteworthy differences between crystals of low molecular weight compounds and crystals of proteins and nucleic acids. Crystals of small molecules exhibit firm lattice forces, are highly ordered, generally physically hard and brittle, easy to manipulate, usually can be exposed to air, have strong optical properties, and diffract X rays intensely. Macromolecular crystals are by comparison usually more limited in size, are very soft and crush easily, disintegrate if allowed to dehydrate, exhibit weak optical properties, and diffract X rays poorly. Macromolecular crystals are temperature sensitive and undergo extensive damage after prolonged exposure to radiation. In some cases many crystals must be analyzed for a structure determination to be successful.

The extent of the diffraction pattern from a crystal is directly correlated with its degree of internal order. The more extensive the patterns, or the higher the resolution to which it extends, the more uniform are the molecules in the crystal and the more precise is their periodic arrangement. The level of detail to which atomic positions can be determined by a crystal structure analysis corresponds closely with the degree of crystalline order. While conventional molecular crystals often diffract almost to their theoretical limit of resolution, protein crystals, by comparison, are characterized by diffraction patterns of limited extent.

The liquid channels and solvent cavities that characterize macromolecular crystals are primarily responsible for the limited resolution of the diffraction patterns. Because of the relatively large spaces between adjacent molecules and the consequent weak lattice forces, every molecule in the crystal may not occupy exactly equivalent orientations and positions in the crystal but may very slightly from lattice point to lattice point. Furthermore, because of their structural complexity and their potential for conformational dynamics, protein molecules in a crystal may exhibit slight variations in the course of their polypeptide chains or the dispositions of side groups.

Although the presence of extensive solvent regions is a major contributor to the poor diffraction quality of protein crystals, it is also responsible for their value to biochemists. Because of the very high solvent content, the individual macromolecules in protein crystals are surrounded by hydration layers that maintain their structure virtually unchanged from that found in bulk solvent. As a consequence, ligand binding, enzymatic and spectroscopic characteristics, and other biochemical features are essentially the same as for the native molecule in solution. In addition the size of the solvent channels is such that conventional chemical compounds, which may be ions, ligands, substrates, coenzymes, inhibitors, drugs, or other effector molecules, may be freely diffused into and out of the crystals (see Chapter 10). Crystalline enzymes, though immobilized, are usually accessible for experimentation through alteration of the surrounding mother liquor. Thus a protein crystal can serve as a veritable ligand binding laboratory (McPherson, 1987).

CRYSTALLIZATION STRATEGY

The strategy employed to bring about crystallization is to guide the system very slowly toward a state of reduced solubility by modifying the properties of the solvent. This is accomplished by increasing the concentration of precipitating agents, or by altering some physical property, such as pH. In this way a limited degree of supersaturation may be

TABLE 2.1 Strategies for creating supersaturation

1. Direct mixing to immediately create a supersaturated condition (batch method)
2. Alter temperature
3. Alter salt concentration (salting in or out)
4. Alter pH
5. Add a ligand that changes the solubility of the macromolecule
6. Alteration of the dielectric of the medium
7. Direct removal of water (evaporation)
8. Addition of a polymer that produces volume exclusion
9. Addition of a cross-bridging agent
10. Concentration of the macromolecule
11. Removal of a solubilizing agent

achieved. Table 2.1 lists a number of strategies. Whatever the procedure used, no effort must be spared in refining the parameters of the system, solvent and solute, to encourage and promote specific bonding interactions between molecules and to stabilize them once they have formed. This latter aspect of the problem generally depends on the chemical and physical properties of the particular protein or nucleic acid being crystallized.

In very concentrated solutions the macromolecules may aggregate as an amorphous precipitate. This result is to be avoided if possible and is indicative that supersaturation has proceeded too extensively or too swiftly. One must endeavor to approach very slowly the point of inadequate solvation and thereby allow the macromolecules sufficient opportunity to order themselves in a crystalline lattice.

For a specific protein, the precipitation points, or solubility minima, are usually dependent on the pH, temperature, the chemical composition of the precipitant, and the properties of both the protein and the solvent. As shown in Figure 2.3, at very low ionic strength a phenomenon known as "salting-in" occurs in which the solubility of the protein rises as the ionic strength increases from zero. The physical effect that diminishes solubility at very low ionic strength is the absence of ions essential for satisfying the electrostatic requirements of the protein molecules. As the ions are removed, and in this region of low ionic strength cations are most important (Cohn and Ferry, 1950; Czok and Buecher, 1960), the protein molecules seek to balance their electrostatic requirements through interactions among themselves. Thus they tend to aggregate and separate from solution. Alternatively, one could say that the chemical activity of the protein is reduced at very low ionic strength. The salting-in effect, when applied in the direction of reduced ionic strength, can itself be used as a crystallization tool. In practice, one extensively dialyzes a protein that is soluble at moderate ionic strength against distilled water. Many proteins have been crystallized by this means.

As ionic strength, in Figure 2.3, is increased, the solution again reaches a point where the solute molecules begin to separate from solvent and preferentially form self-interactions among themselves that result in crystals or precipitate. The explanation for this "salting-out" phenomenon is that the salt ions and macromolecules compete for the attention of solvent molecules, that is, water. Both the salt ions and the protein molecules require hydration layers to maintain their solubility. When competition between ions and proteins becomes sufficiently intense, the protein molecules begin to self-associate in order to satisfy, by intermolecular interactions, their electrostatic requirements. Thus dehydration, or the elimination and perturbation of solvent layers around protein molecules, induces insolubility.

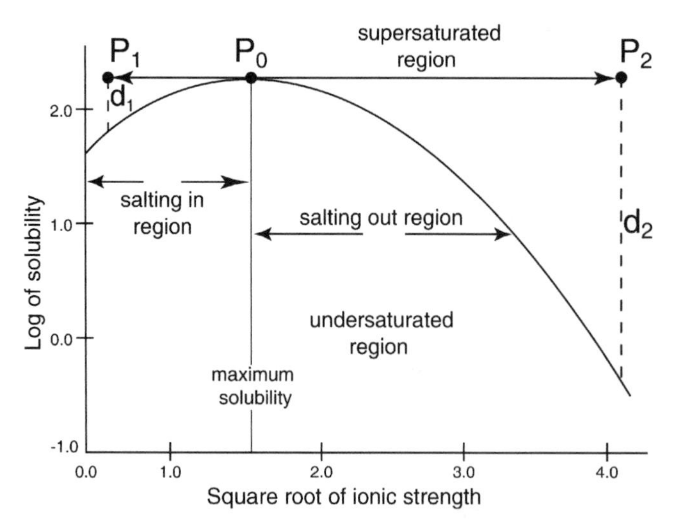

FIGURE 2.3 The solubility of a protein is indicated by the curve, which has a maximum in this case at about 1.5 M ionic strength. To the left of the maximum, at lower ionic strength, is what is termed the "salting-in" region, and to the right, at higher ionic strength, is the "salting-out" region. Supersaturation may be attained by bringing the protein solution to point P_0, clarifying the solution of solid, and then removing ions, thereby moving the solution into the salting in regime to point P_1. Supersaturation may be brought about by adding salt and transforming the protein solution into the salting out region to point P_2. d_1 and d_2 indicate the degree of supersaturation attained in the two cases.

Just as proteins may be driven from solution at constant pH and temperature by the addition or removal of ions, as illustrated in Figure 2.4, they can similarly be crystallized or precipitated at constant ionic strength by changes in pH or temperature. This is because the electrostatic character of the macromolecule, its surface features, or its conformation may change as a function of pH, temperature, and other variables as well. By virtue of its ability to inhabit a range of states, proteins may exhibit a number of different solubility minima as a function of the variables, and each of these minima may afford the opportunity for crystal formation. Thus we may distinguish the separation of protein from solution according to methods based on variation of precipitant at constant pH and temperature from those based on alteration of pH, temperature, or some other variable at constant precipitant concentration. The principles described here for salting-out with a true salt are not appreciably different if polymeric precipitating agents are used. In practice, proteins may equally well be crystallized from solution by increasing the poly(ethylene glycol) concentration at constant pH and temperature, or at constant poly(ethylene glycol) concentration by variation of pH or temperature (McPherson, 1976, 1985).

The most common approach to crystallizing macromolecules, be they proteins or nucleic acids, is to alter gradually the characteristics of a highly concentrated protein solution to achieve a condition of limited supersaturation. As discussed above, this may be achieved by modifying some physical property, such as pH or temperature, or through equilibration with precipitating agents. The precipitating agent may be a salt such as ammonium sulfate, an organic solvent such as ethanol or methylpentanediol, or a highly soluble synthetic polymer such as poly(ethylene glycol). The three types of precipitants act by slightly different mechanisms, although all share some common properties.

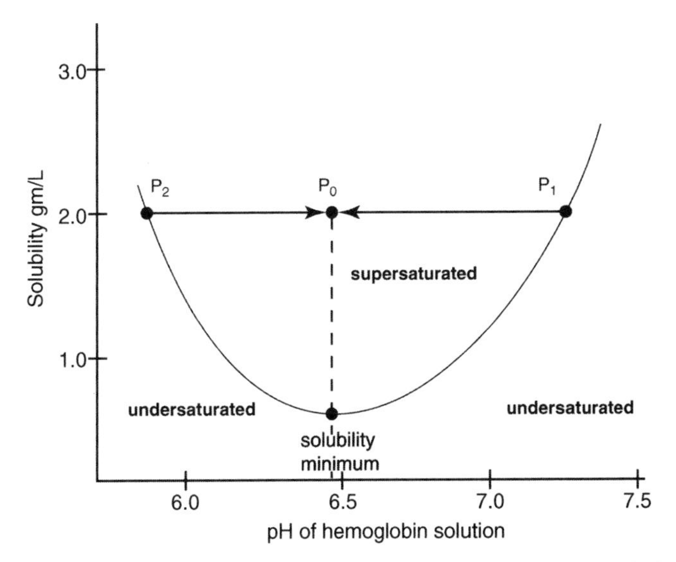

FIGURE 2.4 The solubility of hemoglobin as a function of pH at constant ionic strength and temperature is shown by the curve. Hemoglobin displays a sharp solubility minimum at about pH 6.5, but is freely soluble at both lower and higher pH. A supersaturated hemoglobin solution (or that of many other proteins) may be created by making a saturated protein solution at high (point P_1) or low (point P_2) pH, clarifying the solution by filtration or centrifugation, and then gradually altering the pH so that the solution is transformed to the point P_0. The length of the vertical dashed line is a measure of the degree of supersaturation attained.

In highly concentrated salt solutions competition for water exists between the ions and polyionic protein molecules. The degree of competition will depend on the surface charge distribution of the protein as well as the propensity of the salt ions to bind water. The former is primarily a function of pH. Because protein molecules must bind water to remain solvated, when deprived of sufficient water by ionic competition, they are compelled to associate with other protein molecules. Aggregates may be random in nature and lead to linear and branched oligomers, and eventually to precipitate. When the process proceeds in a systematic fashion and specific chemical interactions are used in a repetitive and periodic manner to give ordered, three-dimensional aggregates, then the nuclei of crystals will form and grow.

The reduction of available solvent by addition of precipitant is, in principle, no different than the crystallization of sea salt from tidal pools as the heat of the sun slowly drives the evaporation of water. It is a form of dehydration but without physical removal of water. A similar effect may be achieved by the slow addition to the mother liquor of certain organic solvents such as ethanol or methyl pentanediol. The only essential requirement for the precipitant is that at the specific temperature and pH of the experiment, the additive does not adversely affect the structure and integrity of the protein. This is often a very stringent requirement and deserves more than a little consideration. Organic solvent competes, to some extent, like salt for water molecules, but it also reduces the dielectric screening capacity of the intervening solvent. Reduction of the bulk dielectric increases the effective strength of the electrostatic forces that allow one protein molecule to be attracted to another.

Polymers such as poly(ethyleneglycol) also serve to dehydrate proteins in solution as do salts, and they alter somewhat the dielectric properties in a manner similar to organic solvents. They produce, however, an additional important effect. Poly(ethyleneglycol) perturbs the natural structure of the solvent and creates a more complex network having both water and itself as structural elements. The underlying basis for the solvent exclusion effect is that polymeric precipitants, such as poly(ethylene glycol), are not like proteins, as they lack any fixed or consistent conformation. They writhe and twist randomly in solution, exhibit a large hydrodynamic radius, and occupy far more space than they otherwise deserve. This results in less solvent available space for the other macromolecules, which then segregate, aggregate, and ultimately form a solid state, often crystals.

Crystallization of macromolecules may also be accomplished by increasing the concentration of a precipitating agent to a point just below supersaturation, and then adjusting the pH or temperature to reduce the solubility of the protein. Modification of pH can be accomplished very well with the vapor diffusion technique (described below) when volatile acids and bases such as acetic acid and ammonium hydroxide are used. This process is analogous to saturating boiling water with sugar and then cooling it to produce rock candy.

SCREENING AND OPTIMIZATION

There are really two phases in the pursuit of protein crystals for an X-ray diffraction investigation, and these are (1) the identification of chemical, biochemical, and physical conditions that yield some crystalline material, though it may be entirely inadequate, and (2) the systematic alteration of those initial conditions by incremental amounts to obtain optimal samples for diffraction analysis. The first of these is fraught with the greater risk, as some proteins simply refuse to form crystals, and any clues as to why are elusive or absent. The latter, however, often proves to be the more demanding, time-consuming, and frustrating.

There are basically two approaches to screening for crystallization conditions. The first is a systematic variation of what are believed to be the most important variables, precipitant type and concentration, pH, temperature, and so forth. An example of this strategy is shown in Figure 2.5. The second is what we might term a shotgun approach, but a shotgun aimed with intelligence, experience, and accumulated wisdom. While far more thorough in scope, and more congenial to the scientific mind, the first method usually does require a significantly greater amount of protein. In those cases where the quantity of material is limiting, it may simply be impractical. The second technique provides much more opportunity for useful conditions to escape discovery, but in general requires less precious material.

The second approach also has, presently at least, one other major advantage, and that is convenience. There is currently on the commercial market, from numerous companies, a wide variety of crystallization screening kits. The availability and ease of use of these relatively modestly priced kits, which may be used in conjunction with a variety of crystallization methods (hanging and sitting drop vapor diffusion, dialysis, etc., see below) make them the first tool of choice in attacking a new crystallization problem. With these kits, nothing more is required than combining a series of potential crystallization solutions with one's protein of interest using a micropipette, sealing the samples, and waiting for success to smile. Often it does, but sometimes not, and this is when the crystal grower must begin using his own intelligence to diagnosis the problem and devise a remedy.

Once some crystals, even if only microcrystals, are observed and shown to be of protein origin (and one ardently hopes for this event) then optimization begins. Every component in the solution yielding crystals must be noted and considered (buffer, salt, ions, etc.), along

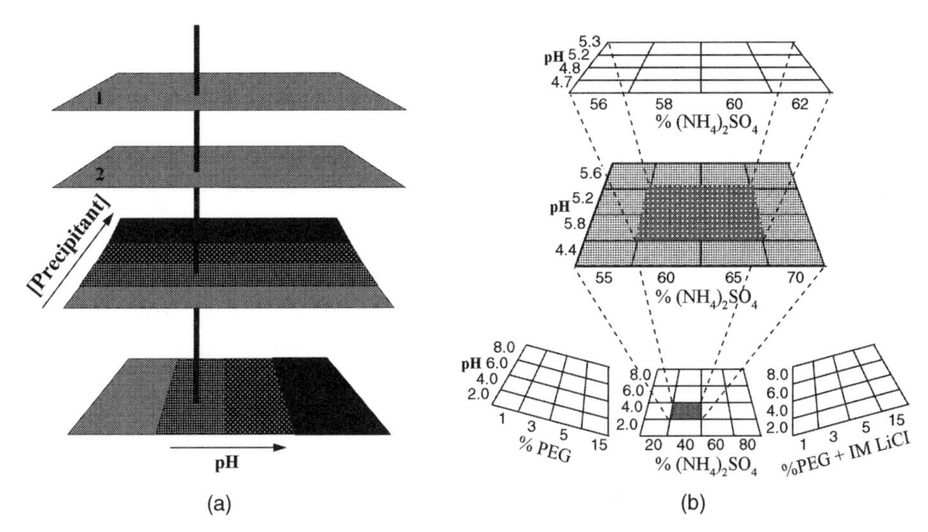

FIGURE 2.5 Diagram of the successive automated grid search strategy for protein crystallization (Cox and Weber, 1988). In (*a*) components of the grid are displayed separately. The bottom square shows the variation in pH across the columns. The square above it shows the variation in precipitant concentration in the rows. The combination of these two layers produces the pH versus precipitant grid that serves as the basis for the two dimensional crystallization strategy. Fixed concentrations of other reagents can be added onto this grid as indicated by the upper squares labeled 1 and 2. The diagram in (*b*) illustrates how solution parameters are chosen using the approach for protein crystallization. Broad screen experiments (shown at the bottom) are set up using three different precipitating agents. Focused ranges of pH and precipitant concentration are centered about droplets containing crystals.

with pH, temperature, and whatever other factors (see below) might have an impact on the quality of the results. Each of these parameters or factors is then carefully incremented in focused trial matrices encompassing a range spanning the conditions which gave the "hit." Because the problem is nonlinear, and one variable may be coupled to another, this process is often more complex and difficult than one might expect (McPherson, 1982, 1999; Ducruix and Giege, 1992; Bergfors, 1999 McPherson). It is here that the amount of protein and the limits of the investigator's patience could prove a formidable constraint.

CREATING THE SUPERSATURATED STATE

Crystallization of a novel protein using any method is unpredictable as a rule. Every macromolecule is unique in its physical and chemical properties because every amino acid or nucleotide sequence produces a unique three-dimensional structure having distinctive surface characteristics. Thus lessons learned by investigation of one protein are only marginally applicable to others. This is compounded by the behavior of macromolecules, which is complex owing to the variety of molecular masses and shapes, aggregate states, and polyvalent surface features that change with pH and temperature, and to their dynamic properties.

Because of the intricacy of the interactions between solute and solvent, and the shifting character of the protein, the methods of crystallization must usually be applied over a broad

TABLE 2.2 Methods for attaining a solubility minimum

1. Bulk crystallization
2. Batch method in vials
3. Evaporation
4. Bulk dialysis
5. Concentration dialysis
6. Microdialysis
7. Liquid bridge
8. Free interface diffusion
9. Vapor diffusion on plates (sitting drop)
10. Vapor diffusion in hanging drops
11. Sequential extraction
12. pH-induced crystallization
13. Temperature-induced crystallization
14. Crystallization by effector addition

set of conditions with the objective of discovering the particular, minimum (or minima) that yield crystals. In practice, one determines the precipitation points of the protein at sequential pH values with a given precipitant, repeats the procedure at different temperatures, and then examines the effects of different precipitating agents.

There are a number of devices, procedures, and methods for bringing about the supersaturation of a protein solution, generally by the slow increase in concentration of some precipitant such as salt or poly(ethylene glycol). The most popular of these are listed in Table 2.2. Many of these same approaches can be used as well for salting-in, modification of pH, and the introduction of ligands that might alter protein solubility. These techniques have been reviewed elsewhere (McPherson, 1976, 1982, 1999, 2004; Bergfors, 1999; Ducruix and Giege, 1992) and will not be dealt with exhaustively here. Only three of these, microdialysis, free interface diffusion, and vapor equilibration will be described as examples of the methods in most common use.

Dialysis is familiar to nearly all biochemists as a means of changing properties of a protein-containing solution. The macromolecule solution is maintained inside a membrane casing or container having a semipermeable membrane partition. The membrane allows, through its pores, the passage of small molecules and ions, but the pore size excludes passage of the much larger protein molecules. The vessel or dialysis tube containing the protein is submerged in a larger volume of liquid having the desired solution properties of pH, ionic strength, ligands, and so forth. With successive changes of the exterior solution and concomitant equilibration of small molecules and ions across the semipermeable membrane, the protein solution gradually acquires the desired properties of the exterior fluid.

Exactly the same procedure, in some manifestation or other, can and has been used to crystallize a number of proteins on a bulk scale. It is generally applicable on a large scale, however, only when substantial amounts of the protein are available. It has the advantage that by liquid–liquid diffusion through a semipermeable membrane, a protein solution can be exposed to a continuum of potential crystal-producing conditions without actually altering directly the mother liquor. Diffusion through the membrane is slow and controlled. Because the rate of change of constituents in the mother liquor is proportional to the gradient of concentrations across the membrane, the nearer the system approaches equilibrium, the more slowly it changes.

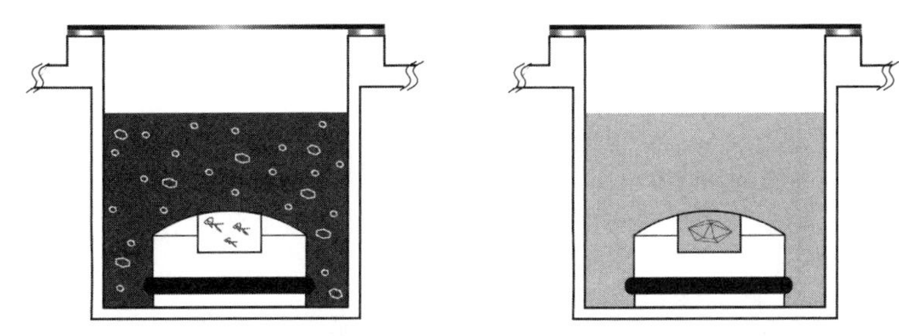

FIGURE 2.6 A small amount of protein solution is confined by dialysis membrane within a **5** to **50 μl** cavity bored into a dome-shaped plastic button. The button is then submerged into a larger volume of external fluid. By exchange of ions and small molecules across the membrane, the protein solution is brought to a state of supersaturation.

This method has been adapted to much smaller amounts of protein by crystallographers who now use almost exclusively microtechniques involving no more than **5** to **50 μl** of protein solution in each trial (Zeppenzauer et al. 1968; Zeppenzauer, 1971). By this method a small amount of protein solution is confined within a glass capillary, or the microcavity of a small plexiglass button. The cavity within the button, or the ends of the microcapillary tube, are then closed off by a semipermeable dialysis membrane. The whole arrangement, charged with protein solution, is then submerged in a much larger volume of an exterior liquid, and the whole system is kept within a closed vessel such as a test tube or vial. If the exterior solution is at an ionic strength or pH that causes the mother liquor to become supersaturated, crystals may grow. If not, the exterior solution may be exchanged for another and the experiment continued.

The dialysis buttons, seen in Figure 2.6 are particularly ingenious. Not only are they compact and easy to examine, but they have a shallow groove about their waist. After a section of wet dialysis membrane is placed over the mother liquor filled cavity, it can be held firmly and precisely in place by simply slipping a common rubber **O** ring over the top of the button and seating it in the groove. The buttons are now in wide use, and have proved themselves successful. Their cavities range in size from **5** to **50 μl** and then can be reused many times.

A modification of the liquid–liquid diffusion method is the free interface diffusion technique (Weber and Goodkin, 1970; Salemme, 1972) illustrated in Figure 2.7. Here, the membrane is dispensed with completely, and the mother liquor is simply layered upon a second precipitating solution in a glass tube or capillary. In some applications the bottom solution is first frozen before the second is layered to ensure a sharp demarcation between the two. In the free-interface diffusion method, direct diffusive and convective mixing at the interface generates concentration gradients that produce regions of local supersaturation. These can in turn yield nuclei that may grow to a size useful for diffraction analysis.

Currently the most widely used method for bringing about supersaturation in microdrops of protein mother liquid is vapor diffusion (Hampel et al., 1968; McPherson, 1976, 1999, 2004). This approach may be divided into those procedures that use a “sitting drop” and those employing a “hanging drop.” Whatever its form, the method relies on the transport of either water or some volatile agent between a microdrop of mother liquor, generally **1** to **10 μl** volume, and a much larger reservoir solution of **0.75** to **25 ml** volume. Through the vapor phase, the droplet and reservoir come to equilibrium. Because the reservoir is of

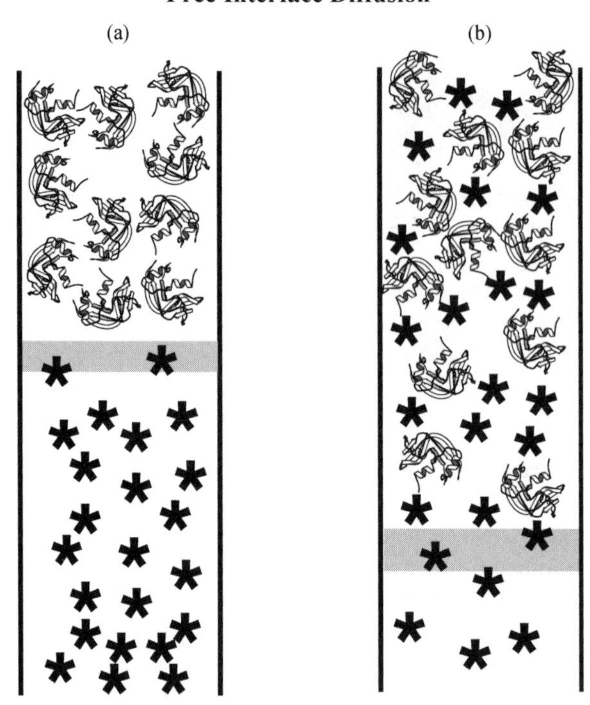

FIGURE 2.7 With the free interface diffusion method illustrated schematically here, the protein sample, in buffer, is simply layered, with care, atop the precipitant solution, which may be either salt or polyethylene glycol. Salt ions diffuse rapidly into the protein solution aided by convective transport, and local concentration gradients are created in the region of the interface. With polymeric precipitants, both the polymer and the protein diffuse into one another, but at a greatly reduced rate.

much larger volume, the final equilibration conditions are essentially those of the initial reservoir state.

Through the vapor phase, then, water is removed slowly from the droplet of mother liquor, its pH may be changed, or volatile solvents such as ethanol may be gradually introduced. As with the liquid–liquid dialysis and free-interface diffusion methods, the procedure may be carried out at a number of different temperatures to gain advantage of that parameter as well.

In the popular "sitting drop" method, illustrated in Figure 2.8 a drop of protein containing mother liquor, **1** to **5 μl** in volume, is dispensed onto a shallow depression in the chamber of a plastic plate. The chamber is contiguous with a reservoir containing precipitant at higher concentration, or at a different pH. Through the vapor phase the concentration of salt or organic solvent in the reservoir equilibrates with that in the sample. In the case of salting out, the droplet of mother liquor must initially contain a level of precipitant lower than the reservoir, and equilibration proceeds by distillation of water out of the droplet and into the reservoir. This holds true for nonvolatile organic solvents, such as methylpentanediol, and for poly(ethylene glycol) as well. In the case of volatile precipitants, none need be added initially to the microdroplet, as distillation and equilibration proceed in the opposite direction.

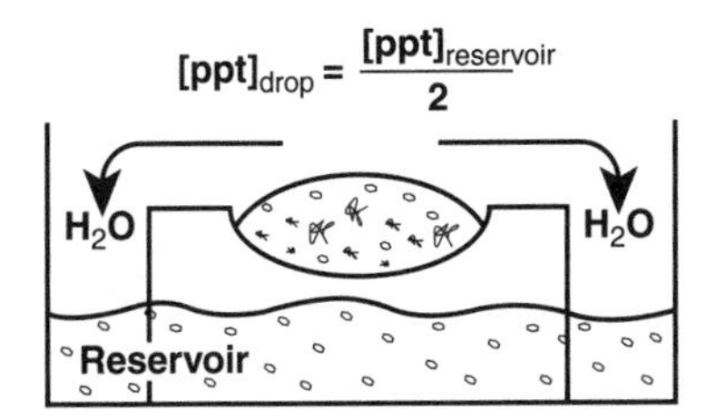

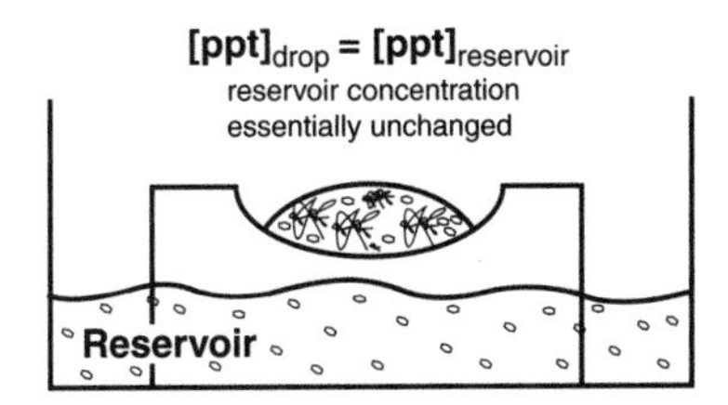

FIGURE 2.8 In sitting drop vapor diffusion, a small volume of protein sample, combined with, generally but not necessarily, an equal volume of reservoir solution to make a drop of **1** to **10μl**. The drop is dispensed onto a small platform, which is contiguous through the vapor phase with a reservoir of much larger volume. Through the vapor phase, the microdroplet establishes equilibrium with the reservoir. By loss of water, as well as by concentration of the protein, the droplet is brought to a state of supersaturation.

In recent years there has been an increased emphasis on large-scale crystallization screening experiments involving both large numbers of test conditions, and often large numbers of proteins. In addition, as the proteins addressed by X-ray crystallography have become increasingly more difficult to produce and purify, a premium is now placed on carrying out the crystallization trials with a small amount of material. In response to those pressures, efficient robotic systems have been designed, and are on the market that efficiently and accurately pipette droplets of mother liquor in the nanoliter range. The type of sitting drop plate used by some robotic systems is shown schematically in Figure 2.9. These systems are generally accompanied by automated photo analysis systems that also speed the

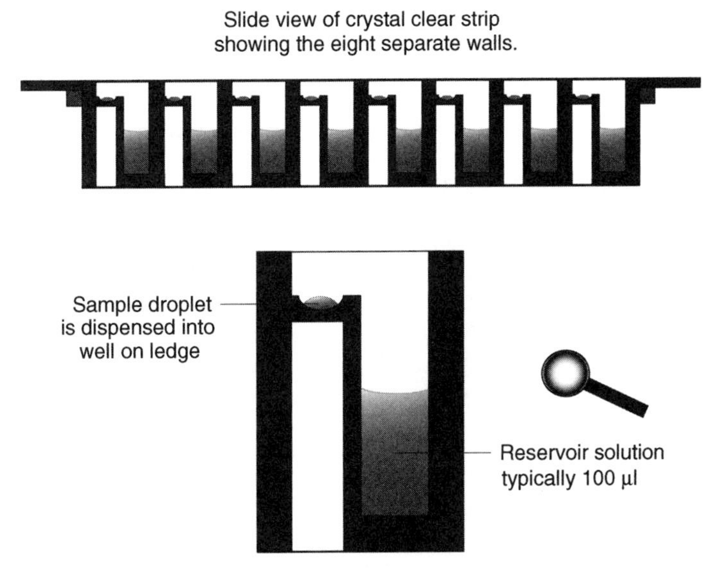

FIGURE 2.9 Robotic systems commonly use 96 well plates for crystal trial arrays. One of these is shown schematically in cross section. The drop, which may be **1 μl** or less, "sits" on the small shelf at the top of the chamber, while the deep well is filled with the precipitant solution. The plates are sealed with clear plastic tape.

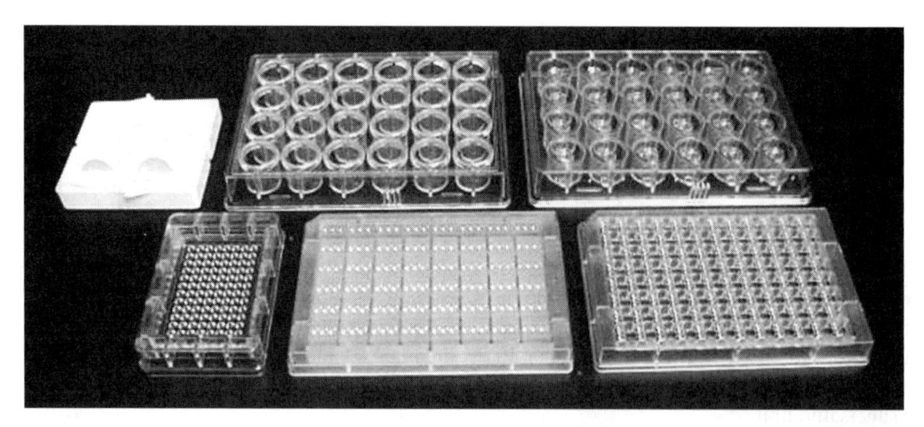

FIGURE 2.10 An array of commercially available and commonly used plastic plates for both sitting and hanging drop vapor diffusion crystallization. Also in the picture is a box of silicone coated cover slips for hanging drops. Courtesy of Hampton Research.

examination and evaluation of trial conditions. With these systems, hundreds of trials per day can be deployed, observed, and recorded with virtually no human intervention. When clear plastic plates like those in Figure 2.9 are used, large numbers of samples can be quickly inspected for crystals under a dissecting microscope and conveniently stored. The plastic plates and instructions for their use are now widely, commercially available. Some of these are shown in Figure 2.10.

The "hanging drop" procedure, illustrated in Figure 2.11, also uses vapor phase equilibration, but with this approach, a micro droplet of mother liquor (as small as **1 μl**) is suspended from the underside of a microscope cover slip, which is then placed over a small well

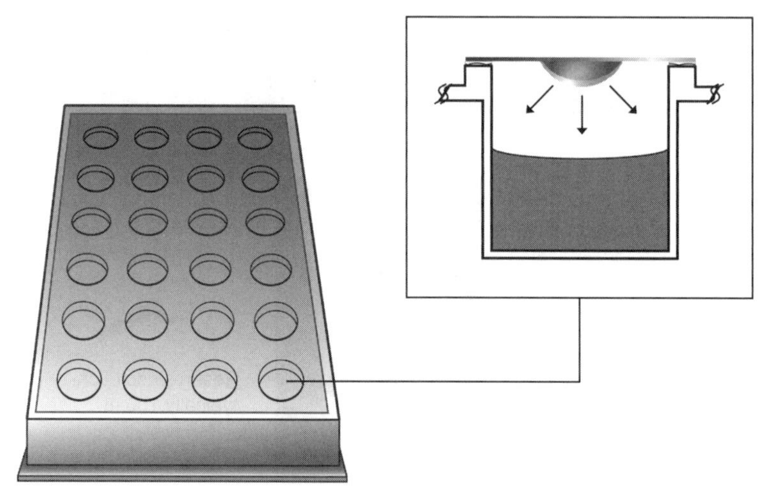

FIGURE 2.11 Vapor diffusion by hanging drop is essentially identical to that using sitting drops, which is illustrated in Figure 2.8. The major difference is that the protein containing droplet hangs from the underside of a silicone coated coverslip, which may be of either glass or plastic. The reservoir chambers are provided by 24 well plastic plates.

containing **0.3** to **1 ml** of the precipitating solution. The wells are most conveniently supplied by disposable plastic plates that have 24 wells with rims that permit sealing by application of silicone vacuum grease around the circumference. These plates provide the further advantages that they can be swiftly and easily examined under a dissecting microscope and they allow compact storage. The hanging drop technique can be used both for the screening of conditions and for the growth of large single crystals.

While the principle of equilibration with both the "sitting drop" and the "hanging drop" are essentially the same, they frequently do not yield the same results even though the reservoir solutions and protein solutions are identical. Because of differences in the apparatus used to achieve equilibration, the path to equilibrium is different even though the end point may be the same. In some cases there are striking differences in the degree of reproducibility, final crystal size, morphology, required time, or degree of order. These observations illustrate the important point that the pathway leading to supersaturation, the kinetics of the process, may be as important as the final point achieved.

As noted earlier, one of the most powerful techniques for producing a supersaturated protein solution is adjustment of the pH to values where the protein is substantially less soluble. This may be done in the presence of a variety of precipitants so that a spectrum of possibilities can be created whereby crystals might form. The gradual alteration of pH is particularly useful because it may be accomplished by a variety of gentle approaches that do not otherwise perturb the system or introduce unwanted effects. Chief among these are dialysis and vapor diffusion.

As with pH, proteins may vary in solubility as a function of temperature, and some are quite sensitive. One can take advantage of this property with both bulk and microtechniques (Jacoby, 1968; McPherson, 1999). Many of the earliest examples of protein crystallization were based on the formation of concentrated solutions at elevated temperatures followed by slow cooling. Osborne in 1892 successfully crystallized over 20 plant seed globulins by cooling relatively crude extracts from **60°C** to room temperature in the presence of varying concentrations of sodium chloride.

Most protein and nucleic acids are conformationally flexible or exist in several conformational equilibrium states. In addition they may assume a substantially different conformation when they have bound coenzyme, substrate, or other ligand. Frequently a protein with bound effector may exhibit appreciably different solubility properties than the native protein. In addition, if many conformational states are available, the presence of effector may be used to select for only one of these, thereby engendering a degree of conformity of structure and system microhomogeneity that would otherwise be absent.

The effect of ligands can be employed to induce supersaturation and crystallization in those cases where its binding to the protein produces solubility differences under a given set of ambient conditions. The effector may be slowly and gently combined with the protein, for example, by dialysis, so that the resulting complex is at a supersaturating level. Complexation of ligands, substrates, and other small molecules has seen widespread use in protein crystallography, since it provides attractive alternatives if the apoenzyme itself cannot be crystallized.

PRECIPITATING AGENTS

Protein precipitants fall into four broad categories: (1) salts, (2) organic solvents, (3) long-chain polymers, and (4) low-molecular-mass polymers and nonvolatile organic compounds.

TABLE 2.3 Precipitants used in macromolecular crystallization

Salts	Volatile Organic Solvents	Polymers	Nonvolatile Organic Solvents
Ammonium phosphate sulfate	Ethanol	Poly(ethylene glycol) 1000, 3350, 6000, 8000, 20,000	2-Methyl-2, 4-pentanediol
Lithium sulfate	Propanel and isopropanol	Jeffamine T, Jeffamine M	2,5-Hexandediol
Sodium or ammonium citrate	1,3-Propanediol	Poly(ethylene glycol) monomethyl ester	Ethylene glycol 400
Sodium or potassium phosphate	2-Methyl-2, 4-pentanediol	Poly(ethylene glycol) monostearate	
Sodium or potassium or ammonium chloride	Dioxane	Polyamine	
Sodium or ammonium acetate	Acetone		
Magnesium or calcium sulfate	Butanol		
Cetyltriethyl ammonium salts	Acetonitrile		
Calcium chloride	Dimethyl sulfoxide		
Ammonium or sodium nitrate			
Sodium or magnesium formate	Methanol		
Sodium or potassium tartrate	1,3-Butyrolactone		
Cadmium sulfate	Ethylene glycol 400		

The first two classes are typified by ammonium sulfate and ethyl alcohol, respectively, and higher polymers such as poly(ethylene glycol) 3350 are characteristic of the third. In the fourth category we might place compounds such as methylpentanediol and low-molecular-mass poly(ethylene glycol). Common members of the four groups are presented in Table 2.3.

As described above, salts exert their effect by dehydrating proteins through competition for water molecules. Their ability to do this is proportional to the square of the valences of the ionic species composing the salt (Hofmeister, 1887; Cohn and Ferry, 1950). Thus multivalent ions, particularly anions are the most efficient precipitants. Sulfates, phosphates, and citrates have traditionally been employed with success. More recently, it was shown (McPherson, 2001) that the salts of organic acids such as malonate, citrate, succinate, and tartrate, because of the potentially hydroden bonding carboxyl groups, as well as their charges, are very effective crystallization salts. They are also, it might be added, good cryo-protective agents at high concentrations.

One might expect little variation between different salts so long as their ionic valences were the same, or that there would be little variation with two different sulfates such as Li_2SO_4 and $(NH_4)_2SO_4$. This, as indicated by Figure 2.12, is often not the case. In addition to salting out, which is a general dehydration effect, or lowering of the chemical activity of

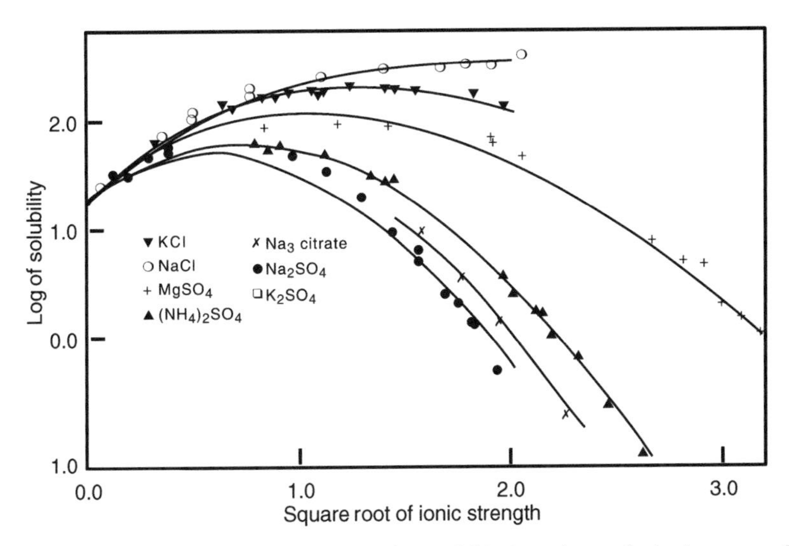

FIGURE 2.12 The solubility of carbonmonoxyhemoglobin in various salts is shown as a function of ionic strength. The regions of the end points of the curve where solubility decreases are called, at low ionic strength, the "salting-in" region, and at high ionic strength, the "salting-out" region. This diagram emphasizes the point that the solubility is not strictly a function of ionic strength but depends as well on the specific ions involved.

water, there are specific protein–ion interactions that could have other consequences. This is particularly true because of the unique polyvalent character of individual proteins, their structural complexity, and the intimate dependence of their physical properties on environmental conditions and interacting molecules. When attempting to crystallize a protein it is therefore never sufficient to examine only one or two salts and ignore a broader range. Changes in salt can sometimes produce crystals of varied quality, morphology, and in some cases diffraction properties.

It is usually not possible to predict the degree of saturation or molarity of a salt required for the crystallization of a particular protein without some prior knowledge of its behavior. In general, however, the concentration is just a small percentage less than that which yields an amorphous precipitate.

The most common organic solvents utilized for crystallization have been ethanol, acetone, butanols, and a few other common laboratory reagents. It seems true that organic solvents have been of more general use for the crystallization of nucleic acids, particularly tRNA and the duplex oligonucleotides. There they have been the primary means for crystal growth. This in part stems from the greater tolerance of polynucleotides to organic solvents and their poly anionic surfaces, which appear to be even more sensitive to dielectric effects than are proteins.

The only general rules are that organic solvents should be used at a low temperature, at or below **0°C**, and they should be added very slowly and with good mixing. Since most are volatile, vapor diffusion techniques are entirely applicable. Ionic strength should, in general, be maintained as low as possible and whatever means are available should be taken to protect against denaturation.

Poly(ethylene glycol) is a polymer produced in various lengths, containing from several to many hundred monomers. It exhibits as its most conspicuous feature a regular alteration of ether oxygens and terminal glycols. In addition to its volume exclusion property, it shares some characteristics with salts that compete for water and produce dehydration, and with organic solvents that reduce the dielectric properties of the medium. Poly(ethylene glycol) is produced in a variety of polymer size ranges. The low-molecular-mass species are oily liquids while those of **Mr** above **1000**, at room temperature, exist as either waxy solids or powders. The size specified by the manufacturer is the mean **Mr** of the polymeric molecules and the distribution about that mean may vary appreciably.

A distinct advantage of poly(ethylene glycol) over other agents is that most proteins crystallize within a fairly narrow range of poly(ethylene glycol) concentration, this being about **4%** to **18%**. The dependence of solubility on poly(ethylene glycol) concentration for a number of proteins is shown in Figure 2.13. In addition the exact poly(ethylene glycol) concentration at which crystals form is rather insensitive and if one is within **5%** to **10%** of the optimal value some success may be achieved. With most crystallizations from high ionic-strength solutions or from organic solvents, one must be within **1%** and **2%** of an optimum lying anywhere between **10%** to **85%** saturation. The advantage of poly(ethylene glycol) is that when conducting a series of initial trials to determine what conditions will give crystals, one can use a fairly coarse selection of concentrations and over a rather narrow total range. This means fewer trials with a corresponding reduction in the amount of protein expended.

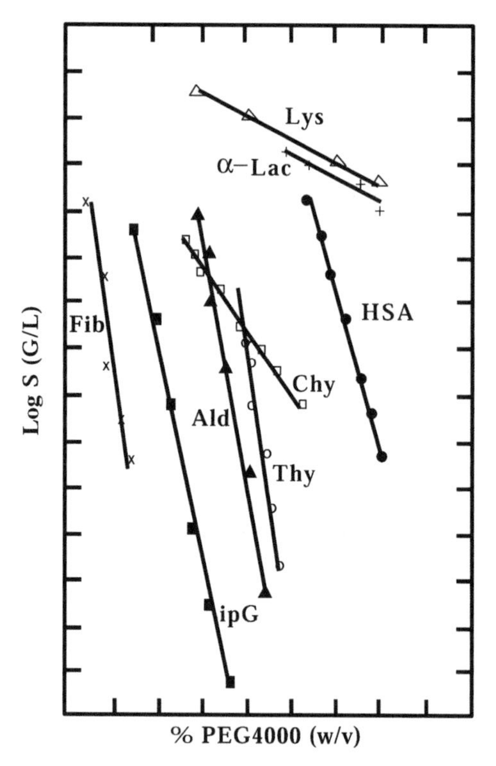

FIGURE 2.13 The solubility of a number of proteins as a function of poly(ethylene glycol) concentration. The dependence is essentially linear, generally steep, and lacks the "salting-in" and "salting-out" regions characteristic of salts. Measurements were made in 0.05 M potassium phosphate, pH 7, containing 0.1 M KCl (Atha and Ingham, 1981).

Since poly(ethylene glycol) solutions are not volatile, this precipitant must be used like salt and equilibrated with the protein by slow mixing or vapor equilibration. This latter approach, utilizing either hanging drops over 0.5 ml reservoirs, or sitting drops in plastic plates, has proved the most popular. When the reservoir concentration is in the range **5%** to **20%**, the protein solution to be equilibrated should be at an initial concentration of about half of that, which is conveniently obtained by adding an equal volume of the reservoir to that of the protein solution.

FACTORS INFLUENCING PROTEIN CRYSTAL GROWTH

Table 2.4 presents some, but probably not all, physical, chemical, and biological variables that may influence to a greater or lesser extent the crystallization of proteins. The difficulty in properly arriving at a just assignment of importance for each factor is substantial for several reasons. Every protein is different in its properties, and surprisingly perhaps, this applies even to proteins that differ by no more than one or just a few amino acids. There are even cases where the identical protein prepared by different procedures or at different times show significant variations. In addition each factor may differ considerably in importance for individual proteins.

Because each protein is unique, there are few means available to predict in advance the specific values of a variable, or sets of conditions that might be most profitably explored. Finally, the various parameters under one's control are not independent of one another and their interrelations may be complex and difficult to discern. It is therefore not easy to elaborate rational guidelines relating to physical factors, or ingredients in the mother liquor

TABLE 2.4 Factors effecting crystallization

Physical	Chemical	Biochemical
1. Temperature/temperature variation	1. pH	1. Purity of the macromolecule/ impurities
2. Surfaces	2. Precipitant type	2. Ligands, inhibitors, effectors
3. Methodology/approach to equilibrium	3. Precipitant concentration	3. Aggregation state of the macromolecule
4. Gravity	4. Ionic strength	4. Posttranslational modifications
5. Pressure	5. Specific ions	5. Source of macromolecule
6. Time	6. Degree of supersaturation	6. Proteolysis/hydrolysis
7. Vibrations/sound/ mechanical perturbations	7. Reductive/oxidative environment	7. Chemical modifications
8. Electrostatic/magnetic fields	8. Concentration of the macromolecules	8. Genetic modifications
9. Dielectric properties of the medium	9. Metal ions	9. Inherent symmetry of the macromolecule
10. Viscosity of the medium	10. Crosslinkers/polyions	10. Stability of the macromolecule
11. Rate of equilibration	11. Detergents/surfactants/ amphophiles	11. Isoelectric point
12. Homogeneous or heterogeneous nucleants	12. Nonmacromolecular impurities	12. History of the sample

that can increase the probability of success in crystallizing a particular protein. The specific components and conditions must be carefully deduced and refined for each individual.

As already noted, temperature may be of crucial importance or it may have little bearing at all. In general, it is wise to duplicate crystallization trials and conduct parallel investigations at **4°C** and at **25°C**. Even if no crystals are observed at either temperature, differences in the solubility behavior of the protein with different precipitants and with various effector molecules, may give some indication as to whether temperature is likely to play an important role. If crystals are observed to grow at one temperature and not, under otherwise identical conditions, at the another, then further refinement of this variable may be justified. This is accomplished by conducting the trials under the previously successful conditions over a range of temperatures centered on the one that initially yielded crystals.

In general, the solubility of proteins as a function of temperature is more sensitive at low ionic strength than at high, explaining why the use of temperature is usually more successful in crystallizing proteins from poly(ethylene glycol) containing solutions, than from those containing high concentrations of salt. Otherwise, the only rules with regard to temperature seem to be that proteins in a high salt solution are usually more soluble at cold than warmer temperatures. Proteins, however, generally precipitate or crystallize from a lower concentration of poly(ethylene glycol), methylpentanediol, or organic solvent at cold than at warmer temperature. One must remember, however, that diffusion rates are less and equilibration occurs more slowly at low than higher temperature, so the times required for precipitation or crystal formation may be longer at colder temperatures.

After precipitant concentration the next most important variable in protein crystal growth appears to be pH. This follows since the charge character of a protein and all of its attendant physical and chemical consequences are intimately dependent on the ionization state of the amino acids or chemical groups that comprise the macromolecule. Not only does the net charge on the protein change with pH, but also the distribution of those charges, the dipole moment of the protein, its conformation, and in many cases its aggregation state. Thus an investigation of the behavior of a specific protein as a function of pH is perhaps the single most important review to carry out when attempting to crystallize a macromolecule.

As with temperature, the procedure is to first conduct multiple crystallization trials at coarse intervals over a broad pH range and then repeat the trials over a finer matrix of values in the neighborhoods of those that initially showed promise. The only limitations on the breadth of the initial range are the points at which the protein begins to show indications of denaturation. In refining the pH for optimal growth, it should be recalled that the difference between amorphous precipitate, microcrystals, and large single crystals may be only a **Δ**pH of less than **0.5**.

In addition to adjusting pH for the optimization of crystal size, it is sometimes also useful to explore the variation of pH as a means of altering the habit or morphology of a crystalline protein. This is occasionally necessary if the initial crystal form is not amenable to analysis because it grows as fine needles or thin plates, or demonstrates some other unfavorable tendency such, as striation or twinning.

SOME USEFUL CONSIDERATIONS

The earliest investigators of protein crystals noted that the concentration of protein in the mother liquor should be as high as possible, **10** to **50 mg/ml**. This is particularly true if one is attempting to grow crystals of a protein for the first time. The probability of obtaining crystals is certainly enhanced by increasing the concentration of protein. Concentration alone is sometimes sufficient to drive the system into a state of supersaturation and into the labile region where stable nuclei can form. This may not, however, be the best approach in growing large, perfect crystals once optimal conditions for all other parameters have been established.

Once conditions for nucleation and growth have been identified and the investigation of variables more or less complete, the concentration of the protein should be gradually reduced in increments to moderate the growth of the crystals. As a rule, the largest and most perfect crystals result when the rate of incorporation of molecules is slow and orderly. Reduction of macromolecule concentration is an effective means for controlling this.

The time required for the appearance and growth of protein crystals is quite variable and may range from a few hours in the best of cases to several months in others. Because no truly systematic investigations have been carried out, how rapidly crystals grow once visible nuclei have formed remains in question. The rate of growth may not be reflected at all in the total amount of time required to obtain crystals adequate for analysis. This includes the time required for solvent equilibration to be achieved, for crystal nuclei to form, and for full growth to occur. When one is screening variables to establish optimal parameters, the practical objective is to promote crystallization at the greatest possible speed to expedite determination of most probable conditions. When optimizing and refining crystallization parameters, time itself becomes an important parameter and longer periods of slow growth are generally desirable.

The most intriguing questions with regard to optimizing crystallization conditions concern what additional components or compounds should comprise the mother liquor in addition to solvent, protein, and precipitating agent (McPherson, 1982, 1999; McPherson and Cudney, 2006). The most probable effectors are those that maintain the protein in a single, homogeneous, and invariant state. Reducing agents such as glutathione or 2-mercaptoethanol are useful to secure sulfhydryl groups and prevent oxidation. EDTA and EGTA are good if one wishes to protect the protein from heavy or transition metal ions, or the alkaline earths. Inclusion of these components may be particularly desirable when crystallization requires a long period of time to reach completion.

When crystallization is carried out at room temperature in poly(ethylene glycol) or low-ionic-strength solutions, attention must be given to preventing the growth of microbes. These generally secrete proteolytic enzymes that may have serious effects on the integrity of the protein under study. Inclusion of sodium azide or thymol at low levels may be necessary to discourage invasive bacteria and fungi.

Substrates, coenzymes, and inhibitors often serve to fix an enzyme in a more compact and stable form. Thus a greater degree of structural homogeneity may be imparted to a population of macromolecules and dynamic behavior suppressed by complexing the protein with a natural ligand before attempting its crystallization. In some cases an apoprotein and its ligand complexes may be significantly different in their physical behavior, and these can, in terms of crystallization, be treated as almost entirely separate problems. This may permit a second or third opportunity for growing crystals if the native protein appears refractile. Thus it is worthwhile, when determining or searching for crystallization conditions, to explore complexes of the macromolecule with substrates, coenzymes, analogues, and inhibitors very early. In many ways such complexes are inherently more interesting in a biochemical sense than the apoprotein when the structure is ultimately determined.

It was noted that microbial growth frequently results in proteolysis of protein samples, something to be avoided. This, however, is not always the case. It has been shown in a number of instances that limited and controlled proteolytic cleavage of a protein can render it crystallizable, while in the native state it was not. In other cases limited proteolysis resulted in a change of crystal form to a more suitable and useful habit. Proteases, it seems, occasionally trim off loose ends or degrade macromolecules to stable, compact domains. These abbreviated proteins are, as a result, more invariant, less conformationally flexible, and they often form crystals more readily than the native precursor. Although one might prefer the intact protein, a partially degraded form sometimes exhibits the activity and physical properties that are of primary interest.

Various metal ions have been observed to induce or contribute to the crystallization of proteins and nucleic acids. In some instances these ions were essential for activity, so it was reasonable to expect that they might aid in maintaining certain structural features of the molecule. In other cases, however, metal ions, particularly divalent metal ions of the transition series, were found that stimulated crystal growth but played no known role in the macromolecules' activity. One of the oldest examples of an animal protein being crystallized is horse spleen ferritin that forms perfect octahedra when a solution containing the protein is exposed to Cd^{++} ions (Laufberger, 1937; Granick, 1941). Metal ions should be included for investigation in that class of additives which might stabilize or engender conformity by specific interaction with the macromolecule.

TYPICAL TRIAL ARRAYS

Initially, the parameters to establish as rapidly as possible are appropriate concentrations for precipitants, optimal pH for solubilization and crystallization, and the effect of temperature. The precipitants that should be examined first are poly(ethylene glycol) 3350 and one or more representatives of salts; probably ammonium sulfate and sodium malonate would be the best choices. They are the two major classes of precipitants in use. If quantity of protein permits than the additional two groups of organic solvents and short chain alcohols (see Table 2.3) should be investigated as well. The best representatives of the latter are ethanol and methylpentanediol, respectively.

Initially, a pH range of **3.5** to **9.0** should be explored in intervals of **0.5**, but the range should be extended, abridged, or modified as appropriate. Generally, it is sufficient to set up two parallel sets of trials and maintain one set at **4°C** and the other at **25°C**. This will provide an indication of the possible influence and value of temperature as a variable. If crystals of any sort are obtained in the first round of trials, then the coarse matrix of conditions is more finely sampled, evaluated, and in successive rounds, the growth of the crystals optimized. If no crystals are obtained, ligand complexes or alternative forms of the protein are explored. If this fails, then effectors such as metal ions and detergents are introduced, and so on.

A consideration in screening crystallization conditions is minimization of the number of trials that must be carried out. Even in those happy cases where the quantity of protein is not a limitation, reduction of trials means less time and effort. Thus one seeks to avoid conditions that are certain to be unprofitable. For example, if the protein is observed to precipitate rapidly at salt concentrations greater than **50%** saturation, or at pH below **5.0**, or at **4°C**, then clearly the trials lying beyond those limits can be eliminated.

The entire strategy of crystallizing proteins is often a process of picking out those areas of variable space that have some chance of yielding success and intuiting those likely to produce failure. A major difficulty in this pursuit is that only a narrow range of conclusions are possible from each crystallization trial. The mother liquor (1) contains some amount of precipitate, (2) it is clear, (3) there is oiling out, or phase separation, (4) large crystals are present, or (5) microcrystals are present. It is always difficult to know how close a trial, or a set of conditions is to success unless crystals are actually present.

Careful examination of precipitates formed in the mother liquor are frequently of some value. Granular precipitate, for example, sometimes is actually microcrystalline when examined under a high-power microscope; a globular or oil-like precipitate often indicates hydrophobic aggregation and suggests the use of detergents; a light, fluffy precipitate is generally a strong negative; a clear trial means a higher precipitation level is needed, or another pH, and so on.

Timing is also important, and when one is carrying out initial trials it is good to examine the crystallization samples frequently, every **12** to **24 h** for the first few days. This way conditions that cause very rapid precipitation or crystal growth can be identified. Once optimal crystal growth conditions have been precisely defined, then that is the time to lay the trials down like fine wine, in a cool, quiet place.

It is wise to pay attention to what might be considered trivial matters. Be certain that the workplace is clean to minimize dust and microbes in the samples. When making a microdroplet, see that it is as hemispherical as possible and does not spread on the glass or plastic to yield a large surface-to-volume ratio. Microfilter protein samples, work quickly to avoid evaporation; do not carry on philosophical conversations while dispensing ingredients. Be alert for unusual events that may later explain anomalous results. Be patient. Examine the volume by McPherson (1999).

THE IMPORTANCE OF PROTEIN PURITY AND HOMOGENEITY

From the perspective of physics there are two important effects to consider about the rate of growth of protein crystals: the transport of molecules to the surface of a growing nucleus, or crystal, and the frequency with which the molecules orient and attach themselves to the growing layer. Crystal growth rates can therefore be considered in terms of transport kinetics and attachment kinetics. For protein crystals, which grow relatively slowly, the transport kinetics are dependent primarily on physical forces and movements in the solution phase, so the transport of molecules is almost certainly the less important of the two, although this effect is sometimes evident. There is not much doubt that at least over most of the period of growth, the predominant limitation on the rate at which protein crystals nucleate and grow is a function of the rate of attachment.

As in any multi-component chemical reaction, the capture of molecules by a growing crystal surface requires, first, that the candidate molecules have the correct orientation when they approach the crystal surface and, second, that they be in the proper chemical state to form interactions essential for coupling to a set of neighbors. Although there may be some things we can do to improve the statistical probability of proper orientation, there is not likely to be very much. On the other hand, we may have opportunities to affect the frequency of attachment by enhancing the number and strength of the interactions between molecules in the lattice. We do this, for example, by optimizing the charge state of the proteins by adjusting pH, providing electrostatic crossbridges, or minimizing the dielectric shielding between potential bonding partners.

Certainly important to promoting periodic bond formation is that the population of molecules be as homogeneous as possible. For this reason any contamination by unwanted species must be suppressed. Within the target population all individuals must be encouraged to assume physical and chemical conformity. Because crystals have, as their essential elements, exact symmetry and periodic translational relationships between molecules in the lattice, non-uniform protein units cannot properly enter the crystal. Imperfect molecules will serve as inhibitors of crystal growth and impose a generally negative effect on the attachment rate. Should imperfect molecules enter the lattice despite their peculiarities, they will introduce imperfections which, by accumulation, will ultimately produce defects, dislocations, and probably termination of crystal growth.

For proteins difficult to crystallize, it is essential to take all possible measures to purify the protein free of contaminants and to do whatever is necessary to engender a state of maximum structural and chemical homogeneity. Frequently we are misled by our standard analytical approaches, such as polyacrylamide gel electrophoresis (PAGE) or isoelectric focusing (IEF), into believing that a specific protein preparation is completely homogeneous. This is frequently illustrated for us by distinctive differences in the crystallizability of several preparations, even when all analyses indicate they are identical. Imperceptible differences may be due to degrees of microheterogeneity within preparations that lie at the margin of our ability to detect them. Many of these differences arise from known posttranslational modifications to denaturation, proteolytic cleavage, and chemical modifications by contaminants such as heavy metals or oxygen.

There are occasions when even the most intense efforts to crystallize a specific protein fail despite the best efforts at ultra-purification and elimination of microheterogeneity. When this occurs, an alternative is to turn to a different source of the protein. Often only very small variations in amino acid sequence, as found for example between different species of organisms, is enough to produce dramatic differences in the crystallization behavior of a protein. Thus, if the protein from one source proves intractable, consider another. With recombinant proteins, of course, one always has the option of producing a vast range of mutants. Variation of sequence in fact provides a powerful approach to crystallizing proteins when the native molecule fails. There is currently much work underway to define effective mutation strategies for protein crystallization (Derewenda, 2004; Dale et al., 2003).

SOLUBILIZATION

The utility of non-ionic detergents is an important, if not crucial factor in the crystallization of membrane proteins and has been treated in detail elsewhere (Michel, 1990; Wiener, 2004). It is useful to point out, however, that detergents may be of value in the crystallization of otherwise soluble proteins as well (McPherson et al., 1986, Larson et al., 2004). Many protein molecules, particularly when they are highly concentrated and in the presence of precipitating agents such as poly(ethylene glycol) or methylpentanediol, tend to form transient and sometimes metastable, nonspecific aggregates. The existence of a spectrum of varying sizes, shapes, and charges presents problems not appreciably different from the crystallization of a protein from a heterogeneous mixture or an impure solution composed of dissimilar macromolecules. An objective in crystallizing proteins is to limit the formation of nonuniform states and reduce the population to a set of standard individuals that can form identical interactions with one another.

Indeed evidence from inelastic light-scattering experiments suggests that the formation of nonspecific or disordered aggregates, particularly linear aggregates, may be a major obstacle to the appearance of crystals. Conditions that tend to produce a preponderance of such aggregates therefore are to be avoided in favor of those yielding ordered three-dimensional arrangements. Laboratories are currently investigating and developing methods to predict, even prior to the observation of microscopic crystals, which conditions favor the latter over the former, and some have proved remarkably successful (George and Wilson, 1993).

Because the key to successfully crystallizing a macromolecule often lies in the procedure, means, or solvent used to solubilize it, some careful consideration should be given to this initial step. This is particularly true of membrane, lipophilic, or other proteins which, for one reason or another, are only marginally soluble in water solutions. In addition to mild detergents there are chaotropic agents that can also be employed for the solubilization of proteins. These include such compounds as urea, guanidinium hydrochloride, and relatively innocuous anions such as SCN^-, ClO_4^-, I^-, Br^-, and NO^- (Hatefi and Hanstein, 1969). These compounds, even at relatively low concentrations, may serve to increase dramatically the solubility of a protein under conditions where it would otherwise be insoluble.

SEEDING

Often it is desirable to reproduce previously grown crystals of a protein where either the formation of nuclei is limiting or spontaneous nucleation occurs at such a profound level of supersaturation that poor growth patterns result. In such cases it is desirable to induce growth in a directed fashion at low levels of supersaturation. This can sometimes be accomplished by seeding a metastable, supersaturated protein solution with crystals from earlier trials. The seeding techniques fall into two categories: those employing microcrystals as seeds and those using larger macro seeds. In both methods the fresh solution to be seeded should be only slightly supersaturated so that controlled slow growth will occur. The two approaches have been described elsewhere in some detail (Fitzgerald and Madsen, 1987; Thaller et al., 1985).

In the method of seeding with microcrystals, the danger is that too many nuclei will be introduced into the fresh supersaturated solution and none of the masses of crystals that result will be suitable for diffraction analysis. To prevent this problem, a stock solution of microcrystals is serially diluted over a very broad range, as shown in Figure 2.14. Some dilution sample in the series will, on average, have no more than one microseed per microliter. Others will have severalfold more, or none at all. **1 μl** of each sample in the series is then added to fresh protein-crystallization trials under what are perceived to be optimal conditions for growth to occur. This empirical test should identify the correct sample to use for seeding by yielding only one or a small number of single crystals when growth is completed. Solutions containing too many seeds will yield showers of microcrystals, and solutions containing too low a concentration of seeds will produce nothing at all. The optimal seeding concentration, as determined by the test, can then be used to seed many additional samples.

The second approach to seeding, illustrated in Figure 2.15, involves crystals large enough to be manipulated and transferred under a microscope. Again, the most important consideration is to eliminate spurious nucleation by transfer of too many seeds. Even if a single large crystal is employed, microcrystals adhering to its surface may be carried across to the fresh solution. To avoid this, it is recommended that the macro seed be thoroughly washed by passing it through a series of intermediate transfer solutions. In so doing, not only are microcrystals removed, but if the wash solutions are chosen properly, some limited

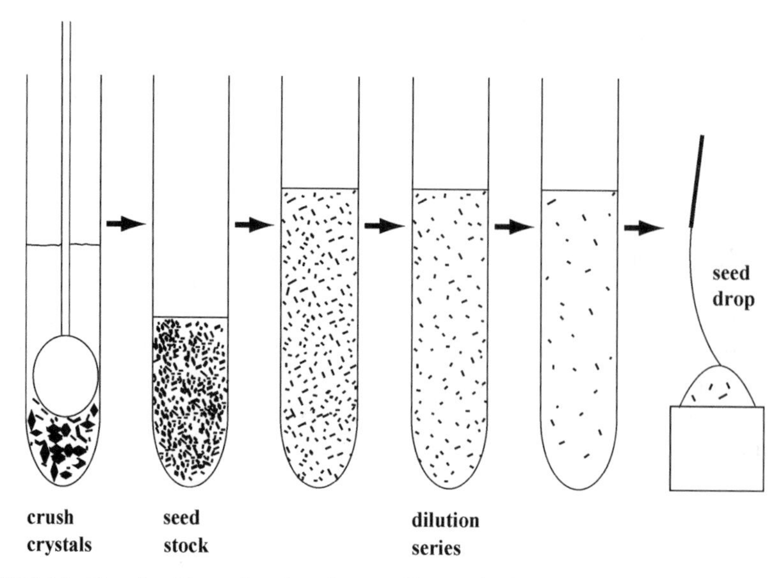

FIGURE 2.14 Drawing illustrating the micro seeding method. Small crystals are crushed in a glass tissue homogenizer to produce a seed stock. The seed stock is serially diluted by a factor of **5** to **10** to produce a dilution series. Each diluted seed solution is used to test optimal seed concentration for growing large single crystals. The diluted seed solutions can be stored for future use.

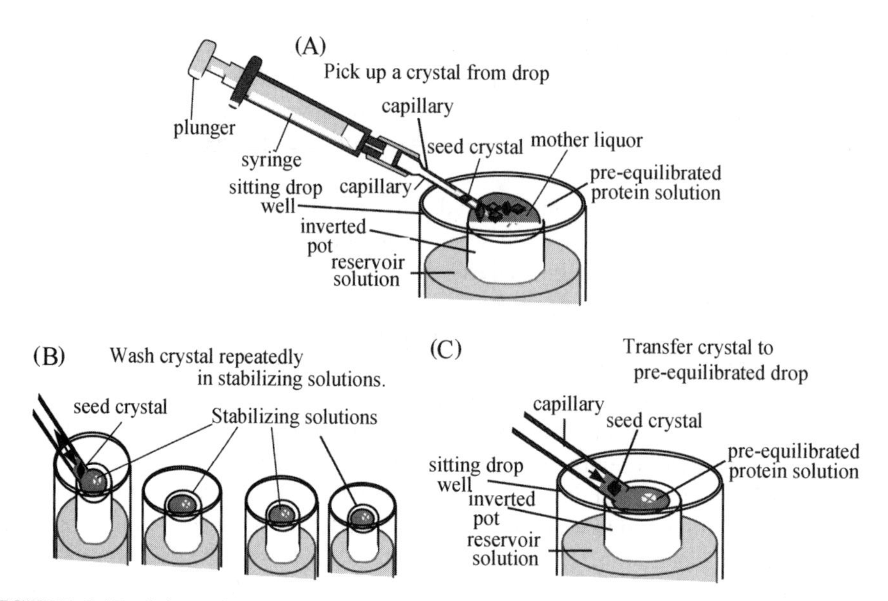

FIGURE 2.15 Schematic illustration of the macro seeding method. In a sitting drop experiment, crystals can be selected directly from the depression where they have been grown using a capillary adapted to a syringe. The crystal is rinsed in a stabilizing solution (typically the precipitant solution in the reservoir) in a series of washes performed in a separate well. The washed crystal is then taken up in a clean capillary (to avoid contamination with crystal fragments) and most of the stabilizing solution is removed from the seed (to avoid diluting the protein solution to be seeded). Solution from the drop to be seeded is then drawn into the capillary, and the seed is transferred to this new protein solution for crystal growth (courtesy of E. Sutra).

dissolution of the seed may take place. This has the effect of freshening the seed crystal surfaces and promoting new growth once it is introduced into the fresh protein solution. Again, the new solution must be supersaturated with respect to protein, but not excessively so, in order to ensure slow and ordered growth. Seeding is frequently a useful technique for initiating the growth of crystals, or inducing nucleation and growth at a lower level of supersaturation than where it might otherwise spontaneously occur. This can only be done, however, where crystals, even poor crystals of the protein under investigation have previously been obtained and can be manipulated to serve as seeds.

A common problem in macromolecular crystallization is inducing crystals to grow that have never previously been observed. The single major obstacle to obtaining any crystals at all is, however, ensuring the formation of stable nuclei of protein crystals. In cases where the immediate problem is growing crystals, attention must be thus directed to the nucleation problem, and any approach that can help promote nucleation should be considered.

One such technique, borrowed in part from classic small molecule crystal growth methodology, is the use of heterogeneous or epitaxial nucleants. In principle, this means the induction of growth of crystals of one substance on crystal faces of another. The classical example is gallium arsenide crystals that nucleate and grow from the faces of crystals of silicon. Because protein molecules possess chemical groups, both charged and neutral, that often readily interact with small molecules, membranes, or other surfaces, the possibility presents itself that the faces of natural and synthetic minerals might help order protein molecules at their surfaces and thereby induce the formation of ordered two-dimensional arrays of the macromolecules. This ordering might occur by mechanical means due to steps and dislocations on the crystal faces, or by chemical means derived from complementarity between groups on the mineral and the protein. Such cooperation between mineral faces and nascent protein crystals might be particularly favored when the lattice dimensions of the protein unit cell are integral multiples of natural spacings in the mineral crystal. McPherson and Shlichta (1988) showed that both heterogeneous nucleation and epitaxial growth of protein crystals from mineral faces do indeed occur.

A second approach to enhancing the formation of crystal nuclei has also been described (Ray and Bracker, 1986). Here, microdroplets of various concentrations of poly(ethylene glycol) were introduced into protein solutions that were also sufficiently high in salt concentration to support crystal growth once stable nuclei were formed. It was shown that protein left the salt-dominated phase of the mixture and concentrated itself in the poly(ethylene glycol)-rich microdroplets, sometimes reaching effective concentrations in these droplets of several hundred milligrams per milliliter. By light microscopy techniques it was demonstrated that crystal nuclei appeared first at the surface of the droplets and then proceeded to grow into the supersaturated salt solution that surrounded them, finally reaching a terminal size appropriate for X-ray analysis.

AUTOMATED CRYSTALLIZATION AND ROBOTICS

When an investigator is focused on one crystallographic objective, the structure solution of a specific macromolecule or macromolecular complex, then it is wise to put as much time into headwork (thinking) as handwork (setting up crystallization samples). Examining and evaluating results, devining insights, and setting new directions are certainly as important as actually dispensing samples into crystallization plates. There are, however, instances where it is more efficient to automate the process by which crystallization conditions are identified and optimized. This is true when there is not one, but many real, or potential targets.

In recent years robots and other automated instruments, and entire integrated systems, have been developed to accelerate the crystallization process (Luft et al., 2003; DeLucas et al., 2003; Hosfield et al., 2003; Bard et al., 2004). They have the capacity to screen thousands of crystallization conditions, and they do so precisely and reliably, with fewer errors and better record keeping than most humans. In many large laboratories these have become essential pieces of equipment. Using standard, usually commercially available screening kits, sometimes supplemented by local favorites, they can often arrive at acceptable crystals in the most expeditious possible manner.

The robotic systems are not only efficient, tireless, and accurate, but they offer another important feature. They can carry out experiments using drop samples of very small volume, drops of a microliter in most cases and nanoliters in some. This produces a requirement for automated, microphotographic visualization systems, and complex storage and handling systems, and the associated expense. On the other hand, a great advantage emerges in that they can perform enormous numbers of crystallization trials using remarkably little biological sample. This in turn relieves the investigator of a great burden in terms of preparing and purifying macromolecules.

Many of the robotic systems are based on reproducing procedures currently used for manual experiments, such as sitting and hanging drops, and microdrops under oil. They are simply carried out on a much smaller scale. More recently, however, even more miniaturized devices have come on the market. These use what is now commonly called nanotechnology to manipulate small amounts of liquids and fluid streams. These devices are only now seeing rigorous evaluation in laboratories, but they clearly show great promise for the future. Another effort is underway to develop crystallization devices that will allow direct X-ray exposure of crystals where they are grown in situ (McPherson, 2000). These would obviate the need for careful mounting, an often problematic aspect of data collection.

TABLE 2.5 Some important principles

1. **Homogeneity**—Begin with as pure and uniform a population of a molecular specie as possible; purify, purify, purify
2. **Solubility**—Dissolve the macromolecule to a high concentration without the formation of aggregates, precipitate, or other phases
3. **Stability**—Do whatever is necessary to maintain the macromolecules as stable and unchanging as possible
4. **Supersaturation**—Alter the properties of the solution to obtain a system that is appropriately supersaturated with respect to the macromolecule
5. **Association**—Try to promote the orderly association of the macromolecules while avoiding precipitate, nonspecific aggregation, or phase separation
6. **Nucleation**—Try to promote the formation of a few critical nuclei in a controlled manner
7. **Variety**—Explore as many possibilities and opportunities as possible in terms of biochemical, chemical, and physical parameters
8. **Control**—Maintain the system at an optimal state, without fluctuations or perturbations, during the course of crystallization
9. **Impurities**—Discourage the presence of impurities in the mother liquor, and the incorporation of impurities and foreign materials into the lattice
10. **Preservation**—Once the crystals are grown, protect them from shock and disruption; maintain their stability

IMPORTANT PRINCIPLES

Table 2.5 provides a concise listing of some important principles that should be born in mind when facing a crystallization problem. Their bases have been discussed, or touched upon in this chapter, but their detailed, specific applications are entirely up to the investigator. Be creative, be imaginative.

A last word of advice regarding success. Once crystals are obtained, then that should not signal the end of the chase. Better crystals for analysis, larger crystals, a more favorable crystallographic symmetry or unit cell, or crystals that diffract to a higher resolution might all be obtained by continued examination of conditions. The ability of a specific protein to form derivatives and ligand complexes is often very much dependent on the crystal lattice interactions. Thus the search for improvements should go forth in parallel as the X-ray analysis commences.

CHAPTER 3

THE NATURE OF CRYSTALS: SYMMETRY AND THE UNIT CELL

Crystals of macromolecules, like those shown in Figure 3.1, are like crystals of all other kinds. They are precisely ordered three-dimensional arrays of molecules that may be characterized by a concise set of determinants that exactly define the disposition and periodicity of the fundamental units of which they are composed. The set of parameters is comprised of three elements. These define the symmetry properties, the repetitive and periodic features, and the distribution of atoms in the repeating unit. The properties may be separated and understood by considering how a crystal can be developed as a three-dimensional form from a basic building block (the asymmetric unit), by the application of symmetry (the space group), and translation (the unit cell, or lattice). As illustrated in Figure 3.2, this can be accomplished in four stages.

THE ASYMMETRIC UNIT

In Figure 3.2 an arbitrary object, here the set of discrete atoms belonging to the backbone structure of a small protein, is chosen as the fundamental unit of construction. This object is termed the asymmetric unit because, in the completed crystal, no part of this object will be systematically related to any other of its parts by crystallographic properties. That is, it has no inherent symmetry or symmetry elements or, if present, they do not coincide with any symmetry operators of the crystal (i.e., the elements generate only internal or local symmetry). In general, the asymmetric unit is one formula unit of a compound, a molecule, or a protein subunit. It can be a small integral number of these, or it may be a fraction such as $\frac{1}{2}$ or $\frac{1}{4}$ if the molecule does posses self-symmetry. The essential property of the asymmetric unit for our purposes is the set of relative x, y, z coordinates of the atoms, which comprise its structure. These are, of course, what we are ultimately interested in.

Introduction to Macromolecular Crystallography, Second Edition By Alexander McPherson

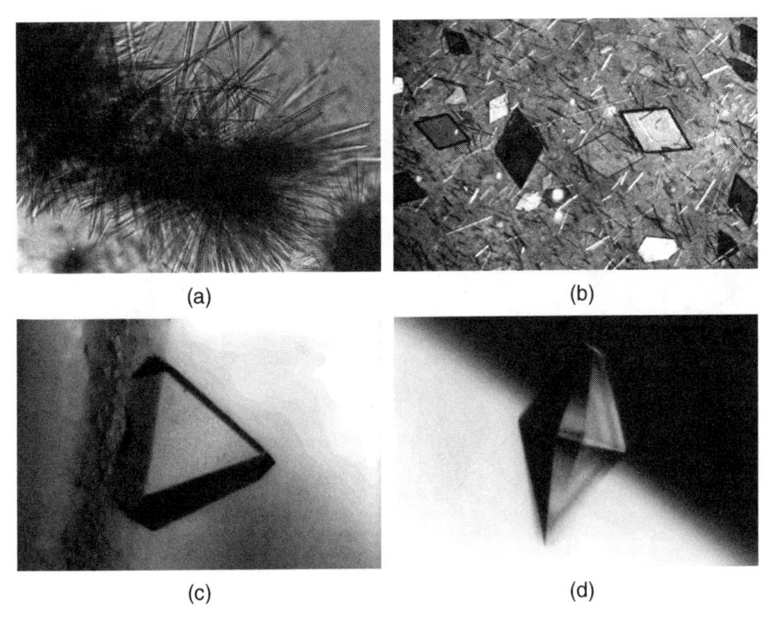

FIGURE 3.1 Crystals of a variety of macromolecules. In (*a*), clusters of needle crystals of bacterial α amylase; in (*b*), crystals of a protease from pineapple, in (*c*), a crystal of satellite tobacco mosaic virus; and in (*d*), a crystal of the sweet protein thaumatin.

THE SPACE GROUP

In the second step, a set of symmetry operations is applied to the asymmetric unit, thereby generating a closely packed, closed set of additional, identical asymmetric units. One type of symmetry operation, rotation about a twofold or dyad axis is shown in Figure 3.3. Operation of this symmetry element results in the rotation of the asymmetric unit by **180°** about a defined axis. The only parameters required to specify the operation are the direction of the symmetry element in space (subject to a defined coordinate system) and its position with respect to the origin of the coordinate system. If, as in Figure 3.3, the axis is parallel with the y direction in space and passes through the origin, and if any point on one object is arbitrarily assigned the coordinates x, y, z, the equivalent point on the symmetrically related object will have coordinates $-x$, y, $-z$. The sets of general coordinates relating identical points on symmetrically related objects are called equivalent positions, and every symmetry operation, or combination of operations yields a unique set. The combination of symmetry operations that characterizes a crystal is called its space group.

In Figure 3.2 a twofold rotation of asymmetric unit **(i)** about the vertical axis results in asymmetric unit **(ii)**. A second twofold operation about the horizontal axis then generates **(iii)** and **(iv)**. The operations could equally well have been performed in reverse order, or starting with, or including the axis not used above, and the same set of four asymmetric objects would have been created; hence the operations are commutative. This is true of the symmetry operations for any space group. Note that as a consequence of the pair of perpendicular and intersecting dyad axes along the vertical and horizontal directions, a set of asymmetric units is produced that possess a third twofold axis that is perpendicular to, and intersecting the first pair. This is a common feature of symmetry operation combinations and

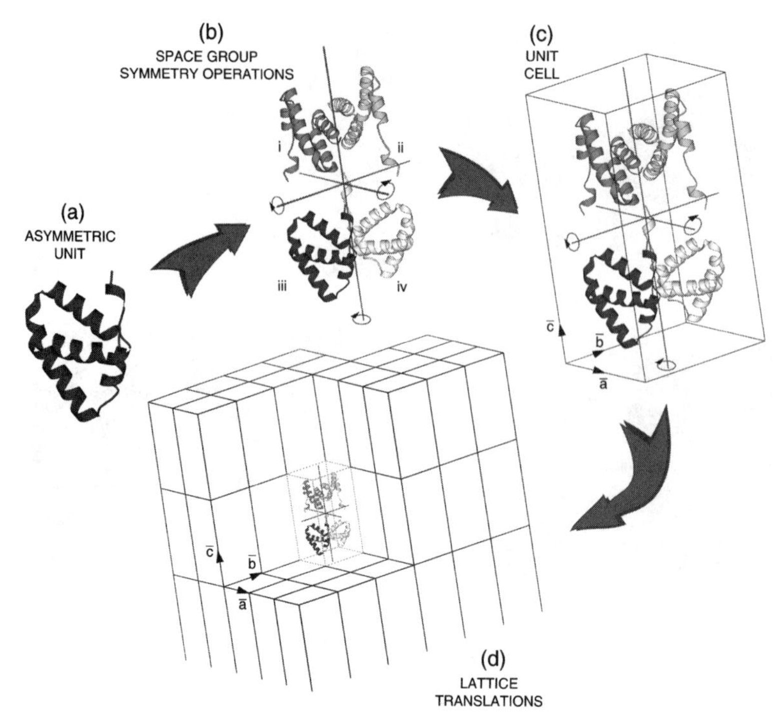

FIGURE 3.2 A crystal may be created from a fundamental asymmetric unit, illustrated by the small protein in (*a*), according to the sequence of steps shown in (*b*) through (*d*). In (*b*), a set of symmetry operators, here three mutually perpendicular twofold axes, called the space group, is applied to the asymmetric unit to produce a small, closed set of asymmetric units. In (*c*), a parallelpiped of minimum volume, called the unit cell, is chosen so that it encloses a full compliment of the asymmetric units, and reflects the symmetry properties of the space group. The space group shown here is called an orthorhombic unit cell. The final stage is shown in (*d*) where the unit cell and its contents are repeated in a periodic manner along the unit cell axes. These translations are called the lattice translations, or lattice vectors, and the corners of the unit cells, define a point lattice.

one that makes it possible to define or describe a complex system of symmetry relationships in terms of only a few basic operations.

There are a limited number of symmetry elements, some of which are illustrated in Figures 3.4, 3.5, and 3.6, that are permitted in a crystal. We will see why only these in a moment. These include twofold, threefold, fourfold, and sixfold rotation and screw axes, mirror planes, glide planes, and centers of symmetry. All allowed types of crystallographic symmetry elements, and their descriptions are given in the *International Tables for X-ray Crystallography*, Vol. I. Combination of these elements in all possible ways yields a total of **230** unique three-dimensional space groups of symmetry operations. The space group of a crystal therefore is a description of the means by which the fundamental set of asymmetric units is self-related. The space group of the arrangement shown in Figure 3.2 is specified as **2 2 2**, since the asymmetric units of the set are related to one another by three, intersecting, and mutually perpendicular twofold axes.

A significant simplification arises when one is dealing with biological molecules (and there are not many, so treasure this one). Many natural products occur only in one

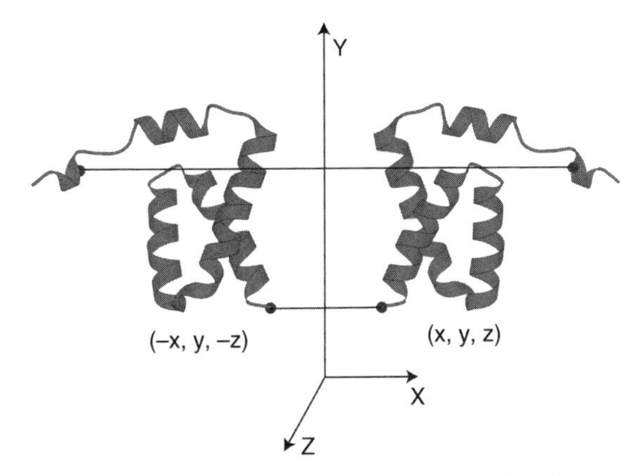

FIGURE 3.3 A twofold axis of symmetry along y (denoted **2**) relates two identical protein molecules. Alternatively, one molecule may be generated from the other by **180°** rotation about the axis. Additional **180°** rotations simply superimpose the existing molecules upon themselves, creating no new asymmetric units (a closed set). A symmetry axis like this, or of any kind, may be specified in terms of a coordinate system and its origin. That shown here is coincident with the y axis and passes through the origin; thus any arbitrary point x, y, z on one protein molecule is related to the identical point on the other by $-x, y, -z$. All points in this system having the relationship x, y, z and $-x, y, -z$ are called equivalent positions.

stereoisomer. For example, proteins are composed only of L-amino acids, nucleic acids and nucleotides are composed of D-ribose or D-deoxyribose, and most polysaccharides have only one chiral form of monomer. These molecules therefore can crystallize only in space groups that do not posses inversion symmetry (i.e., symmetry that does not require a change in the hand of the molecule). Thus all space groups that contain a center of symmetry, a mirror plane, or a glide plane, which for completeness are illustrated in Figure 3.7, are eliminated from consideration. For biological macromolecules only **65** distinct space groups are available rather than **230**, and these allowable space groups are listed in Table 3.1. They are generated only from combinations of strict rotation and screw symmetry operations, plus fractional unit cell translations that give rise to centered unit cells (see below).

THE UNIT CELL

The third stage in our crystal synthesis exercise is to construct the smallest parallelepiped possible whose edges are parallel or coincident with the primary symmetry elements relating the set of asymmetric units, and chosen in such a way that it encloses a full complement of the set. The parallelepiped is called the unit cell of the crystal, and it may have a number of shapes, depending on the angles between the cell edges and the relative lengths of the edges. The particular kind of unit cell chosen will be determined by the symmetry elements relating the asymmetric units in the cell. It too must support the same elements. That is, the unit cell cannot have lower symmetry than the aggregate of asymmetric units, nor will it (except for mirror planes or by coincidence) have higher symmetry. If the space group contains a fourfold axis, then the unit cell parallelpiped must have a fourfold axis. If the space group relating the asymmetric units has a threefold axis, then a threefold axis is required to be present in the unit cell, and so on. The unit cell in Figure 3.2 is an orthorhombic

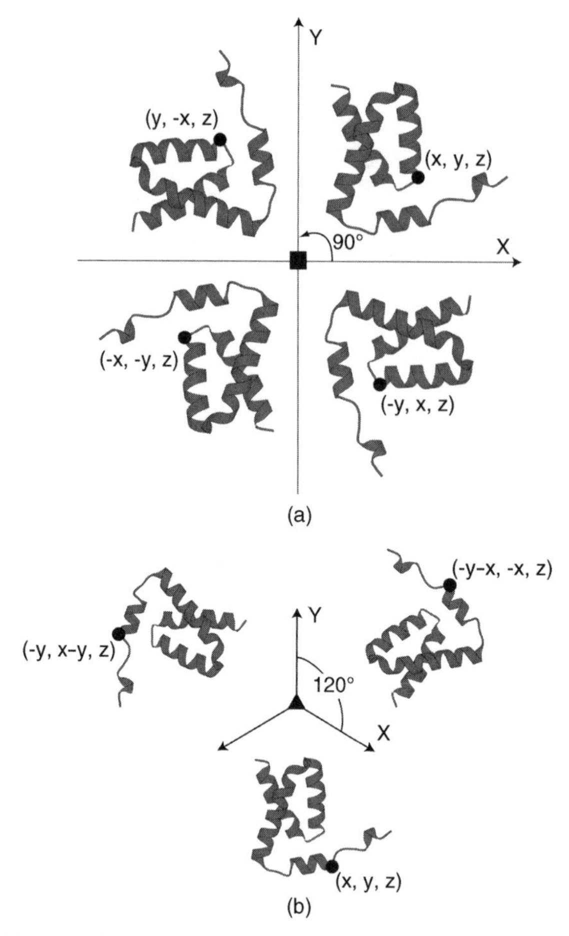

FIGURE 3.4 In (*a*), a fourfold axis of symmetry (denoted **4**) along z is defined mathematically by the four equivalent positions indicated. Application of the equivalent position relationships to the set of atomic coordinates $\boldsymbol{x},\ \boldsymbol{y},\ z$ of a single asymmetric unit yields the positions of all atoms in the unit cell. In (*b*), three protein asymmetric units are related to one another by a threefold axis of symmetry (denoted **3**), perpendicular to the $\boldsymbol{xy}$ plane along z. An orthogonal coordinate system would here obscure the mathematical symmetry of the arrangement. This is preserved, however, if a hexagonal coordinate system is chosen with an angle of **120°** between $\boldsymbol{x}$ and $\boldsymbol{y}$. In the hexagonal coordinate system, the three equivalent positions are $\boldsymbol{x},\ \boldsymbol{y},\ z$; $\boldsymbol{y}-\boldsymbol{x},\ -\boldsymbol{x},\ z$ and $-\boldsymbol{y},\ \boldsymbol{x}-\boldsymbol{y},\ z$. If the threefold axis were instead a $\mathbf{3_1}$ axis along z, then the symmetry equivalent positions would be $\boldsymbol{x},\ \boldsymbol{y},\ z$; $\boldsymbol{y}-\boldsymbol{x},\ -\boldsymbol{x},\ z+\frac{1}{3}$ and $-\boldsymbol{y},\ \boldsymbol{x}-\boldsymbol{y},\ z+\frac{2}{3}$.

unit cell. No cell edge length is equal to any other (i.e., $\mathbf{a} \neq \mathbf{b} \neq \mathbf{c}$), but all interaxial angles are equal to **90°**. It is a rectangular box having three mutually perpendicular twofold axes. Table 3.1 includes the allowable unit cell types found in crystals and their distinguishing characteristics.

The unit cell is generally chosen so as to contain only one complete complement of the asymmetric units, in which case it is called primitive and designated by a capital *P*. Occasionally, however, the parallelepiped may be defined to include more than one full complement in order to preserve higher symmetry. This point will be addressed more fully

TABLE 3.1 Unit cells and allowed space groups for biological macromolecules

Crystal System	Types of Lattices	Symmetry of Unit Cell	Unit Cell Edges and Angles	Diffraction Symmetry	Permissible Space Groups
Triclinic	P	None	$a \neq b \neq c$ $\alpha \neq \beta \neq \gamma$	$\bar{1}$	$P1$
Monoclinic	P C	A single twofold axis	$a \neq b \neq c$ $\alpha = \gamma = 90^\circ$ $\beta \neq 90^\circ$	$2/m$	$P2$, $P2_1$, $C2$
Orthorhombic	P C I F	Three mutually perpendicular twofold axes	$a \neq b \neq c$ $\alpha = \beta = \gamma = 90^\circ$	mmm	$P222$, $P2_12_12_1$, $P222_1$, $P2_12_12$, $C222$, $C222_1$, $I222$, $I2_12_12_1$, $F222$
Tetragonal	P I	A single fourfold axis and perpendicular twofold axes	$a = b \neq c$ $\alpha = \beta = \gamma = 90^\circ$	$4/m$ $4/mmm$	$P4$, $P4_1$, $P4_3$, $P4_2$ $I4$, $I4_1$ $P422$, $P4_122$, $P4_322$, $P4_222$, $P42_12$, $P4_12_12$, $P4_32_12$, $P4_22_12$, $I422$, $I4_122$
Rhombohedral	R	A single threefold axis and perpendicular	$a = b = c$ $\alpha = \beta = \gamma \neq 90^\circ$	3	$R3$, $R32$
Trigonal	P	twofold axes	$a = b \neq c$ $\alpha = \beta = 90^\circ$ $\gamma = 120^\circ$	$3m$	$P3$, $P3_1$, $P3_2$, $P321$, $P312$, $P3_121$, $P3_221$, $P3_112$, $P3_212$
Hexagonal	P	A single sixfold axis and perpendicular twofold axes	$a = b \neq c$ $\alpha = \beta = 90^\circ$ $\gamma = 120^\circ$	$6/m$ $6/mmm$	$P6$, $P6_1$, $P6_5$, $P6_3$, $P6_2$, $P6_4$ $P622$, $P6_122$, $P6_522$, $P6_322$, $P6_222$, $P6_422$
Cubic	P I F	Threefold axes on body diagonals fourfold axes on face normals, and twofold axes along face diagonals	$a = b = c$ $\alpha = \beta = \gamma = 90^\circ$	$m3$ $m3m$	$P23$, $P2_13$, $I23$, $I2_13$, $F23$, $P432$, $P4_132$, $P4_332$, $P4_232$, $I432$, $I4_143$, $F432$, $F4_132$

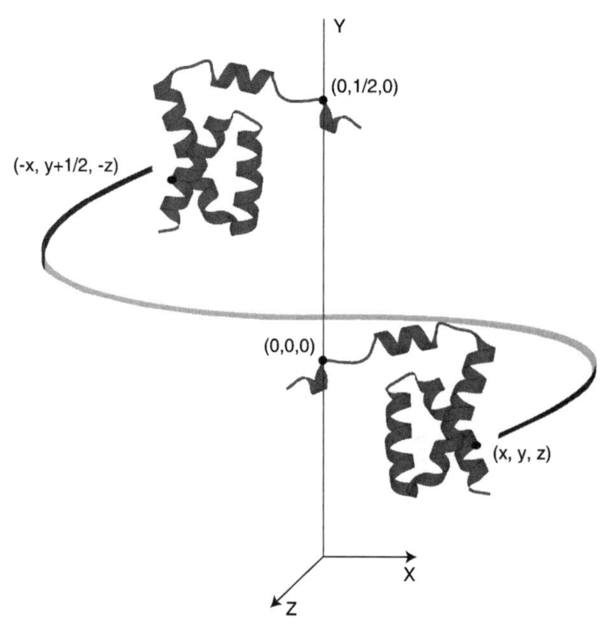

FIGURE 3.5 Two identical protein molecules in a unit cell are related by a twofold screw axis, denoted $\mathbf{2_1}$. One molecule may be generated from the other by rotating it by **180°** about the axis (here along y), and then translating the molecule by $\frac{1}{2}$ the unit cell length, and parallel with the axis. A screw axis of any sort is composed of a rotational component and a translational component. The angle of rotation between asymmetric units is **360°** divided by the large number, and the fraction of the unit cell translated is the subscript divided by the large number. The symmetry equivalent positions of identical points on $\mathbf{2_1}$ related asymmetric units (for $\mathbf{2_1}$ along y) are $x,\ y,\ z$ and $-x,\ y+\frac{1}{2},\ -z$.

below. Unit cells that contain an asymmetric unit complement greater than one set are called centered, or nonprimitive unit cells. Table 3.1 describes those allowed in crystals. In all instances the additional asymmetric unit sets are related to the first by simple fractions of the unit cell edges, such as $(\frac{1}{2},\ \frac{1}{2},\ \frac{1}{2})$ for the body centered **I** cell, or $(\frac{1}{2},\ \frac{1}{2},\ \mathbf{0})$ for the single-face-centered cell (***C*** face in this example).

The unit cell of a crystal can be completely specified by three vectors $\bar{a}, \bar{b}, \bar{c}$, which are coincident with the unit cell edges and of those lengths. The lengths of $\bar{a}$, $\bar{b}$, and $\bar{c}$, (generally given in angstroms) are called the unit cell dimensions, and their directions define the major crystallographic axes. A unit cell may be defined by specifying $\bar{a}$, $\bar{b}$ and $\bar{c}$, or alternately by specifying the lengths $|\bar{a}|,\ |\bar{b}|,\ |\bar{c}|$ and the angles between the vectors, α, β, and γ

In virtually all cases, X-ray crystallography is concerned only with the contents of an individual unit cell, and the distribution (or coordinates) of the atoms within a single unit cell. A neighboring unit cell is simply a repetition. Thus all of the mathematical formalism, except where periodicity is an explicit consideration, is based on what pertains within a single cell. To accommodate this reality each of the three vectors are conveniently assigned a length of one unit (these are what mathematicians call base vectors). This provides a coordinate system that permits identification of every point within a unit cell by an ordered set of three fractional numbers, as illustrated in Figure 3.8. It further provides a very useful means for defining spatial relationships within the crystal and, in particular, equivalent positions of symmetry-related objects or points (see Figures 3.3 through 3.6). It also

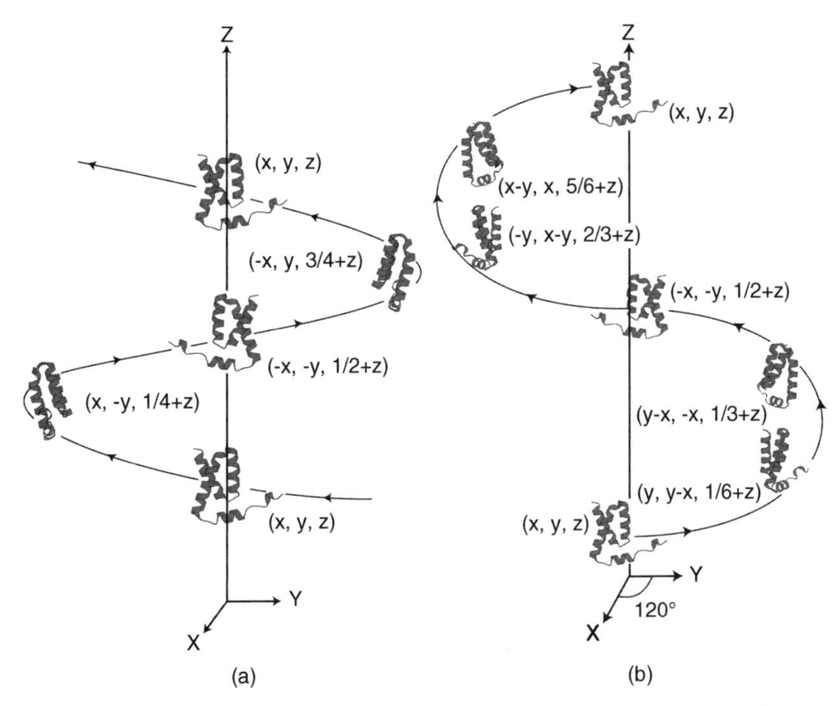

FIGURE 3.6 In (*a*), the disposition of asymmetric units related by a $\mathbf{4_1}$ screw axis is illustrated. As with the $\mathbf{2_1}$ screw axis, continued application of the symmetry operator in a crystal simply generates asymmetric units in adjacent unit cells which were already present due to unit cell translations. In (*b*), a $\mathbf{6_1}$ screw axis produces six identical asymmetric units whose equivalent positions are specified according to a hexagonal coordinate system. It follows that such a symmetry axis could only be compatible with a unit cell having a hexagonal face (i.e., a hexagonal prism).

greatly simplifies equations and calculations, which are always cast in terms of fractional coordinates.

THE LATTICE TRANSLATIONS

The final stage in the creation of a crystal is to sequentially translate the unit cell and its contents along $\bar{a}, \bar{b}, \bar{c}$, by distances $\boldsymbol{a}, \boldsymbol{b}, \boldsymbol{c}$, respectively, many times, to generate a contiguous three-dimensional array of periodically repeated unit cells. The continuous solid so formed constitutes a crystal and exhibits all of its properties. In a sense, this final step of translation of the unit cell contents along $\bar{a}$, $\bar{b}$, and $\bar{c}$ is conceptually redundant with the previous definition of the unit cell parameters, which defines the directions and magnitudes of the translations. It is nonetheless necessary to actually carry out the operations in order to generate the physical crystal.

There is one other feature of the crystal that needs to be described, one that is not essential to Figure 3.2 but that is crucial in realizing some of its properties. If we replace all of the molecules in each primitive unit cell in the entire crystal with a single point, always choosing the identical location for it in every unit cell, as in Figure 3.9, then we generate what is called the crystal lattice. It does not matter where we place the point in the unit cell because,

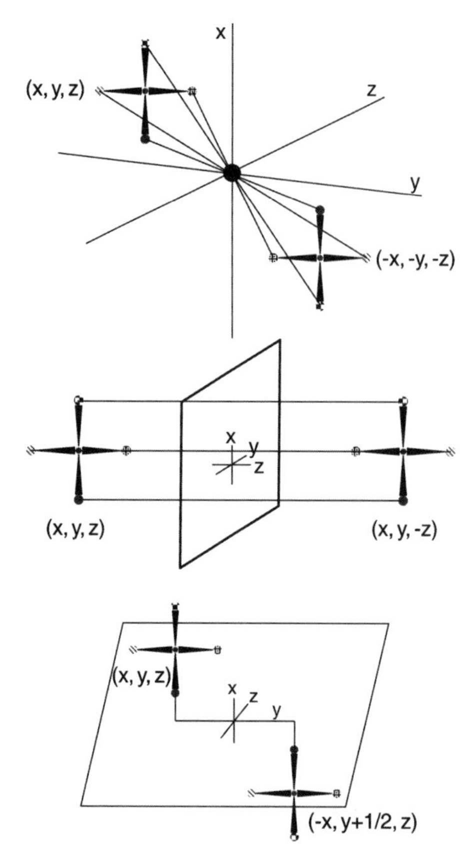

FIGURE 3.7 Illustrated here are three types of inversion symmetry operators applied to an asymmetric carbon atom having four unique substituents. In (*a*), is a center of symmetry, or inversion center. The asymmetric unit at $x,\ y,\ z$ produces that at $-x,\ -y,\ -z$ by inversion through the origin. Note particularly, that in carrying out such an operation, the hand of the asymmetric unit is inverted so that a right-handed enantiomorph becomes left-handed, and vice versa. In (*b*), the asymmetric unit at left is reflected through a mirror plane to produce that at the right, again with inversion of hand. The glide plane operation in (*c*) is somewhat similar to a $\mathbf{2_1}$ screw operator in that it contains a translational component. Instead of rotation, however, it combines translation with reflection through a plane, and again inverts the hand of the asymmetric unit. Because all of these inversion and reflection operators require both right- and left-handed enantiomers to be present in the set of asymmetric units, they cannot be present in crystals of biological macromolecules, which are produced as only one stereoisomer by their biosynthesis.

as long as we choose the same place in all cells, we will obtain the same lattice. Another way to look at this is to take a single point that represents the contents of an entire primitive unit cell and reproduce it in a periodic manner at the ends of vectors $\boldsymbol{m} \times \bar{a}$, $\boldsymbol{n} \times \bar{b}$, and $\boldsymbol{p} \times \bar{c}$, where $\boldsymbol{m}$, $\boldsymbol{n}$, and $\boldsymbol{p}$ take on all integer values.

The idea of a lattice, which expresses the translational periodicity within a crystal as the systematic repetition of the molecular contents of a unit cell, is a salient concept in X-ray diffraction analysis. A lattice, mathematically, is a discrete, discontinuous function. A lattice is absolutely zero everywhere except at very specific, predictable, periodically distributed points where it takes on a value of one. We can begin to see, from the discussion

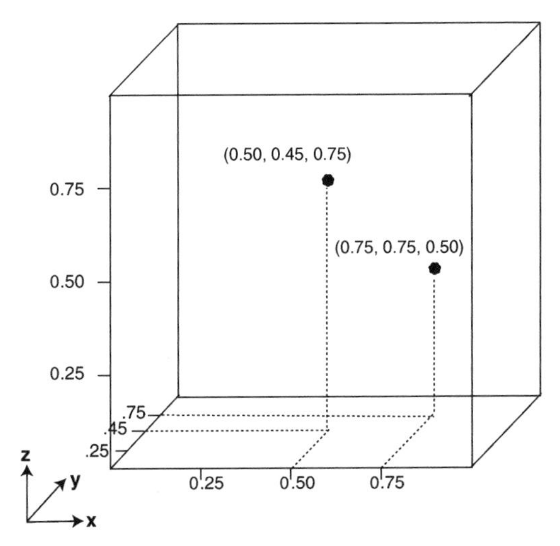

FIGURE 3.8 Because of translational periodicity of unit cells in a crystal, the unit cell edges may be defined as having lengths of 1 (period along $\bar{a}$, 1 period along $\bar{b}$, and 1 period along $\bar{c}$). That is, the unit cell edges define the unitary vectors of the crystal lattice. This is consistent with the idea of the unit cell contents constituting a periodic, three-dimensional electron density wave traveling along the directions of the unit cell vectors $\bar{a}$, $\bar{b}$, and $\bar{c}$. Points in the unit cell, such as the two indicated here, can then be defined in terms of three fractions (of the three periods). Addition of 1 to a fractional coordinate simply yields the identical point in an adjacent unit cell. Addition of integers $\boldsymbol{n}, \boldsymbol{m}, \boldsymbol{p}$ to any fractional coordinate generates the identical point $\boldsymbol{n}, \boldsymbol{m}$ and $\boldsymbol{p}$ unit cells away along $\bar{a}$, $\bar{b}$, and $\bar{c}$, respectively. The mathematical equations and expressions dealing with crystals will always be formulated in terms of this fractional coordinate system. Indeed any atom in any unit cell, anywhere in the crystal, can always be defined in terms of this fractional coordinate system.

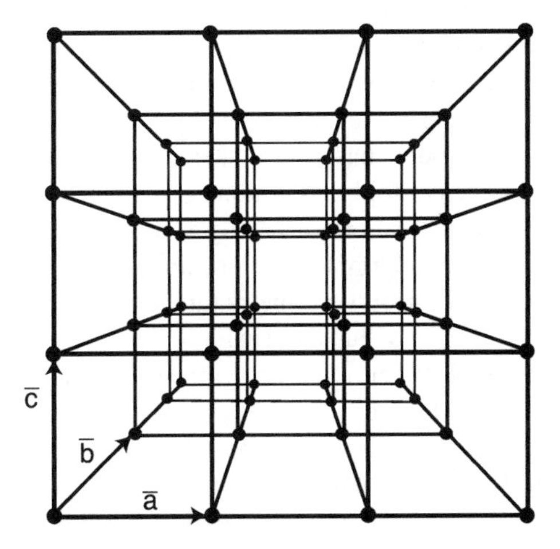

FIGURE 3.9 If any arbitrary point is sequentially translated along direction $\bar{a}$ by $\boldsymbol{m}|\bar{a}|$, where $\boldsymbol{m}$ is integral, then a periodic, linear array of points is generated. If that line of points is then sequentially translated along $\bar{b}$ by $\boldsymbol{n}|\bar{b}|$, a periodic two-dimensional plane of points results. If that plane of points is then translated along $\bar{c}$, by $\boldsymbol{p}|\bar{c}|$, a three-dimensional point lattice is the result. A point lattice, mathematically speaking, is everywhere zero except at discrete points corresponding to all integral multiples of $\bar{a}$, $\bar{b}$, and $\bar{c}$ where it has a value of 1.

above that a crystal may also be thought of as the combination of two things, a cluster of real molecules whose atoms are related by space group symmetry and an imaginary point lattice that describes their periodic placement in three-dimensional space. The lattice, notice, defines the unit cell, just as the unit cell defines the lattice.

SYMMETRY AND EQUIVALENT POSITIONS

We tend to think of symmetry elements as operators (in a mechanical sense) that take one asymmetric object, such as a molecule, and, treating it like a template, reproduces it in space according to a specific scheme. Alternately, we can think of symmetry elements as representing systematic relationships among asymmetric objects in a set. There is, however, another way of considering symmetry elements, and that is as mathematical relationships between points in a defined coordinate system, or space. To define the necessary coordinate system, we must have an origin and three non co-linear axes. The unit cell vectors drawn from some common point in the unit cell generally supplies such a spatial reference system.

In general, we don't choose the origin arbitrarily but in a way that simplifies all subsequent formulations and calculations (crystallography is hard enough, we need every simplification we can get). This is done by choosing the origin at some spatially unique point in the cell, on a symmetry axis, for example, at $\frac{1}{4}$ or $\frac{1}{2}$ of a unit cell translation, at the intersection point of symmetry axes (the **222** symmetry point in Figure 3.2), and so forth. Usually there is some particular point that optimizes the simplification and this is chosen. Don't worry about finding it. Crystallographic forefathers have already done so for every possible space group and unit cell, and kindly recorded them in the *International Tables of X-ray Crystallography*, Vol. I.

Once we have defined the origin and the coordinate axes of the unit cell, let us examine each of the asymmetric units in a unit cell, any unit cell, and mark the identical point (e.g., any arbitrary point we wish, say atom C1 of amino acid 46) on each. Any two of these identical points will be related by a space group symmetry element, a rotation or screw axis. If you measure the x, y, z coordinates of the two points, you will find that they share a simple relationship. For example, if they are related by a twofold axis along the y direction, as in Figure 3.3, then if one point has coordinates x, y, z, the second point will have coordinates $-x, y, -z$. If the twofold axis is along z, then the relationship will be x, y, z and $-x, -y, z$. If the axis is a 2_1 screw axis along x, then the relationship will be x, y, z and $x + \frac{1}{2}, -y, -z$. Some sort of simple relationship of this kind will exist between all identical points related by space group symmetry (and centering operations as well) in the unit cell. For the symmetry elements illustrated in Figures 3.3 through 3.6, the corresponding equivalent positions are noted. When multiple symmetry elements are combined, as they are in most space groups, and Figure 3.10 is an example, the equivalent positions for the entire arrangement are simply the collection of those for the individual elements. The listing of all coordinate relationships for all of the space group related asymmetric units are known as the equivalent positions, and these too are found in the *International Tables of X-ray Crystallography*, Vol. I.

The sensitive reader might be troubled by screw axes, which, because of their translational component, seem to generate a never-ending series of asymmetric units in space. Thus the set of equivalent positions for a screw axis would be infinite as well. When incorporated into a crystal, which is translationally periodic along all three axes, and therefore along any permissible screw axis, we see, however, that this is of no concern. Screw axes, having

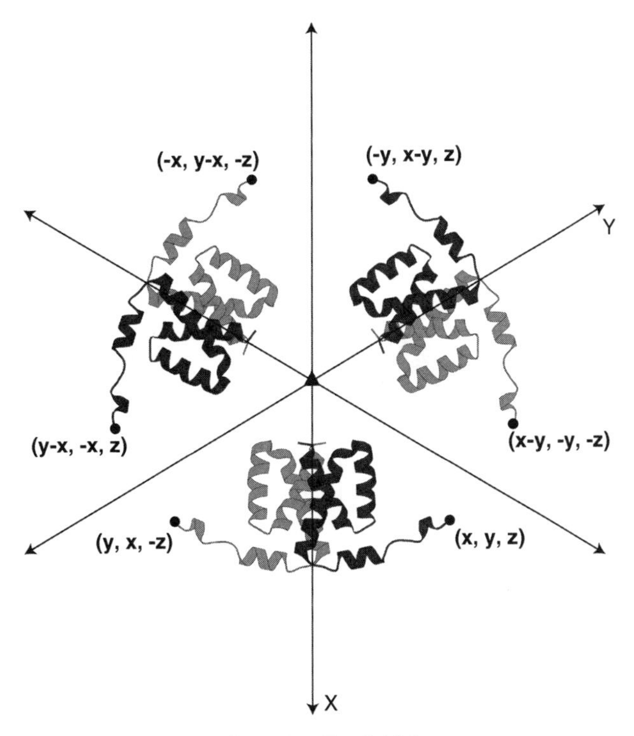

FIGURE 3.10 Symmetry elements may be combined with one another as shown in this example. The heavy and light molecules of any pair are related by a twofold axis in the xy plane. The other pairs in the set of six may then be produced by application of a threefold axis, through the origin, along z. Note that application of the threefold axis to any one of the twofold axes generates the other dyads; hence only one twofold axis is required in the definition of the symmetry, which is denoted **32** symmetry. We might equally well have operated on any one of the molecules with the threefold axis to produce a set of three, and then applied the twofold operators, and the same final arrangement would have been obtained.

integral rotation and translation components, produce a unique set of asymmetric units, and therefore equivalent positions, only within the confines of an individual unit cell. Beyond that they simply reproduce translationally equivalent asymmetric units in neighboring unit cells. This is illustrated in Figure 3.11 for a $\mathbf{2_1}$ screw axis. The same is true of glide planes in those crystals where they are permitted.

The ensemble of all equivalent positions for a space group is unique and may be considered the mathematical definition of the space group. It provides the basis for manipulating objects and points related by symmetry in a digital computer. Equivalent positions are another way of stating both the space group and the Bravais lattice of a crystal.

If the coordinates of the points comprising a single asymmetric unit are known, those of the equivalent points of all the asymmetric units in the unit cell are then known, or may be generated from the space group symmetry (i.e., by equivalent positions). Any point then in any asymmetric unit anywhere in the crystal, even if thousands of unit cells away, can be found by applying the unit cell translations. The coordinates of the atoms in a single asymmetric unit, plus space group symmetry, plus the unit cell vectors, completely specify

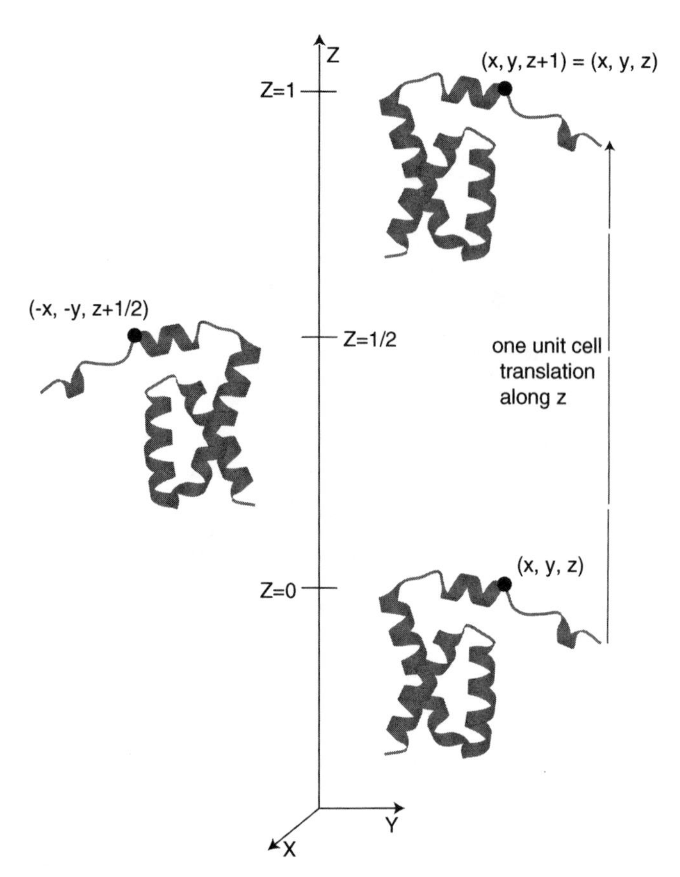

FIGURE 3.11 The asymmetric unit at x, y, z creates, point by point, an equivalent asymmetric unit at $-x, -y, z + \frac{1}{2}$ by application of the $\mathbf{2_1}$ screw axis along z. Further application of the $\mathbf{2_1}$ axis then creates the asymmetric unit at $x, y, z + 1$, which is the same as x, y, z. Because of the periodicity of the unit cells in a crystal, the asymmetric unit at $x, y, z + 1$, was already specified by translation of the first asymmetric unit at x, y, z along z. Thus a $\mathbf{2_1}$ screw axis, like a twofold axis, is described by only two equivalent positions.

every point, that is, every atom in the entire crystal. Since the latter two properties can be deduced in a straightforward manner, even in the course of a preliminary X-ray analysis, the solution of "a crystal structure" implies the determination of the relative coordinates x, y, z of all the atoms that comprise a single asymmetric unit.

A point that deserves emphasis is that, because of periodicity (application of the lattice translations), symmetry elements that define the arrangement of asymmetric units within the unit cell become continuous throughout the crystal and therefore relate all points in the crystal; that is, space group operators are not local relationships confined to a single unit cell. By virtue of the translational periodicity, space group operators are global symmetry operators and extend throughout the crystal. It should also be observed that by virtue of the repetition of the cell along the crystallographic axes, new symmetry elements appear at the corners or edges of the unit cells. None, however, are of higher order than the space group symmetry of the cells. These are also universal and relate the atoms in the molecules occupying adjacent unit cells. Thus all symmetry present in the crystal is explicitly stated or implied by a maximum of three space group operations plus the unit cell translations.

WHY SO FEW KINDS OF UNIT CELLS

Initially it might appear that there could be a virtually limitless number of symmetry elements that we could use in a crystal to relate asymmetric units, and a seemingly unlimited number of kinds of unit cells we could design to contain these sets of asymmetric units. This is, however, not the case. The limitation comes really from two conditions. First, the symmetry elements must generate a closed (finite) set of asymmetric objects. This means only rotational symmetry that uses a rotation angle θ for which $360^{\circ}/\theta = n$ would be permissible, since the $n + 1$ rotational increment would just superimpose the $n + 1$ object upon the first. The same kind of argument can be made for screw symmetry operators taking into account the unit cell translations in a crystal. Nonetheless, this condition of a closed set might still allow for the existence of 11-fold rotation axes or 47-fold screw axes.

The dominant constraint comes from a condition pertaining to the unit cell. Remember, the unit cell must support, or possess, the rotational symmetry of the space group. Thus, if there is a sixfold axis, or a sixfold screw axis of some sort in the space group, then the unit cell must also have at least one sixfold axis. If the space group has three perpendicular twofold axes, then so must the unit cell. If we had a space group containing an 11-fold rotation axis (which we can not), then the corresponding unit cell would have to have one too. But another condition that we have more or less ignored to this point is that when we assemble the unit cells into a three-dimensional array, (i.e., we make a crystal out of them), there can be no gaps or spaces or voids anywhere. The unit cells in a sense have to fit together perfectly, be completely contiguous with one another, and be absolutely continuous and periodic in their disposition throughout space.

This poses more of a problem then we might think for a unit cell designer. In fact it imposes a severe limitation. A sevenfold rotation axis might be employed to relate asymmetric units, but it would require a unit cell having sevenfold symmetry as well. One face of this cell would have to be a septagon. Have you ever seen a floor tiled with seven sided tiles? No, you haven't, for the same reason you've never seen floors, or bathrooms covered with five sided or eleven sided, or most other polygonal tiles. They don't fit together in a perfectly contiguous manner on a two-dimensional surface. If you lay them out in the best way possible, they will never fit together without gaps and spaces, and they will never completely cover the floor. If you can't accomplish the task in two dimensions, you certainly can't do it in three. Unit cells having such faces, then, clearly cannot fill three-dimensional space. The only tiles, and therefore the only shapes permissible for the sides, or faces of unit cells that can satisfy the necessary criteria are (1) parallelograms, (2) rectangles, (3) triangles, (4) squares, and (5) hexagons. Any other shapes having more sides on a face are impossible in crystals.

A further word is required here to explain why the trigonal unit cell in Table 3.1 is assigned $\gamma = 120^{\circ}$ rather than what intuition would suggest, $\gamma = 60^{\circ}$. If an equilateral triangle were translated along unit cell vectors having an angle of 60° between them, as in Figure 3.12*a*, the resultant array of unit cells would cover only half of the total plane, leaving triangular "holes" in between. This would be incompatible with our definition of a unit cell. On the other hand, we can clearly see that the plane could indeed be covered

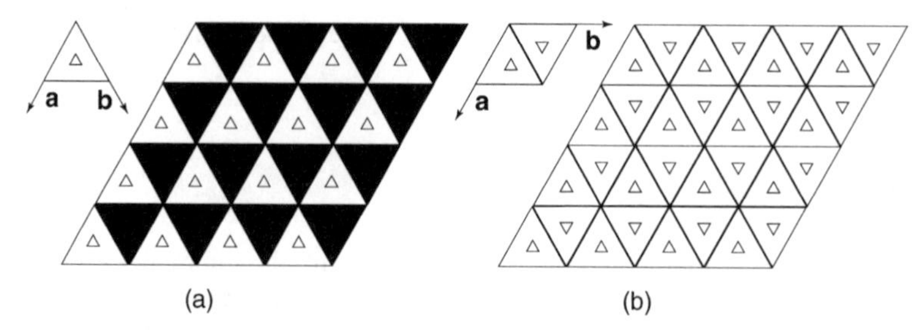

FIGURE 3.12 If the triangular unit cell in (*a*), having an angle of **60°** between ***a*** and ***b*** is translated and repeated sequentially along ***a*** and ***b***, the resultant triangles will cover only half of the plane. The dark triangles remain empty. If two triangles are taken as a rhombus having an angle of **120°** between ***a*** and ***b***, however, and the two triangles translated and repeated along ***a*** and ***b***, then the entire plane is filled. This is the basis for our choice of two triangular unit cells in (***b***) to define the unit cell of a trigonal crystal.

completely with triangles. If, however, we choose two contiguous triangular unit cells, which form a rhombus with unit cell vectors $\bar{\mathbf{a}}$ and $\bar{\mathbf{b}}$ related by **120°**, as in Figure 3.12*b*, then the problem is solved. The double-unit cell, the rhombus, still contains the requisite threefold symmetry elements, but now translation along the rhombic unit cell vectors results in an array that completely covers the plane. With a third axis, $\bar{c}$, perpendicular to $\bar{a}$ and $\bar{b}$, the unit cells can then fill three-dimensional space by pure translation along unit cell vectors.

Now that we know what shapes are possible, we can ask how these various shapes can be put together to make unit cells, and what are the symmetries of those cells. This is important, remember, because the space groups, or symmetry arrangements that fill the parallelpipeds, have to be compatible with the symmetries of the unit cells, and vice versa. Without belaboring the point any further, let us simply assert that the unit cells shown and described in Table 3.1 are the only ones permissible in crystals, and consequently only the 230 space groups compatible with their symmetries can exist.

PRIMITIVE AND CENTERED LATTICES

From a previous discussion it was seen that if an arbitrary point was chosen in one unit cell, this point and the identical points in all other unit cells formed a lattice with neighboring points related by the unit cell vectors. Any type of unit cell gives rise by repetition to a specific kind of general lattice, and the lattice, unique to a particular crystal, is exactly specified by the lengths of, and angles between, the unit cell axes. A lattice representation, illustrated by Figure 3.9 and further by Figure 3.13, is conceptually useful because it completely characterizes the distribution of transitionally identical points and unit cells throughout the crystal without regard to unit cell contents. Furthermore it reflects the symmetry properties of the type of unit cell it relates.

Our discussion of a crystal, and how one would create it from an asymmetric unit (see Figure 3.2), tacitly assumed that in assembling our unit cells into a solid form the only way that we could suitably put them together was by strict translation along the three unit cell vectors. This is not entirely true. There are in fact some alternative ways of stacking the same set of cells together without creating any discontinuity between

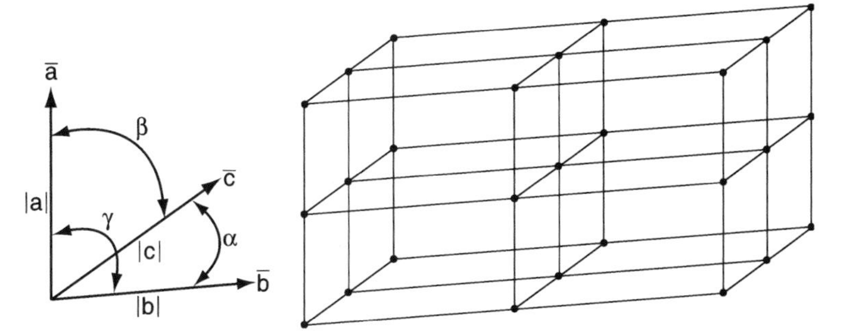

FIGURE 3.13 Example of a point lattice, where the "base vectors," $\bar{a}$, $\bar{b}$, and $\bar{c}$ are nonorthogonal and create a completely general triclinic lattice in which there are no angles of **90°**. The lattice nonetheless is periodic in all directions and has a value of zero except at the specified points.

them. Masons long ago discovered that there is more than one way to lay bricks. When we do this, we may use periodic translations of $|\bar{a}|/2$, $|\bar{b}|/2$, and $|\bar{c}|/2$ along some directions, combined with entire unit cell translations $|\bar{a}|$, $|\bar{b}|$, and $|\bar{c}|$ along others. For example, we could stack the rectangular cells of Figure 3.14*a* in a different manner to create the array in Figure 3.14*b*. Both are completely solid and allow the creation of a crystal.

There is a problem, however. If we did use the alternate packing arrangement shown in Figure 3.14, then we might feel obligated to redefine the unit cell as the monoclinic unit cell indicated, since that would be the unit cell of smallest volume that enclosed an entire set of asymmetric units. But we should be reluctant to do this because the redefined unit cell does not express as high a degree of symmetry as was produced by the space group operators applied to the asymmetric units in the unit cells; that is, the monoclinic unit cells are of lower symmetry. We want to preserve symmetry relationships whenever we can in crystallography because, as we will see, symmetry greatly simplifies and expedites the analysis at virtually every stage.

The solution, which has been adopted by crystallographers, is to choose unit cells that have some integral multiple of the volume of the primitive unit cells, and that enclose multiple full sets of asymmetric units. Although larger by integral multiples, they maintain the underlying symmetry of the space group. These we refer to as centered, nonprimitive unit cells. Just as primitive unit cells gave rise to lattices with one lattice point associated with each unit cell, crystals having centered lattices arise from unit cells containing additional lattice points. The additional lattice points are always simply related to a lattice point at $(\mathbf{0}, \mathbf{0}, \mathbf{0})$ by half unit cell translations, such as $(\mathbf{0}, \frac{1}{2}, \frac{1}{2})$ or $(\frac{1}{2}, \frac{1}{2}, \frac{1}{2})$, depending on the type of centering. There are three types of centering that are permissible in crystals, single-face centered lattices denoted ***C***, body centered, ***I***, and those centered on all faces, ***F***. All of the possible kinds of lattices that can exist for crystals, including both primitive and centered are shown in Figure 3.15. These are referred to as the Bravais lattices. As we will see later, there is neither more nor less information about the contents of the unit cell in the diffraction pattern as a consequence of centering, and furthermore centering presents no significant complication for an X-ray diffraction analysis. Centered unit cells do, however, produce broad classes of what are termed systematic absences in their diffraction patterns (see Chapter 6) that allow us to immediately recognize their presence in a crystal.

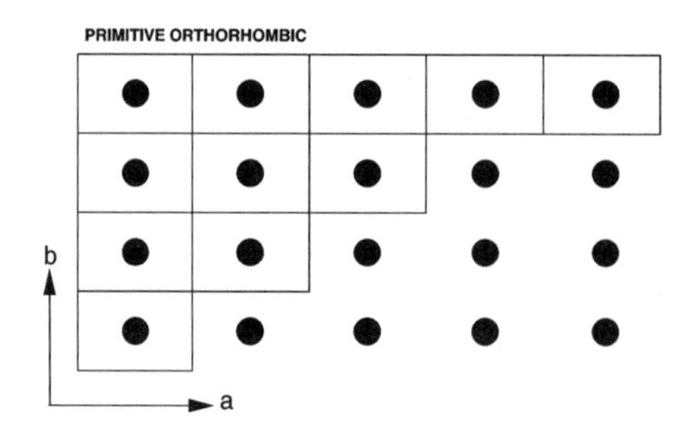

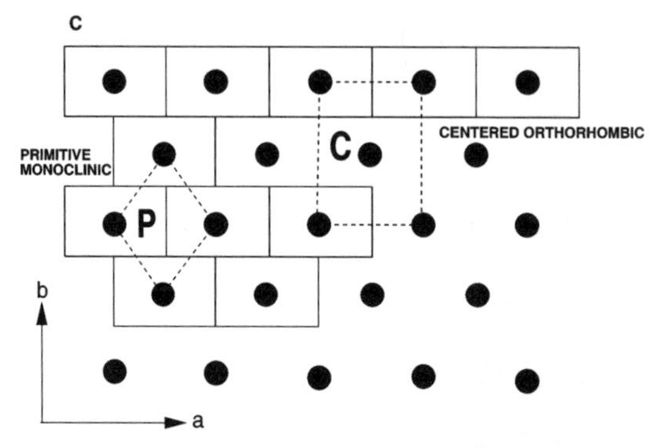

FIGURE 3.14 In (*a*) at top, unit cells are related by full unit cell translations along $\bar{a}$ and $\bar{b}$, yielding a primitive lattice characterized by a primitive unit cell ***P*** containing one lattice point. The periodicity of the unit cells generating the point lattice in (*b*) below is somewhat more intricate because of the $\bar{a}/2$ translation of alternate rows. Two different unit cells (shown with dotted lines) could be drawn that contain at least one full lattice point (one full compliment of asymmetric units), the primitive monoclinic cell (***P***) that contains a single lattice point, or the centered orthorhombic ***C*** unit cell which contains two. Because the ***C*** centered cell preserves the underlying orthogonal symmetry of the lattice, while the monoclinic unit cell does not, the ***C*** centered cell whose axes are along $\bar{a}$ and $\bar{b}$ is chosen. Note, however, that this choice implies a unit cell dimension along $\bar{b}$ that is twice that of the primitive orthorhombic unit cell.

PLANES, MILLER INDEXES, AND CONVOLUTIONS

Ultimately we want to know how a crystal diffracts X rays and produces the diffraction pattern that it does, and conversely, how the diffraction pattern can be used to reconstruct the crystal. It will be found useful in this regard to consider the crystal as the combination, or product of two distinct components, or functions. The first of these is the contents of a unit cell, characterized mathematically by the coordinates of the atoms in an asymmetric unit along with their space group symmetry equivalent positions. The second is a point lattice that describes the periodic distribution of the unit cell contents, and is characterized by $\bar{a}$, $\bar{b}$, and $\bar{c}$. A crystal may then be concisely defined as the first component, or function, repeated in identically the same way at every nonzero point of the second. This physical process of repetitive superposition is termed a convolution. It can be formulated mathematically as the product of the two components, or functions as

$$g(x, y, z) \times f(x, y, z) = g[f(x, y, z)].$$

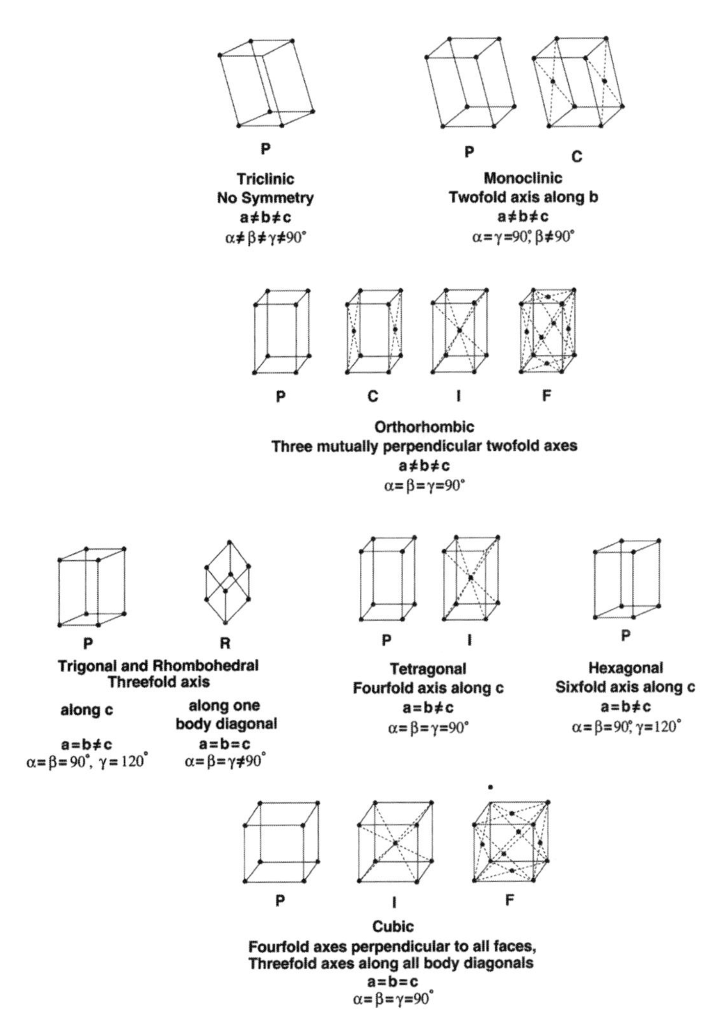

FIGURE 3.15 The types of unit cells that form the basis for the allowable lattices of all crystals (known as the Bravais lattices). There are 15 unique lattices (see *International Tables, Volume I*, for further descriptions). All primitive (***P***) cells may be considered to contain a single lattice point (one-eighth of a point contributed by each of those at the corners of the cell), face-centered (***C***) and body-centered (***I***) cells contain two full points, and face-centered (***F***) cells contain four complete lattice points.

Here $f(x, y, z)$ is the distribution of atoms in a unit cell, and $g(x, y, z)$ is the distribution of points in the lattice. Abstract mathematical formulations such as this are frequently incomprehensible to many of us, so the illustration in Figure 3.16 is offered as an aid.

When we consider diffraction from a crystal, we will further find it useful and convenient to consider a point lattice in even more simplified terms. We can introduce this simplification by organizing the lattice points, distributed in a periodic manner through three-dimensional space, into easily characterized families of two-dimensional planes. This is accomplished by defining all families of planes of equal interplanar spacing that include all of the points in the lattice. Examples are shown in Figure 3.17. Every plane of the family need not contain

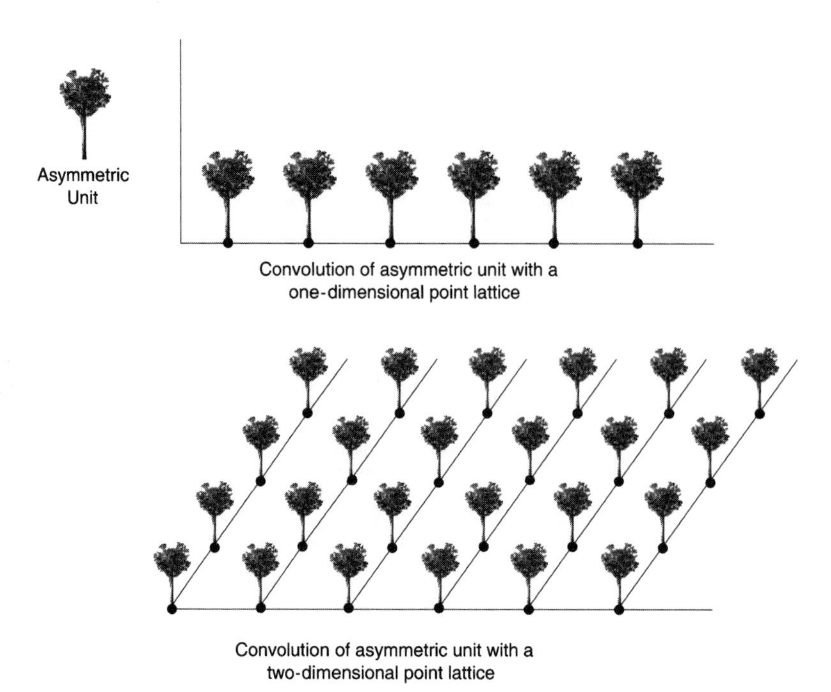

FIGURE 3.16 The concept of convolution of a continuous spatial function (an object, e.g., an asymmetric unit) with a discontinuous, periodic function (a lattice) is shown for one dimension in (*a*) and two dimensions in (*b*). In (*a*), the tree is the object, a continuous distribution of points. Along the horizontal axis to the right is a periodic sequence of points. The convolution is carried out by planting a tree, in an identical manner, at each point in the periodic sequence. In (*b*), an orchard (similar to a two-dimensional crystal with its rows and columns) is created by the convolution of the scattering function, the tree, at positions specified by the lattice function, the gird of points on the property. The idea shown here is no different than that by which we create a crystal by convoluting the atoms within a single unit cell with a lattice specifying the periodicity of the unit cells.

a lattice point, but every lattice point must lie on some plane of the family. Such families of planes have special, and particularly valuable properties.

Note that we are not requiring any molecules, atoms, or even electrons to lie on the planes. We are not interested in them right now, only in simplifying the point lattice, which defines their periodic distribution. We might observe, first of all, that there appear to be an infinite number of such families of planes because we could always subdivide any family even further. There are, but do not be concerned with this. We will see that only some are pertinent to our objective, those with interplanar spacings d greater than a certain constant value. More important, we note that any set of planes, or family, can be uniquely characterized by the number of times it would intercept each cell edge $\bar{a}$, $\bar{b}$, and $\bar{c}$, over the course of one unit cell. Figure 3.18 illustrates several of these families of planes for an individual unit cell. The number of intercepts on each unit cell axis over the course of one unit cell, if we have defined the families as above, must be integral (for rigorous proof, see Buerger, 1942, 1960). The three integers that uniquely define a given family of planes (number of times it intersects $\bar{a}$, intersects $\bar{b}$, intersects $\bar{c}$) are known as the Miller indexes. These are designated $\boldsymbol{hk\ell}$. If the unit cell dimensions and interaxial angles are known for a particular crystal, the interplanar spacing $\boldsymbol{d}$ for any set of Miller indexes $\boldsymbol{hk\ell}$ can be directly

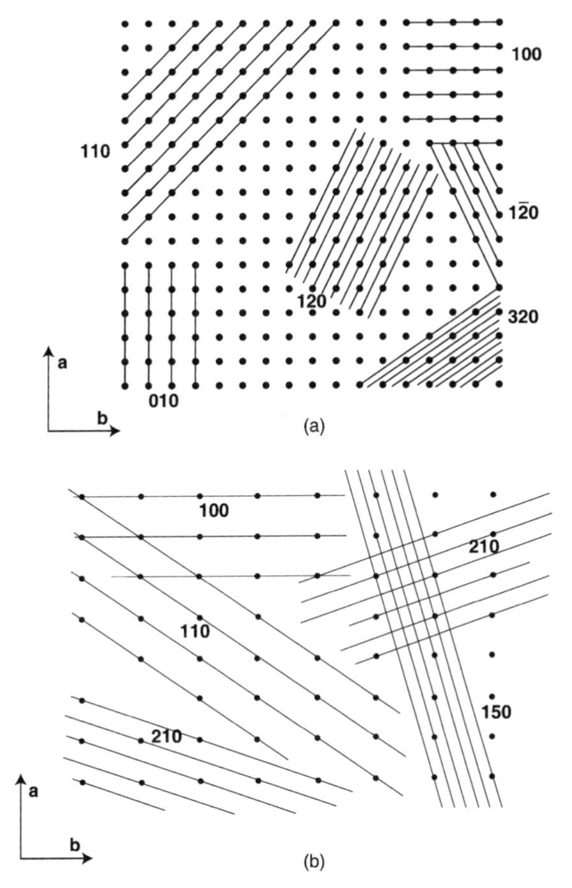

FIGURE 3.17 In (*a*) and (*b*) are two different lattices. In each lattice, the points have been organized as families of planes. The planes, like the points, are periodic throughout the crystal and can be precisely specified by their interplanar spacing and a plane normal (a vector perpendicular to the planes). The families are designated by three integers $hk\ell$ called Miller indexes, and these specify the slopes of the planes in a family.

calculated. The formulas for these relations are given in Table 3.2. Similarly the directions of the normals to each family of planes can also be readily calculated. The normals define the orientations of the planes in the crystal.

The families of planes also may be thought of as sampling the contents of the unit cells at regular intervals, like serial cross sections in microscopy. By analogy, if one knew what lay on each plane, by interpolation one could reconstruct the contents of the unit cell. For families of large interplanar spacings (low Miller indexes), this would yield rather poor results; that is, it would give a very low resolution image of the unit cell contents. For families of small interplanar spacings (with high Miller indexes), this interpolation would be much better.

If one knew how atoms were disposed about many families of planes of different spacings and orientations, the contents of the unit cell could, in principle, be mapped in considerable detail by recombination of all the planes in a suitable fashion. The level of detail that resulted (i.e., the resolution) would be a function of the smallest interplanar spacing, or

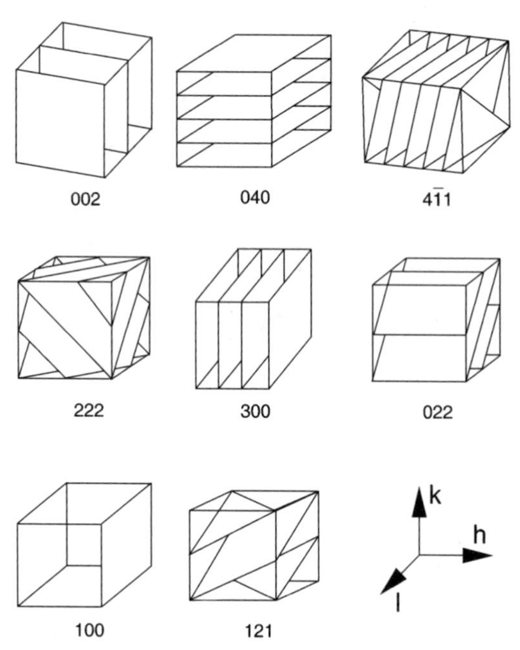

FIGURE 3.18 Families of planes making rational intercepts with the unit cell edges are identified by a set of three integers ***hkℓ*** known as Miller indexes. The values of these indexes are equal to the number of times the family intercepts each unit cell edge. Each family of planes has associated with it a vector ***hkℓ***, called the reciprocal lattice vector, which is normal to the planes and has a length inversely proportional to the interplanar spacings. The various families of planes may be considered to be the spectral or harmonic components of the electron density in the unit cell, each having a period equal to the interplanar spacing, or a frequency given by ***hkℓ***. By mathematical recombination of many sets of planes, each having a characteristic sampling frequency, the electron density of the cell can be reconstructed. It is these families of planes that give rise to the intensity maxima that comprise the X-ray diffraction pattern.

TABLE 3.2 Interplanar distance formulas for the seven crystal systems

Crystal System	Interplanar Spacing of the (**hkl**) Plane
Cubic	$\frac{1}{d^2} = \frac{h^2 + k^2 + l^2}{a^2}$
Tetragonal	$\frac{1}{d^2} = \frac{h^2 + k^2}{a^2} + \frac{l^2}{c^2}$
Orthorhombic	$\frac{1}{d^2} = \frac{h^2}{a^2} + \frac{k^2}{b^2} + \frac{l^2}{c^2}$
Hexagonal	$\frac{1}{d^2} = \frac{4(h^2 + hk + k^2)}{3a^2} + \frac{l^2}{c^2}$
Trigonal	$\frac{1}{d^2} = \frac{4(h^2 + hk + k^2)}{3a^2} + \frac{l^2}{c^2}$
Rhombohedral	$\frac{1}{d^2} = \frac{(1 + \cos\gamma)[(h^2 + k^2 + l^2) - (1 - \tan^2\gamma/2)(hk + kl + hl)]}{a^2(1 + \cos\gamma - 2\cos^2\gamma)}$
Monoclinic	$\frac{1}{d^2} = \frac{(h^2)}{a^2(\sin^2\beta)} + \frac{k^2}{b^2} + \frac{l^2}{c^2(\sin^2\beta)} - \frac{2hl\cos\beta}{ac\sin^2\beta}$

fineness of the sampling that contributed to the synthesis. This process, described here in real space, is analogous to what is done in diffraction space in the course of an X-ray structure determination. We cannot directly visualize the distribution of atoms about the planes, but we can measure how the $\boldsymbol{hk\ell}$ families of planes diffract X rays as a direct consequence of the atomic distribution about them.

THE RECIPROCAL LATTICE

Initially it may not be obvious why we proceed to describe the following operations, which define what is called the reciprocal lattice, but bear up for awhile and hopefully your patience will be rewarded.

For every family of planes having integral Miller indexes $\boldsymbol{hk\ell}$, a vector can be drawn from a common origin having the direction of the plane normal and a length $\mathbf{1}/\boldsymbol{d}$, where $\boldsymbol{d}$ is the perpendicular distance between the planes. The coordinate space in which these vectors are gathered, as in Figure 3.19, is called reciprocal space, and the end points of the vectors for all of the families of planes form a lattice that is termed the reciprocal lattice.

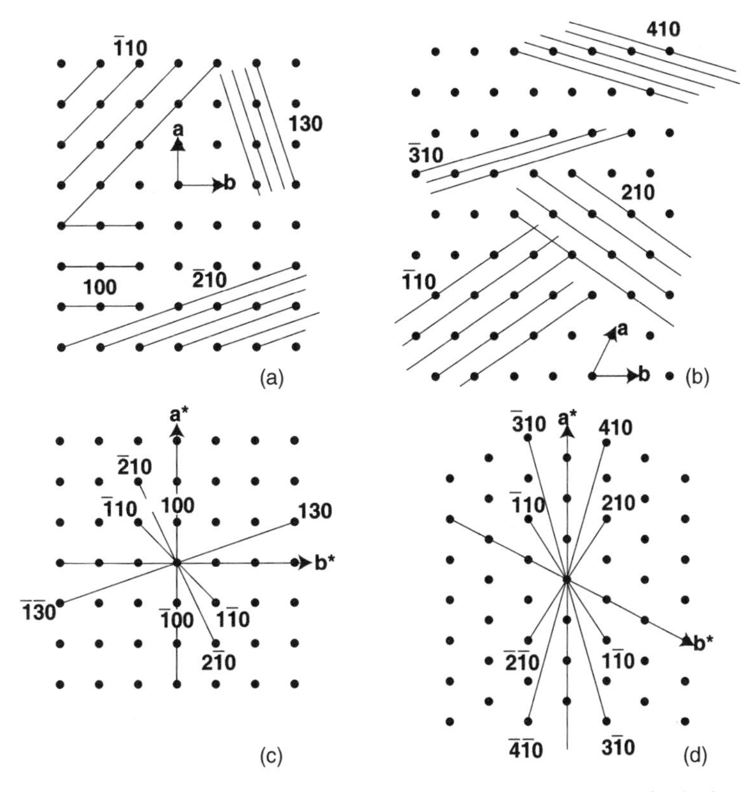

FIGURE 3.19 In (*a*) is a rectangular point lattice, and in (*b*) a hexagonal point lattice containing several families of planes of various orientations and interplanar spacings, denoted by their Miller indexes. In order to include every lattice point, only those families that divide the unit cell edges into an integral number of intervals (have rational intercepts) are permitted. In (*c*) and (*d*) are two-dimensional reciprocal lattices and the reciprocal lattice vectors corresponding to each of the families of planes shown in (*a*) and (*b*), respectively.

Any reciprocal lattice vector, or reciprocal lattice point is uniquely specified by the set of three integers, $hk\ell$, which are the Miller indexes of the family of planes it represents in the crystal. Thus there is a one-to-one correspondence between reciprocal lattice points and families of planes in a crystal. It will be seen shortly that the reciprocal lattice is the Fourier transform of the real lattice, and vice versa. This was in fact demonstrated experimentally in Figure 1.7 of Chapter 1 by optical diffraction. As such, reciprocal space is intimately related to the distribution of diffracted rays and the positions at which they can be observed. Reciprocal space, in a sense, is the coordinate system of diffraction space.

Like a crystal lattice, a reciprocal lattice is also specified completely by three unit vectors denoted by $\bar{a}^*$, $\bar{b}^*$, and $\bar{c}^*$. When the real lattice vectors are orthogonal, the reciprocal lattice vectors are also orthogonal and have lengths of $1/|\bar{a}|$, $1/|\bar{b}|$, and $1/|\bar{c}|$. The relationships for triclinic, monoclinic, and hexagonal systems having non 90° interaxial angles are not so direct. They contain some trigonometric terms, but they still may be readily derived from the real lattice parameters. Note that lattice points far from the origin of reciprocal space arise from families of planes of very small spacings, and those close to the origin from planes of low Miller indexes corresponding to large interplanar spacings. Thus it is possible to correlate directly the distance of a particular lattice point from the origin of reciprocal space with a measure of the level of detail sampled by its corresponding family of planes in the crystal. Points farther from the origin bear higher resolution information than those closer in.

The reciprocal lattice, like the lattice of the crystal, may also be divided into unit cells with the reciprocal unit vectors $\bar{a}^*$, $\bar{b}^*$, and $\bar{c}^*$ as edges. Since reciprocal space of a crystal is zero everywhere except at lattice points, however, the interiors of the reciprocal unit cells will be vacant. The relation between orthorhombic and monoclinic unit cells, and the corresponding reciprocal unit cells derived from them are shown in Figures 3.20 and 3.21. The type of reciprocal unit cell will be the same as the real cell from which it arises, and the reciprocal unit cell, hence the reciprocal lattice, will manifest the symmetry and centering properties of the real crystal lattice.

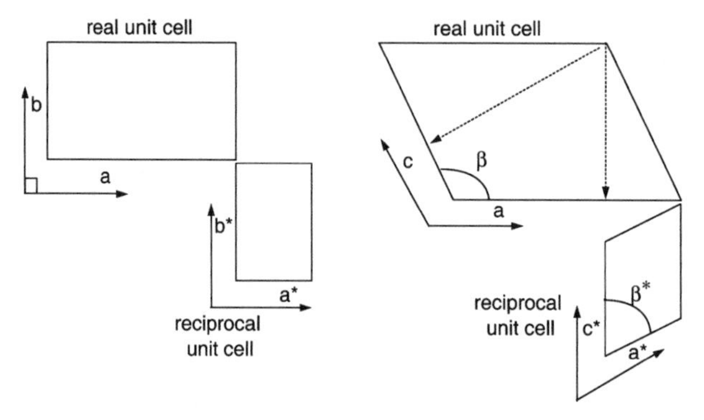

FIGURE 3.20 The orthorhombic crystallographic unit cell on the left has a corresponding orthorhombic reciprocal unit cell with $\bar{a}^*$, $\bar{b}^*$, and $\bar{c}^*$ parallel with $\bar{a}$, $\bar{b}$, and $\bar{c}$, respectively. The monoclinic unit cell at right has a corresponding monoclinic reciprocal unit cell in which the β^* angle is equal to $180^o - \beta$. In this system, $\bar{a}^*$ and $\bar{c}^*$ are not parallel with $\bar{a}$ and $\bar{b}$ but have the relationship that a^* and c^* are perpendicular to $\bar{c}$ and $\bar{a}$, respectively. A good point to remember is that whenever you are looking along any reciprocal lattice axis, you must then be looking perpendicular to a real unit cell face, and vice versa. This is true for any crystal, and for all crystal classes.

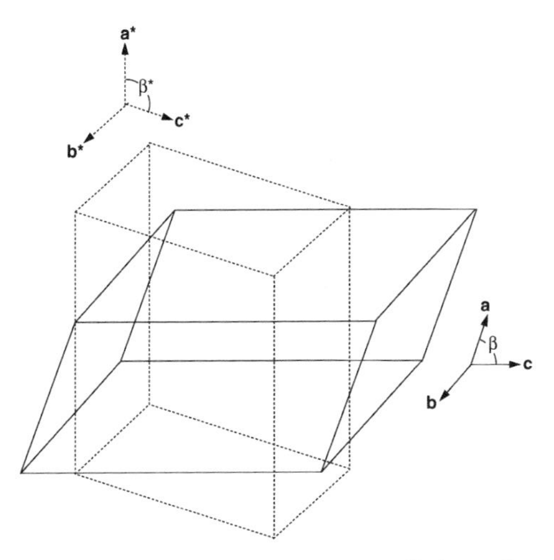

FIGURE 3.21 A real, three-dimensional monoclinic unit cell (*heavy lines*) and its corresponding reciprocal unit call (*thin lines*). The unique $\bar{b}$ and $\bar{b}^*$ axes (unique because of the single twofold axis along $\boldsymbol{b}$) are parallel for the monoclinic system, but the relationship of the $\bar{a}$ and $\bar{a}^*$ to the $\bar{c}$ and $\bar{c}^*$ axes depends on the angle between $\bar{a}$ and $\bar{c}$. Note the inverse relationship between the real and reciprocal edge lengths.

Since any family of planes $\boldsymbol{hk\ell}$ has a plane normal that is both positive and negative, that is, in opposite directions in space, every family of planes also gives rise to a second reciprocal lattice point, $-\boldsymbol{h} - \boldsymbol{k} - \boldsymbol{\ell}$. Thus reciprocal space will always contain a center of symmetry (see Figure 3.7) at its origin independent of the crystal from which it is derived.

CRYSTALS AS WAVES OF ELECTRONS IN THREE-DIMENSIONAL SPACE

From Figure 3.2, a crystal emerges as a virtually infinite array of identical unit cells that repeat in three-dimensional space in a completely periodic manner. Like a simple sine wave in one dimension, it repeats itself identically after a period of $\bar{a}$, $\bar{b}$, or $\bar{c}$ along each of the three axes. A crystal is in fact a three-dimensional periodic function in space, a three-dimensional wave. The period of the wave in each direction is one unit cell translation, and the value of the function at any point x_j, y_j, z_j within the cell, or period, is the density of electrons at that point, which we designate $\rho(x_j, y_j, z_j)$.

Because electrons are concentrated around atomic nuclei, knowing $\rho(x_j, y_j, z_j)$ for all points $\boldsymbol{j}$ is essentially the same as knowing the distribution of atoms in the unit cell, which in turn means the structure of the molecules which inhabit the unit cell. This is illustrated in Figures 3.22 and 3.23. Thus another way of looking at a crystal is that it is a three-dimensional, periodic, electron density wave that repeats in a perfectly regular manner in space. This is important because several hundred years of physics and mathematics have been focused on periodic waves and their properties, and many clever mathematical tools exist that allow us to manipulate and analyze them.

We need not discard the now familiar representation of a crystal seen schematically in Figure 3.2 in favor of a more abstract wave concept. The two are complimentary. Both describe different aspects of the same thing, and both can be held in mind at the same time.

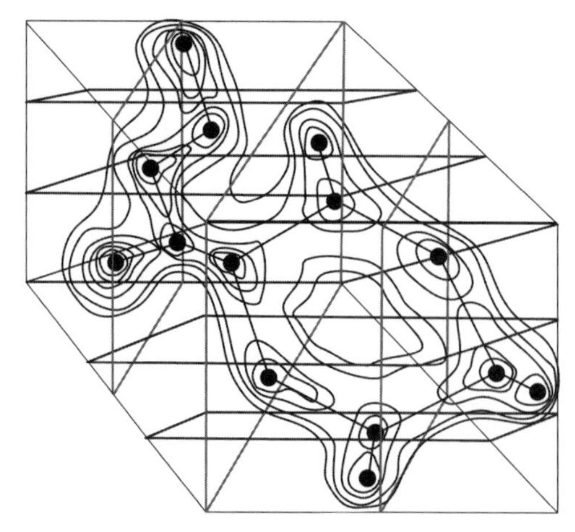

FIGURE 3.22 We can think of the distribution of atoms in a unit cell (here a single molecule, but it could just as well be a set of molecules related by space group symmetry) as a continuous electron density function that rises near atomic nuclei, falls to lower values between, and is otherwise zero. The electron density of one unit cell is contiguous with identical electron density in adjacent unit cells, and therefore throughout the crystal. Viewed in this way, it is evident that the electron density in a crystal is equivalent to a three-dimensional, periodic electron density wave that varies in amplitude over the course of one unit cell, namely over the course of $\bar{a}$, $\bar{b}$, and $\bar{c}$, but repeats identically along the three unit cell directions. $\bar{a}$, $\bar{b}$, and $\bar{c}$ are equivalent to the period, or wavelength, of the electron density wave function. Families of planes, like those shown, section or sample the electron density wave at regular intervals (i.e., with distinct frequencies, and identically for all unit cells of the crystal. The families of planes define the harmonic components of the electron density waves, that is, its spectrum (or Fourier components).

The important thing is that what is true about one concept must be true, and have some analogue, for the other.

The reason for introducing the idea of the crystal as a periodic electron density wave $\rho(x, y, z)$ is that any periodic wave, no matter how elaborate its form over a single period, no matter how complex the distribution of atoms, can be synthesized by adding together sine waves of all frequencies that have the correct phases and amplitudes (see Chapter 4). The electron density that describes a molecule, even a complex macromolecule, can therefore be synthesized by properly adding together a large number of component waves. In other fields we might refer to these contributing waves as harmonics or the spectral components of the complex wave.

The converse of what has just been described must also be true. If a complex wave can be synthesized from its harmonic components, then one must also be able to break down, or analyze, a complex wave into spectral components. That is, given the electron density within a unit cell, one should be able to derive in some fashion its component waves. What, in the real crystal, corresponds to the harmonic components we have been discussing? The families of planes, each characterized by a set of Miller indexes, are the spectral components of the unit cell and its contents, the electron density.

We might look at this in yet another way. If you systematically cut, or sectioned a unit cell in a manner corresponding to a particular family of planes, say the $h = 3, k = 5, l = 7$ family, then you could examine what lay on each of the planes in the family and then have

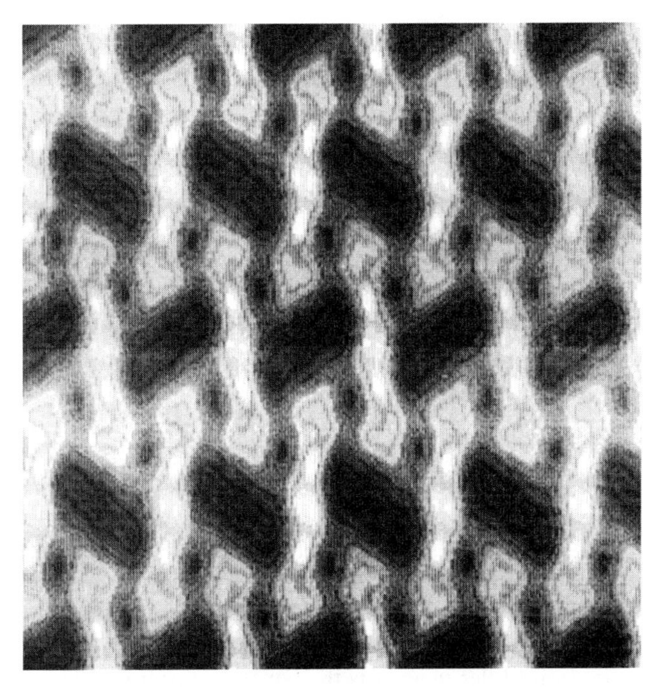

FIGURE 3.23 The digitized image of an electron micrograph of a negatively stained microcrystal of B. subtilis α-amylase. Here one gets the visual impression of an electron density wave function periodically repeating in space. Indeed a comparable image is obtained by Fourier transform of the low resolution X-ray diffraction data recorded from larger crystals.

at least some idea of what the contents of the unit cell looked like. It wouldn't be a very good idea because the planes are separated from one another by the interplanar spacing $\boldsymbol{d}$, and the electron density could change appreciably from one plane to another, but it would provide what we might call a low-resolution image.

Now let us section the unit cell again, only more finely, say we cut it to reveal the $\boldsymbol{h = 6, k = 10, l = 14}$ or even the $\boldsymbol{h = 12, k = 20, l = 28}$ planes. Clearly, as we sliced the cell more and more finely, we would get a progressively better idea of what was inside. We could do even better. Instead of using families of planes parallel to one another but of increasingly finer spacings, as we did here, we could also section the cell along many other directions and sample the electron density. Families of planes having high indexes (and small d spacings) would provide what we term high-resolution information, while families of planes of lower indexes would provide lower resolution information.

In the end we could obtain the best possible image of the contents of the unit cell, the molecules, by systematically combining together the information, or electron density images, yielded by each of the families of planes that can be drawn through the unit cell. Now we can't, of course, really do that because we don't have knives so sharp or eyes as keen as we would need to section a single unit cell. But we can measure the diffraction of X rays by the families of planes, and as we will later see, those diffracted rays carry much the same kind of information.

The family of planes within a single unit cell cannot, of course, provide a diffracted ray intense enough to be measured, but the family of planes cutting through all of the unit cells in the crystal can. As noted already, as long as they have integral Miller indexes, then every family of planes will cut through every unit cell in exactly the same way. This being true, then all the unit cells throughout the crystal will produce exactly the same diffracted ray for a particular family of planes, and these will sum together (when Bragg's law is satisfied; see Chapter 5) to give an observable reflection.

Clearly, the atoms, or electron densities in the unit cells, do not lie exactly on planes, any family of planes. In reality, they lie mostly in between. Why then should imaginary planes, which these planes are, scatter X rays at all? The answer is that they don't. We can, however, make the assumption that all atoms do lie on nearby planes of each family, and then compensate for the fact that they are really displaced from the planes when we calculate the diffracted wave from those planes. This is somewhat analogous to assuming all of the mass of an object is concentrated at its center of mass, or representing a complicated charge distribution by a dipole moment.

In summary then, a crystal can be conceived of as an electron density wave in three-dimensional space, which can be resolved into a spectrum of components. The spectral components of the crystal correspond to families of planes having integral, Miller indexes, and these can, as we will see, give rise to diffracted rays. The atoms in the unit cell don't really lie on the planes, but we can adjust for that when we calculate the intensity and phase with which each family of planes scatter X rays. The diffracted ray from a single family of planes (which produces a single diffraction spot on a detector) is the Fourier transform of that family of planes. The set of all diffracted rays scattered by all of the possible families of planes having integral Miller indexes is the Fourier transform of the crystal. Thus the diffraction pattern of a crystal is its Fourier transform, and it is composed of the individual Fourier transforms of each of the families of planes that sample the unit cells.

CHAPTER 4

WAVES AND THEIR PROPERTIES

In a mathematical sense, a periodic wave is any function $f(x)$ whose value varies in a repetitive and perfectly predictable manner over discrete intervals of some variable x. A physical way of describing waves is that they are some property of the medium in which they exist that changes in a regular and periodic manner as a function of the distance from some point, or as a function of time if one stands at a fixed point in space and measures the unique property. For sound, the property may be pressure; for waves in the water, it may be height above or below the surface; for light or X rays, the properties are electromagnetic.

THE PROPERTIES OF WAVES

Waves have quantifiable properties. In our physical world of time and space, the maximum value that a wave can attain is its amplitude. The distance it travels before it repeats is its wavelength, the time required for it to travel one wavelength, or to complete one full oscillation is its period. The number of periods it completes per unit time is its frequency. Period and frequency are inverses of one another. We normally say that the period of a wave corresponds to **360°** (or 2π), and that any point within the period corresponds to some angle between zero and **360°**.

A simple wave with which most of us are familiar, seen in Figure 4.1*a*, is the sine wave. It is like the ripples on smooth water when you cast a stone. As the wave progresses from its point of origin it rises continuously to a maximum, decreases continuously through zero to its negative minimum, and then returns in the same way to its starting point at zero. We can have more complex waves like that in Figure 4.1*b*, that experience complicated variations over their period, and we can, as in Figures 3.22 and 3.23 from the previous chapter, have waves in two or three dimensions. The important point, however, is that any wave, no matter how complex, begins again and repeats its course after completing a period or traversing a

Introduction to Macromolecular Crystallography, Second Edition By Alexander McPherson

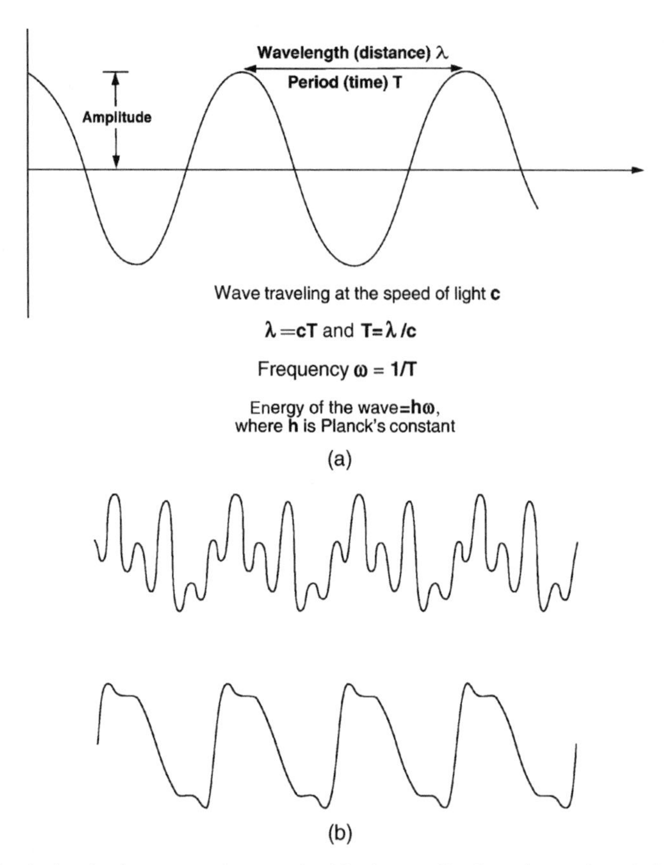

FIGURE 4.1 A simple sine wave, characterized by its amplitude and wavelength, is shown in (*a*). If it is an electromagnetic wave such as light or X rays, then it travels with the speed of light and has associated with it a distinctive energy. In (*b*) are shown two examples of more complex waves, which nonetheless have distinct frequencies, or wavelengths. These are waveforms produced by a violin (top) and a piano (below) playing concert **A**. As Fourier demonstrated, complex, but periodic, waves such as these can be constructed by adding together simple sine waves like those in (*a*), but each having a different amplitude, phase, and frequency. Think of these, if you will, as one-dimensional electron density waves in a crystal.

wavelength. Aside from its complexity, it is otherwise no different than a sine wave. It follows that if we measure $n(\lambda)$ along the direction of propagation from any point on a periodic wave, we find ourselves with exactly the same value of $f(x)$, but n wavelengths ahead.

Now a subtle complication arises, for all kinds of waves, if we choose a fixed point in space or time and call that the origin of our system. Usually we can take the origin anywhere we choose, but we have to pick somewhere. If there is only a single wave in our system, then we logically would choose the point on that wave where it is zero as our origin. After one period of 2π it returns to that value. But what about a collection of independent waves, each originating from a different starting point? Again, we can choose a point of zero amplitude on one of the waves as the origin; that means the corresponding starting points for all of the other waves are shifted positively or negatively by some distance x with respect to this origin. They will be displaced some fraction of their whole wavelength x/λ, or some angle $\phi = 2\pi x/\lambda$ between 0 and 2π. We call ϕ the phase angles relative to the first wave.

Given any set of waves, no matter how large, we are seldom interested in finding some absolute origin; we will leave that to cosmologists. What we are concerned with in X-ray diffraction is knowing the relative phase angles of every wave in our system with respect to some single, arbitrary wave. In general, it doesn't matter which of the waves we choose to define our origin, the relative phases of the other waves will always be the same. The question of relative phase is, as we will see, the central problem of X-ray crystallography and the key to solving macromolecular structures.

When we have a large number of individual waves, like those produced by the scattering of X-rays from families of planes, or from all of the unit cells in a crystal, or from all of the atoms within a unit cell, we are ultimately interested in knowing how all of the waves add together to yield a resultant wave that we can observe, characterize, and use. Waves are more complicated to sum than simple quantities like mass or temperature because they have not only an amplitude, a scaler, but also a phase angle ϕ with respect to one another. This must be taken into account when waves are combined. As will be seen below, waves share identical mathematical properties with vectors (and with complex numbers, which are really nothing but vectors in two dimensions).

Before beginning a discussion of the mathematics of waves diffracted by atoms in a crystal, and how they can be added, it is important to note at least one profound simplification in X-ray crystallography. The physical waves we will be dealing with, our X-rays, because they are experimentally generated, all have the same frequency (and wavelength). That frequency does not change as a consequence of being scattered, or diffracted, by atoms. Thus we can, in understanding how a crystal diffracts X rays, essentially neglect features of the mathematics that are frequency dependent. Frequency will, however, return as an important property when we consider how the contents of a crystal unit cell, its electron density, are reconstructed from the diffraction pattern.

WAVES AS VECTORS AND COMPLEX NUMBERS

Any sinusoidal wave such as sound, light, electrons, or X rays traveling through space and varying with time may be described by an expression of the form

$$\bar{K} = K \sin(\omega t + \phi),$$

where $\bar{K}$ represents the wave at some point of observation, K is the amplitude or maximum value that $\bar{K}$ can attain, ω is the wave frequency in degrees or radians per unit time, t is time, and ϕ is the phase angle that relates the origin of the wave to a common origin point for all such waves.

X rays are electromagnetic waves and like all such waves, they travel at the speed of light and they contain energy (we will not demonstrate all of that here, but we refer the reader to Maxwell, 1878). The energy of an electromagnetic wave is $h\omega$, where ω is the frequency and h is Planck's constant. In X-ray crystallography we are usually not consciously concerned with energies, but in a sense, we do have to pay attention to how it is distributed in the wave.

An electromagnetic wave has two components, electric E and magnetic H. The two components vary in a coordinated fashion with time, and they are orthogonal in space with respect to one another as shown in Figure 4.2. As the wave proceeds through space (and time) the electrical component E varies sinusoidally, and the magnetic component of equal amplitude H does the same, but the two sinusoidal waves are out of phase with one another by $90°$, or $\pi/2$.

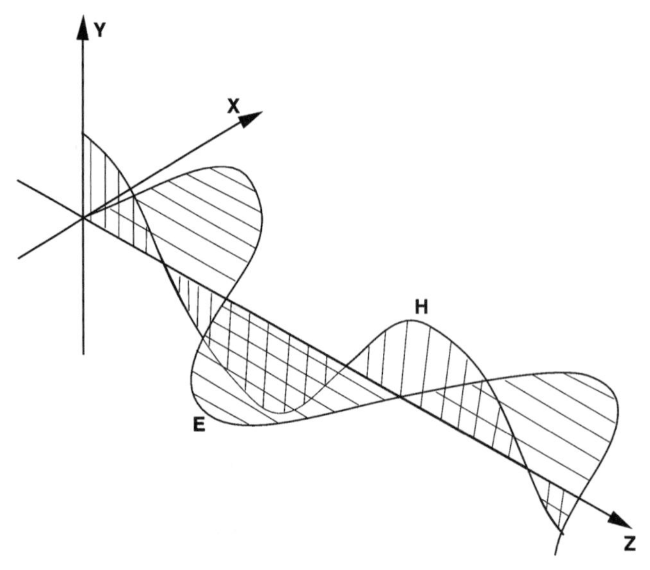

FIGURE 4.2 An X ray is an electromagnetic wave traveling at the speed of light, which distributes its total energy between an electric field ***E*** and a magnetic field ***H***. The energy oscillates in a sinusoidal fashion between the two fields, which are orthogonal to one another in space. When the energy is a maximum in ***E***, it is zero in ***H***, and vice versa. In X-ray diffraction we are interested in the electrical field component because that is what is scattered by electrons, and what can be detected in practice. We can therefore set the magnetic field aside and consider the X ray to be an oscillating electrical wave that travels through space and varies between zero and its maximum value in a sinusoidal fashion.

When the electric component is at its maximum or minimum and contains all of the wave's energy, the magnetic component is zero, and vice versa. At any intermediate point the energy is distributed between the two components but always sums to the constant energy of the wave $h\omega$. Because the ***E*** and ***H*** components are out of phase by **90°**, we can assign one to be a sine wave and the other a cosine wave and recall that $\sin^2 + \cos^2 = 1$. Thus the total energy of the wave at any point or time must be the square root of the amplitude of ***E*** squared plus the amplitude of ***H*** squared.

In X-ray diffraction investigations we are seldom interested in the magnetic component of the waves because it is the electric component that interacts with electrons around atoms and produces diffraction effects. It does, however, soothe the mind to remember that the magnetic component is where the extra energy is found when ***E*** is not at its maximum (in fact all of it when ***E*** = zero). If we consider the wave as a function of distance rather than time, we may express an X ray as a simple sine wave of electric energy ***E*** oscillating between its maximum, zero, and its minimum, as a wave represented mathematically as

$$\begin{aligned}\bar{K} &= K \sin([\omega t] + \phi) \\ &= K \sin\left(\left[\frac{2\pi x}{\lambda}\right] + \phi\right),\end{aligned}$$

where ω (in radians/sec) is the frequency of the wave, t is time, λ (in **Å** or nm) its wavelength, and x is the distance from the origin at which we observe or measure it, and ϕ is the phase angle of the wave with respect to the origin of our coordinate system. Why we use K instead of E in X-ray diffraction, I do not know.

Note that if a number of waves having the same λ and ω are traveling in the same direction, and we are interested in adding them together, it really does not matter at what t or x we choose to do the summing, because the waves' relative phases ϕ remain the same. Thus, when we are concerned only with adding scattered waves together, as we usually are in X-ray diffraction, we can essentially ignore ωt and $2\pi x/\lambda$.

ADDITION OF WAVES

As illustrated in Figure 4.3, a series of waves, in this case three, can be combined by direct scaler addition of their respective amplitudes at any given point along the time or distance axis to give the amplitude of the resultant wave at that point. This can be done for many points, thereby describing the entire resultant wave. The synthesized, or resultant wave depends not only on the amplitudes of the component waves but also on their relative phases. Note that the resultant wave has the same λ and ω as the waves that produced it. The combination, or addition of the waves in this manner illustrates the concept of interference. The addition of many waves, point by point, is cumbersome and impractical. When all of the waves are of the same frequency and wavelength, but of varying phases and amplitudes, which is the case with scattered monochromatic X radiation, summation of waves can be expedited by employing vectors.

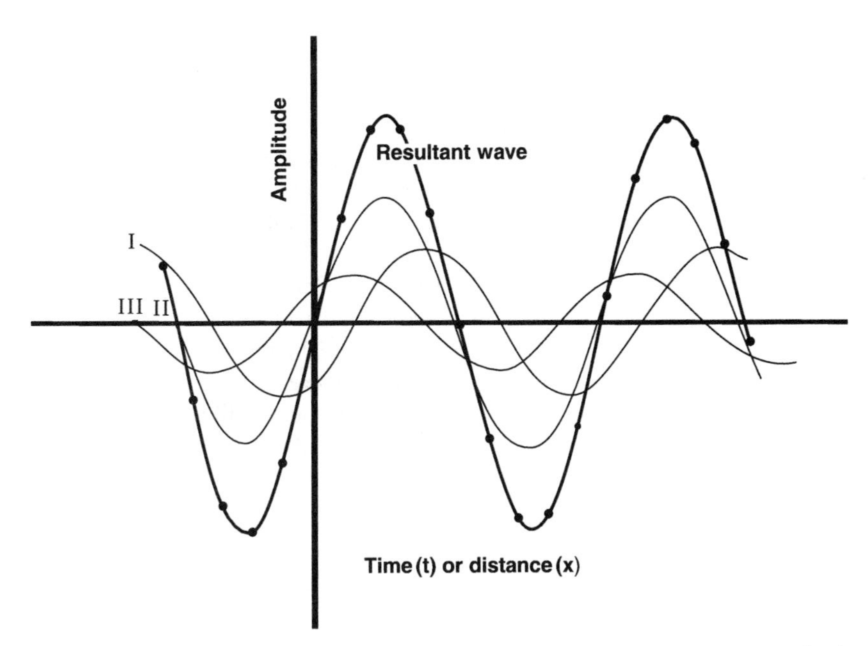

FIGURE 4.3 Three waves of different amplitudes and different phases (starting points), but having common angular velocity (equivalent to frequency ω or wavelength λ), may be added graphically, point by point along the horizontal axis, to give a resultant wave. We say that the three wave components, **I**, **II**, and **III** constitute the spectrum of the resultant wave. Their addition in this manner is an example of wave interference, or diffraction.

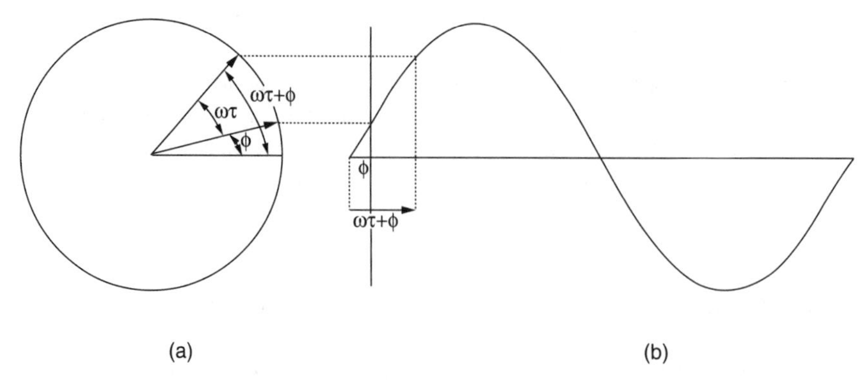

FIGURE 4.4 The horizontal and vertical axes here define an origin in space. Any wave beginning at the origin is said to have a phase of zero. The sinusoidal wave at right begins with a phase of ϕ and travels with an angular velocity or frequency ω, oscillations per second or **360°/s**, or **2π/s** so that at any time t (in seconds) the wave will have traveled through an angle $\phi + \omega t$ with respect to a wave beginning at the origin. The wave at right then has the same mathematical properties as the vector to the left rotating in a plane, from a starting angle ϕ, also with an angular velocity ω. The vector may be thought of as the projection of the wave amplitude at any time t onto the circle, which the vector describes. Simultaneous with the wave completing a full oscillation, or cycle, the vector completes a full rotation.

A wave $\bar{K}$ has an equivalent vector representation, which is illustrated in Figure 4.4. A wave can be considered a vector in the complex plane of length equal to the amplitude of the wave, K, starting with some phase angle ϕ and rotating around the origin with a frequency ω. If we have a collection of waves, as in Figure 4.5*a*, then at time zero they can all be displayed in the complex plane as in Figure 4.5*b*.

The vectors are all rotating with frequency ω, hence they are rotating in lockstep with one another. At any later time, a million rotations later, for example, their relative phase angles will be identically the same. Thus, if we wish to add them together, we can simply consider them stationary vectors at time zero (or any other time for that matter) and add them together as we would any vectors in the complex plane. Because the vectors are defined by length (amplitude) and phase, as are the corresponding waves, addition of the vectors as in Figure 4.5*c*, is equivalent to adding together the waves. The resultant vector rotating with frequency ω is an exact representation of the resultant wave.

As indicated in Figure 4.6, a vector $\bar{K}$ in the complex plane will have a real component $K\cos\phi$, and an imaginary component $iK\sin\phi$. The end of the vector $\bar{K}$ defines a point in the plane, and the coordinates of that point are $K\cos\phi$, $iK\sin\phi$. Thus we see that a wave in space can be defined not only as a vector in the complex plane, but as a complex number

$$a + ib = K(\cos\phi + i\sin\phi).$$

The sum of a large number of waves having the same frequency, then, may be obtained by adding their corresponding vector representatives in the complex plane, which is still awkward, or by simply adding up a list of complex numbers of the form $a + ib$. We can always recover the resultant amplitude K and phase ϕ of the wave from the corresponding

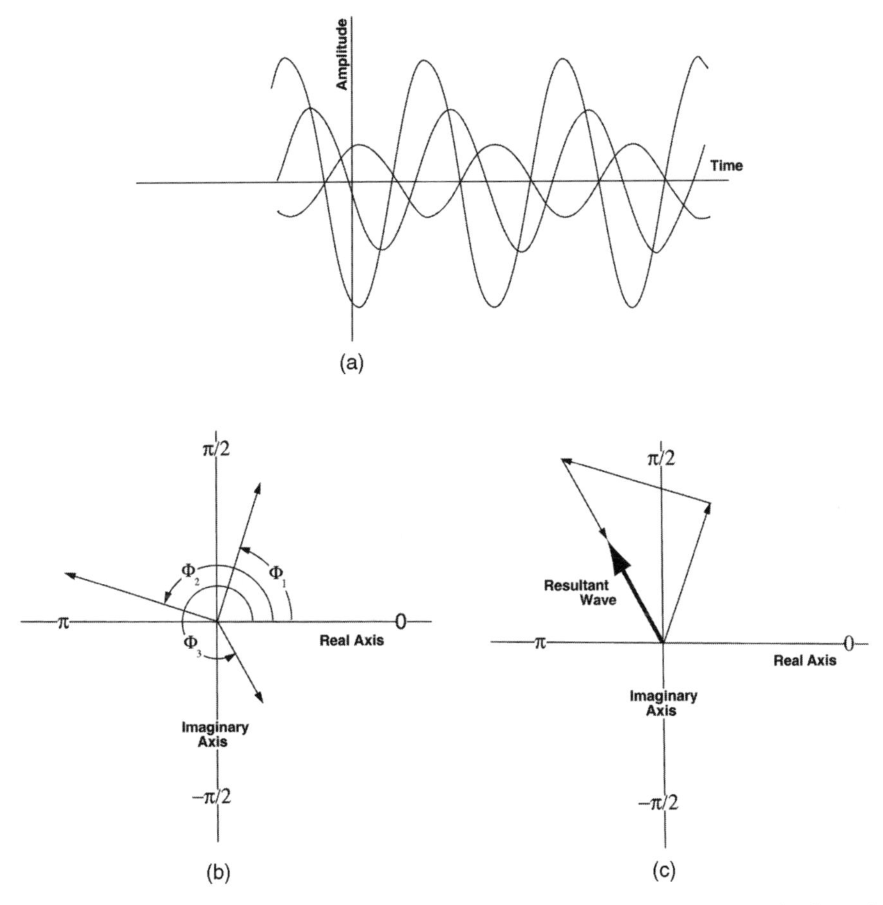

FIGURE 4.5 In (*a*) are three waves of different amplitudes and different phases, ϕ_1, ϕ_2, and ϕ_3 traveling with the same angular velocity ω (frequency). As shown in Figure 4.3, the three waves will add together to produce some resultant wave having the same angular velocity ω and wavelength λ. The three waves are equivalent to vectors rotating with frequency in the complex plane as shown in (b). The vectors have lengths proportional to the amplitudes of their respective waves, and the same phases. Because the vectors rotate with identically the same angular velocity, the phases with respect to one another remain the same; that is, the relative phases are time and distance independent. Thus the resultant vector in (*c*), obtained simply by vector addition of the three vectors in (*b*), is also time independent. We can therefore neglect ωt in adding waves of the same frequency, or wavelength, since their resultant wave depends only on their relative phases and amplitudes.

complex number $a + ib$ by remembering two properties of complex numbers. First, the magnitude of a complex number is equal to its absolute value, which is[1]

$$|\bar{K}| = \sqrt{(a + ib)(a - ib)}.$$

[1] $(a + ib)$ and $(a - ib)$ are called complex conjugates of one another.

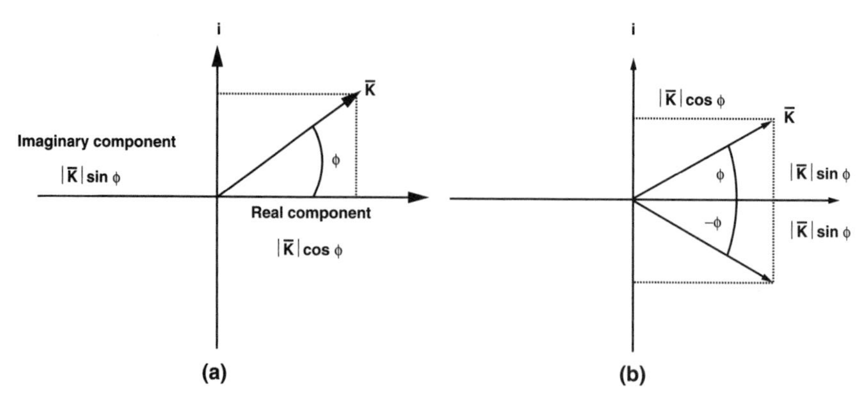

FIGURE 4.6 A vector in the complex plane is composed of a real component, which is its projection on the real axis, and an imaginary component, which is its projection on the imaginary axis. If the vector $\bar{K}$ has length K, and phase ϕ, then the real component is $K \cos \phi$, and the imaginary component is $iK \sin \phi$. Note in (*b*) that for any vector, or wave, the real components are identical for phase angles ϕ and $-\phi$, but that the imaginary components are of opposite arithmetic sign.

Second, looking again at the vector representation of $\bar{K}$, we see that

$$\tan \phi = \frac{\sin \phi}{\cos \phi},$$

which is the same as b/a.

Thus

$$\phi = \tan^{-1} \frac{b}{a}.$$

Complex numbers, which incorporate the imaginary number i, generally strike fear in the hearts of nonmathematicians. The concept of a number, part of which is real and part of which is imaginary, seems to have no reality, no correlate in the physical world. Complex numbers, though, are really just a clever accounting scheme. The total value of a complex number (its absolute value) never changes, but its total value is distributed between two accounts, which we call real and (unfortunately named) imaginary. The distribution of value may flow from one account into the other, from real to imaginary, and vice versa, but the total assets remain constant. We may add the real accounts to other real accounts but not to imaginary, and vice versa. This is exactly analogous to the flow of energy in an electromagnetic wave between the electrical and magnetic components.

The vector representation of waves is often useful in illustrating diffraction phenomena because it allows one to express the radiation scattered by an atom as a single vector, thus permitting visualization of the total scattering from a large array of atoms in terms of a single resultant vector. The complex number representation is advantageous where large numbers of waves must be manipulated, for example, in a digital computer. In addition to the trigonometric, complex, and vector representations of waves, there is an additional relationship ascribed to Euler that establishes an identity between complex numbers and exponential functions such that any wave may also be expressed as

$$\bar{K} = K(\cos \phi + i \sin \phi) = K \exp(i\phi).$$

This last equation provides yet another means of representing and manipulating waves. Because of the compactness of its notation, it is frequently used for expressing diffraction relationships. It is also employed when waves must be multiplied, since the product of two exponentials is obtained simply by adding their exponents. This is important in terms of defining convolutions, which are the products of wave functions. We will encounter those in Chapters 5 and 9.

MANIPULATING VECTORS

Along the way, we will need to carry out a few vector operations, and it may be appropriate here to review some fundamentals. First, two vectors are added by placing the tail of one vector at the head of the other, in either order, and drawing the resultant vector from the free tail to the free head. Two vectors are subtracted from one another by either of two operations. The sense of one vector is reversed, and the two vectors then added as above. Alternatively, the tails of the two vectors are placed at a common point and the resultant vector connecting their two free heads is the difference vector. The sum or the difference of two vectors is always another vector.

Because vectors have three scaler components in space, $\bar{x} = (x, y, z)$, then they can be added and subtracted arithmetically as well. The sum, or resultant, of two vectors $\bar{x}_1 = (x_1, y_1, z_1)$ and $\bar{x}_2 = (x_2, y_2, z_2)$ is $\bar{x}_1 + \bar{x}_2 = (x_1 + x_2, y_1 + y_2, z_1 + z_2)$. Similarly the difference between the two vectors is $\bar{x}_1 - \bar{x}_2 = (x_1 - x_2, y_1 - y_2, z_1 - z_2)$.

Vectors may be multiplied in two ways. These are called the scaler product (or dot product) symbolized by the two vectors separated by a dot, namely $\bar{a} \cdot \bar{b}$, and the vector product (or cross product) symbolized by the two vectors separated by a multiplication sign, $\bar{a} \times \bar{b}$. The dot product is always a scaler quantity, and the cross product is a vector.

The dot product $\bar{x}_1 \cdot \bar{x}_2$ is the number obtained by multiplying the length (or magnitude) of one vector times that of the other, times the cosine of the angle between them when the tails are placed at a common point. The length of a vector multiplied by the cosine is the projection of that vector on the other. The dot product may also be written in component form as $\bar{x}_1 \cdot \bar{x}_2 = (x_1x_2 + y_1y_2 + z_1z_2)$.

The cross product $\bar{x}_1 \times \bar{x}_2$ is again the product of the two lengths, or magnitudes, multiplied here by the sine of the angle between them when their tails are placed at a common point. The result in this case is not, however, a scaler, but a vector that is defined as having direction perpendicular to the plane formed by the two component vectors, and length equal to the product of their lengths times the sine of the angle. The plus or minus direction of the cross product vector is determined by the right-hand rule when the first vector is rotated into the second. Thus the order of multiplication of the vectors forming the cross product is important, as it determines the sign of the direction of the product vector.

SOME USEFUL WAVE RELATIONSHIPS

There are some simple wave relationships that are useful to bear in mind, corollaries so to speak, that can often reduce otherwise imposing problems to trivial exercises. First, we must often define the relative phases at some point of two or more waves traveling through space in order to know how they combine. We can do this in a straightforward manner if we know the relative distance, measured in units of λ, that the waves have traveled from their original source to the point of observation (detector). Remember that a wave traversing a distance of λ corresponds to its equivalent vector rotating through an angle of $\mathbf{360°}$ or $\mathbf{2\pi}$. In Figure 4.7, for example, waves begin at $\boldsymbol{x} = \mathbf{0}, \boldsymbol{\lambda}/\mathbf{4}, \boldsymbol{\lambda}/\mathbf{2}, \mathbf{3}\boldsymbol{\lambda}/\mathbf{4},$ and $\boldsymbol{\lambda}$. Obvious from the diagram is that the two waves beginning at $\boldsymbol{x} = \mathbf{0}$ and $\boldsymbol{x} = \boldsymbol{\lambda}$ have exactly the same phase. This would be true of a wave starting at $\mathbf{2\lambda}$ or $\mathbf{50\lambda}$ or $\boldsymbol{n\lambda}$. Waves starting at $\boldsymbol{x} = \boldsymbol{\lambda}/\mathbf{4}$ would have a phase angle (with respect to the wave beginning at $\boldsymbol{x} = \mathbf{0}$) of $-\boldsymbol{\pi}/\mathbf{2}$, that from $\boldsymbol{x} = \boldsymbol{\lambda}/\mathbf{2}$ of $-\boldsymbol{\pi}$, etc. Thus we can always determine the relative phases of two waves by

$$\phi = \frac{\textbf{path length difference}}{\textbf{wavelength } \lambda} \times 2\pi \text{ (or) } \mathbf{360°},$$

where $\boldsymbol{\phi}$ is the phase difference. Since we can always set the phase of the first wave, or reference wave equal to $\mathbf{0}$, then the phase of the next wave will be $\boldsymbol{\phi_1}$, the next $\boldsymbol{\phi_2}$, and so on.

If two waves, as in Figure 4.8*a*, have the same amplitude and they have only a very small phase difference, that is $\boldsymbol{\phi} \cong \mathbf{0}$ or $\mathbf{2}\boldsymbol{n\pi}$, where $\boldsymbol{n}$ is an integer, then they will add together to give a resultant wave of almost twice the amplitude of either. That is, they give a resultant wave of nearly the greatest possible amplitude, and we say that they constructively interfere. If there are $\boldsymbol{N}$ waves having the exact condition of $\boldsymbol{\phi} = \mathbf{0}$, or $\mathbf{2}\boldsymbol{n\pi}$, then they produce a resultant wave with amplitude $\boldsymbol{N}$ times that of the individuals. All $\boldsymbol{N}$ waves perfectly, constructively interfere. Conversely, if two waves of equal amplitude have a phase difference very close to $\boldsymbol{\phi} = \boldsymbol{\pi}$, or $\mathbf{180°}$, they almost cancel as in Figure 4.8*b*, and they are said to destructively interfere.

Now some more interesting cases may be considered. If we have a very large number of waves *N* of equal, or even approximately equal amplitude but random phase, what will their sum be? Statistically it will tend to be zero. Why? Because for every wave having

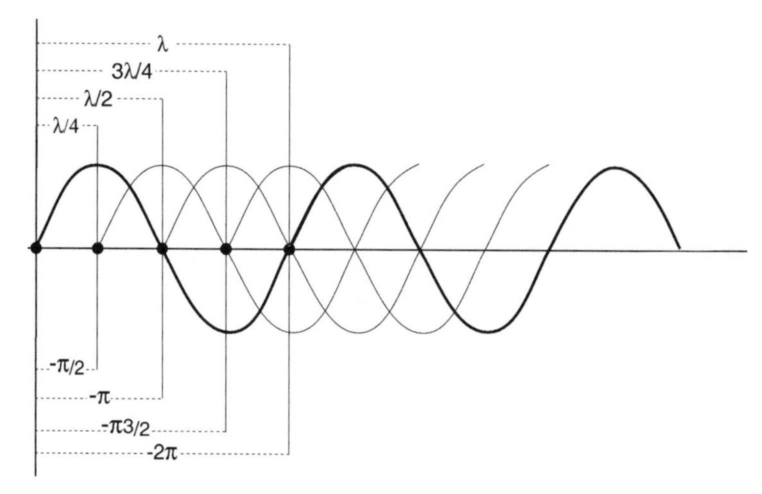

FIGURE 4.7 Shown are four sine waves that begin after a reference sine wave initiates at the origin and has a phase $\boldsymbol{\phi} = \mathbf{0}$. The waves begin $\boldsymbol{\lambda}/\mathbf{4}, \boldsymbol{\lambda}/\mathbf{2}, \mathbf{3}\boldsymbol{\lambda}/\mathbf{4},$ and $\boldsymbol{\lambda}$ behind the reference wave and have corresponding phases of $-\mathbf{360°}/\mathbf{4} = -\mathbf{90°}$, $-\mathbf{360°}/\mathbf{2} = -\mathbf{180°}$, and so on. Alternately, the phases of these waves may be written as $-\mathbf{2}\boldsymbol{\pi}/\mathbf{4}$, $-\boldsymbol{\pi}$, and so on. Thus there is a direct linear relationship between distance traveled and phase in terms of fractions and multiples of wavelength.

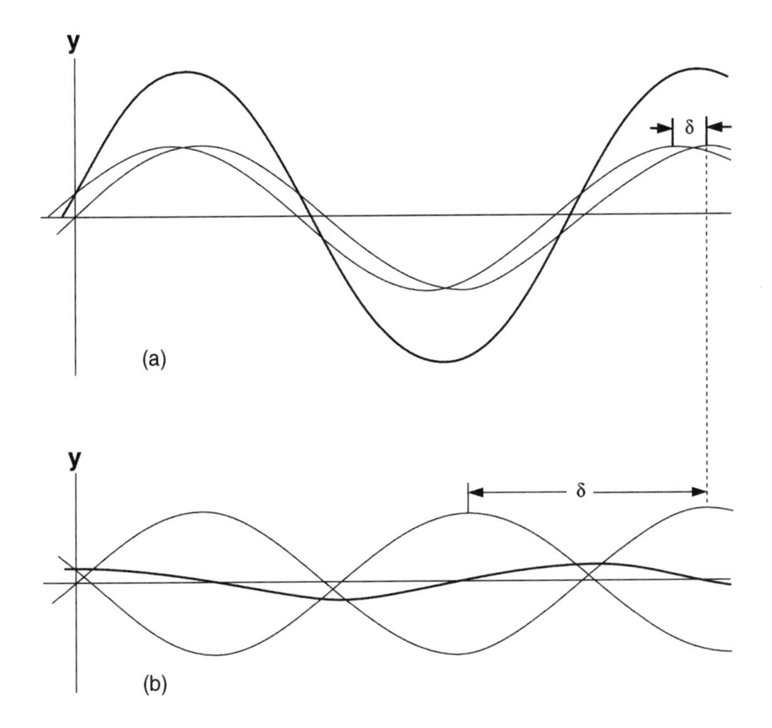

FIGURE 4.8 In (*a*) two waves of equal amplitude and wavelength have a small phase difference δ. Their resultant, shown as the darker wave, has a large amplitude that is very nearly the simple arithmetic sum of the amplitudes of the two components. Thus the two waves constructively interfere. As δ tends to zero, the reinforcement increases, and when $\delta = 0$, there is perfect constructive interference. At this condition of $\delta = 0$ the amplitude of the resultant wave exactly equals the sum of the component amplitudes. In (*b*) two equivalent waves have a phase difference close to $180°$ and destructively interfere to yield a resultant wave of low amplitude. When $\delta = 180°$, the two waves completely cancel and no resultant wave is produced.

phase ϕ we can find some other wave in the set of phase $\phi + \pi$, and the two will cancel. Thus nearly all waves cancel in pairs and the result will tend to zero. While this is exactly true when N is extremely large and approaches ∞, it is virtually true whenever N is any large number.

Another particularly useful case to consider, because it can cut short numerous proofs and arguments to follow, is that of an infinite, or at least extremely large set of waves of the same, or similar, amplitudes, but differing in their phases by some constant amount Δ. That is, wave two differs in phase from wave one by some small increment Δ, wave three from wave two by Δ, and so on. What will be the sum of this series of waves?

All the waves have similar amplitudes, and for every wave with a phase of ϕ there will again be another somewhere in the series having a phase $\phi + 180°$. These two waves will destructively interfere, and they will sum to zero. This will be true for all waves in the series. Thus the series must sum to zero. The only case where this will not be true is when the phase difference Δ is exactly 0 or $2n\pi$, that is, when all the waves are perfectly in phase.

In crystallographic problems this situation arises frequently because diffraction from a crystal can be thought of as involving waves scattered from an almost infinite number of planes in a family, or the almost infinite number of unit cells in a crystal. If those waves are consistently out of phase with one another by a fixed increment, whether that increment be large or small, the resultant wave will be zero. But, when the increment becomes exactly zero, all waves scattered by every plane in a family or every unit cell in the crystal sum together in a perfectly constructive manner. What we have just described can be formulated as

$$\text{resultant wave}\,(\bar{K}) = \sum_{n} \exp\, i2n\pi(\bar{a} \cdot (\bar{k} - \bar{k}')),$$

using Euler's notation for the individual waves that go into the summation. We use this ominous looking formulation of an essentially simple idea not to frighten, but because it is exactly the same formulation that will appear, hopefully in more comprehensible form in the next chapter when we consider the diffraction of X rays by atoms and by families of planes. Here the Δ phase difference is variable and is equal to $(\bar{a} \cdot (\bar{k} \cdot \bar{k}'))2n\pi$. Notice that $\bar{a} \cdot (\bar{k} \cdot \bar{k}')$ will, in general, not be integral. Thus the phase difference will be something other than 2π. If $\bar{a} \cdot (\bar{k} \cdot \bar{k}')$ assumes integral value, however, the phase difference between successive waves in the series disappears, and the waves will sum exactly. All of their amplitudes will be additive. For a large number of waves N, with an average amplitude of I_{avg}, the sum of the waves will simply be $N(I_{avg})$.

The formulation above is known in mathematics as a Dirac delta function. The delta function is everywhere zero except when some condition is satisfied (here, $\bar{a} \cdot (k - k')$ being integral), and then it suddenly attains its maximum possible value. This is equivalent to the physical phenomenon we know of in diffraction analysis and crystallography as the constructive interference of waves. As we will see in the next chapter, Bragg's law is simply one expression of this function.

THE FOURIER SYNTHESIS, PLANES, AND THE ELECTRON DENSITY

To this point we have been interested in the scattered waves, or X rays from atoms that combine to yield the observed diffraction from a crystal. Because the waves all have the same wavelength, we could ignore frequency in our discussions. In X-ray crystallography, however, we are equally interested in understanding how the waves diffracted by a crystal can be transformed and summed, in a symmetrical process, to produce the electron density in a unit cell.

In this process, as we will see in the next chapter, we transform and combine waves arising from families of planes in crystals. The families of planes, remember from Chapter 3, are the harmonic components (the spectrum of component waves) that sum to yield the entire electron density, or electron density wave, within a crystal. A specific family of planes uniquely cuts through a single unit cell, or the entire point lattice, with a frequency given by $hk\ell$, a wavelength equivalent to $1/d$, and a direction corresponding to the normal to the planes. That is, appropriately weighted families of planes, denoted by $hk\ell$, are the component waves of the contents of the unit cells in the crystal. We will show in the next chapter that the weights we assign to each family of planes, or wave, reflect the arrangement of atoms around those planes, and those weights we can consider to be the amplitudes of the plane waves. Note particularly though, that the properties of the Miller planes correspond to waves of differing frequencies, namely $hk\ell$, which is a definition of frequency in three dimensional space.

The idea of frequency has an analog in reciprocal space, which is the same as diffraction space or Fourier transform space. In reciprocal space, the families of planes ***hkℓ***, have corresponding reciprocal lattice points ***hkℓ***. Hence every reciprocal lattice point also has associated with it the feature of frequency. The intensities and their associated phases found at those reciprocal lattice points are not direct measures of the electron density surrounding the corresponding planes ***hkℓ***, but their Fourier transforms. They are the intensities and phases of the resultant waves of X rays diffracted by the families of planes having frequencies ***hkℓ***. Thus in reconstructing the electron density from component waves, in either real space or reciprocal space, we combine and transform waves of different frequency.

Fourier demonstrated that any periodic function, or wave, in any dimension, could always be reconstructed from an infinite series of simple sine waves consisting of integral multiples of the wave's own frequency, its spectrum. The trick is to know, or be able to find, the amplitude and phase of each of the sine wave components. Conversely, he showed that any periodic function could be decomposed into a spectrum of sine waves, each having a specific amplitude and phase. The former process has come to be known as a Fourier synthesis, and the latter as a Fourier analysis. The methods he proposed for doing this proved so powerful that he was rewarded by his mathematical colleagues with accusations of witchcraft. This reflects attitudes which once prevailed in academia, and often still do.

As remarked previously, a crystal acts to decompose the continuous Fourier transform of the electron density in the unit cells into a discrete spectrum, the diffraction pattern, which we also call the weighted reciprocal lattice. Thus a crystal performs a Fourier analysis in producing its diffraction pattern. It remains to the X-ray crystallographer to provide the Fourier synthesis from this spectrum of waves and to recreate the electron density.

Figure 4.9 illustrates this process. At the top is a complex periodic waveform, and below are some of its spectral components. As their number increases and higher frequency waves are included in the synthesis, the resultant wave increasingly resembles the wave at the top. This is in one dimension, but it would be the same in two or three dimensions. Furthermore, it is exactly what is done in X-ray crystallography, where the complex waveform is the electron density that repeats from unit cell to unit cell throughout the three dimensions of the crystal. The component waves of the electron density are the electron density planes passing through the unit cell. Higher frequencies correspond here to the families of planes of smaller and smaller spacings, higher and higher Miller indexes, and higher frequency of sampling of the unit cell contents. Figure 4.10 is another example of a Fourier synthesis where a saw tooth wave is produced by simple sine waves of common phase but having frequencies of increasing integral multiples.

Some points are noteworthy. According to Fourier, formally, the series must be summed over all integral frequencies from $-\infty$ to $+\infty$ to be mathematically exact. In practice of course, this is never possible. As the number of terms increases, however, as higher frequency terms are included, the approximation to the exact resultant wave function becomes more nearly correct. As shown in Figure 4.9, it often doesn't require all that many terms before a quite acceptable result is obtained. The difference between the exact waveform and the one we obtain from summing a limited series of Fourier terms is known as "series termination error." As illustrated by the two-dimensional case in Figure 4.11, the phases of the component waves in the synthesis play a crucial role in determining the form of the resultant wave.

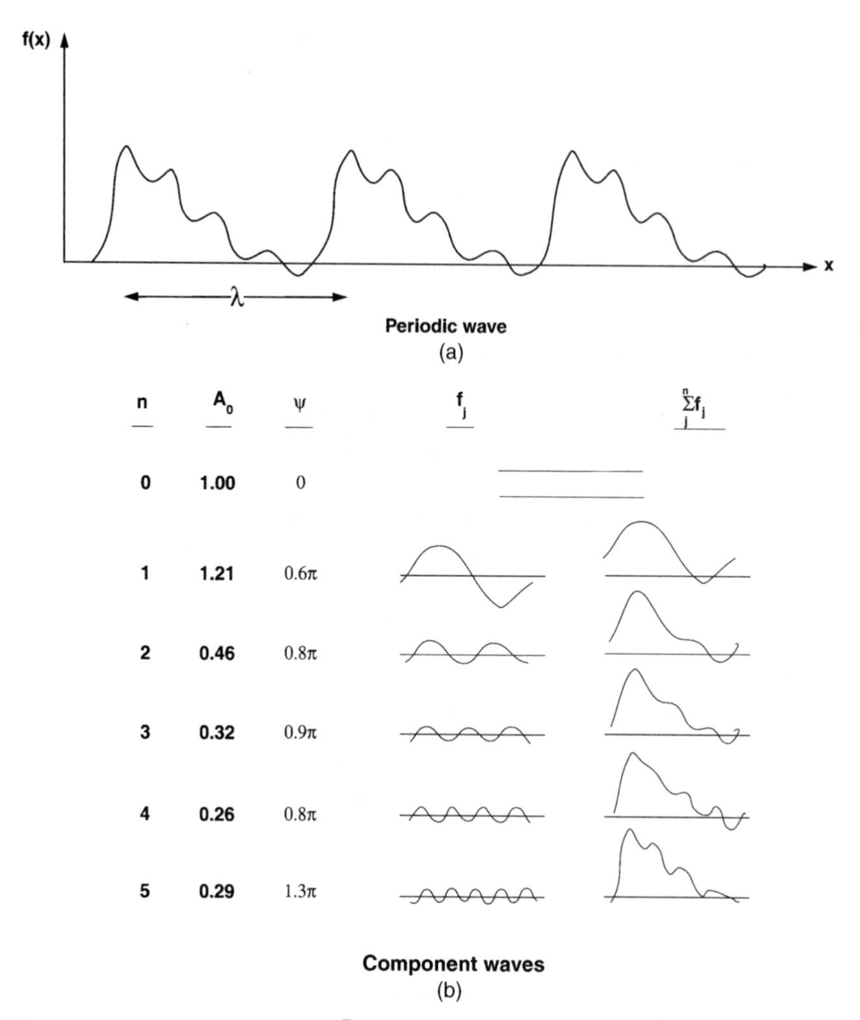

FIGURE 4.9 In (*a*) is a periodic wave $\bar{f}(x)$, which could just as well be $\bar{f}(t)$, where x is distance from the origin and $|\bar{f}(x)|$ is the amplitude of the wave at that point. As demonstrated by Fourier, the wave $\bar{f}(x)$, can be constructed by addition of a series of component waves having diverse frequencies, amplitudes and phases. In (*b*) is shown the zero order (a constant), and first five components of $\bar{f}(x)$. As each wave $\bar{f}_j$ is included in the Fourier synthesis, the resultant wave at the right more closely approximates the waveform of $\bar{f}(x)$. Were the series of component waves infinite, the Fourier series would match the waveform of $\bar{f}(x)$ perfectly. Just as $\bar{f}(x)$ can be synthesized by adding together a series of component waves $\bar{f}_j$, $\bar{f}(x)$ can also be decomposed in a symmetrical process called Fourier analysis to yield the component waves, its spectrum. In X-ray crystallography the electron density in the unit cells of a crystal (the molecules) is a wave function that is periodic throughout the crystal. It may be decomposed into its component waves, the families of planes $\boldsymbol{hk\ell}$, which also may be viewed as electron density waves of wavelength varying according to their interplanar spacings.

Note in Figure 4.9 that the low frequency terms of the series, which could correspond to the low resolution X-ray reflections arising from families of planes of large spacings and low Miller indexes, fill out the broad outlines of the synthetic wave. The higher frequency terms, corresponding to families of planes of fine spacings, provide the details. The same

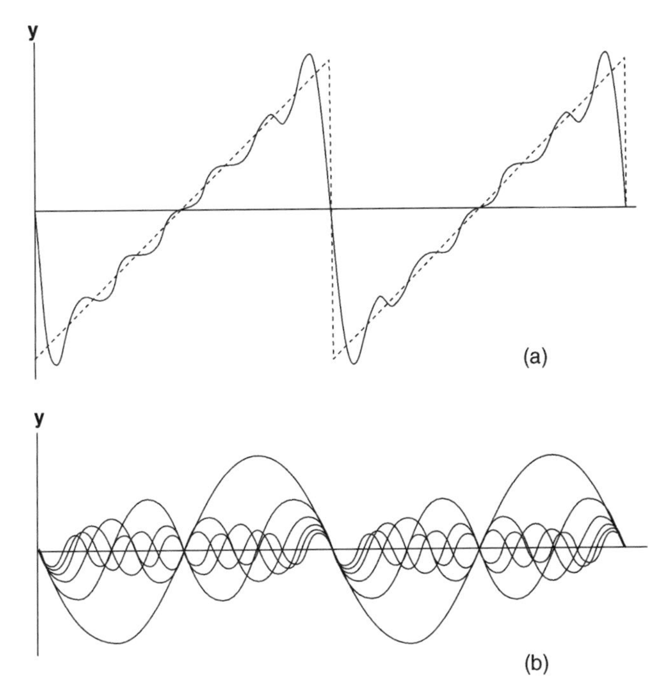

FIGURE 4.10 The sawtooth curve in (*a*) is created by adding together the component waves in (*b*). The lowest frequency component wave, which has the greatest amplitude, has exactly the same frequency, or wavelength, as the resultant wave in (*a*). All the component waves have frequencies of twice, three times, four times, and so on. The more high-frequency waves that are added into the summation, the more the resultant wave resembles the sawtooth curve.

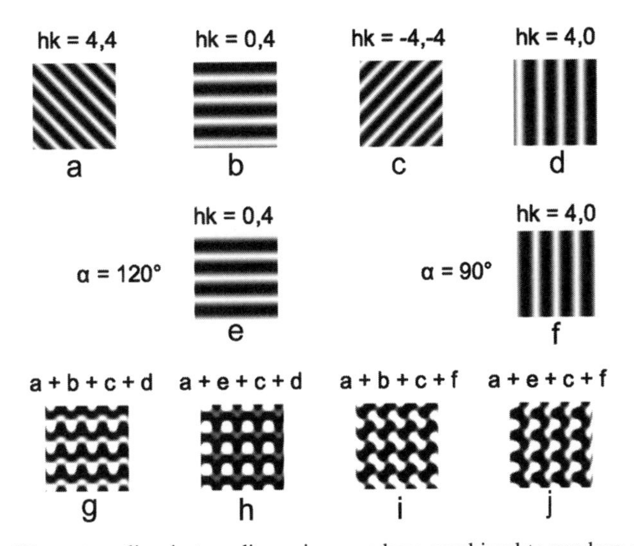

FIGURE 4.11 Waves traveling in two dimensions are here combined to produce resultant waves. These syntheses depend on both the frequencies of the waves, their amplitudes, and their phases with respect to one another. The four waves $\boldsymbol{hk} = \mathbf{4, 4}$; $\boldsymbol{hk} = \mathbf{0, 4}$; $\boldsymbol{hk} = \mathbf{-4, -4}$; $\boldsymbol{hk} = \mathbf{4, 0}$ in (*a*) through (*d*) all have phases of **0**. They combine to give the resultant wave pattern in (*g*). If the phase of only wave $\boldsymbol{hk} = \mathbf{0, 4}$ in (*b*) is set to **120°**, the resultant wave in (*h*) is obtained. If only the phase of the wave in (*d*), $\boldsymbol{hk} = \mathbf{4, 0}$ is set to **90°**, then the resultant wave in (*i*) is obtained. If the phases of both (*b*) and (*d*) are shifted to **120°** and **90°**, respectively, the synthesis in (*j*) is observed.

is true in X-ray crystallography. The low-resolution terms arising from families of planes widely spaced give us the gross shape of the molecule and suggest its undulations and protrusions, but high-resolution terms from families of planes of narrow spacings are essential to show us the locations and orientations of the amino acid side chains.

Finally, one could ask just how accurate the amplitudes and phases of the individual wave components have to be before the synthesis accurately resembles the true waveform. Well, infinity is a lot of terms. With that many terms, every one doesn't have to be right on. In fact an individual term only has to be sort of close, but if enough of these rather poorly determined components are available, the final result still looks pretty good. Fortunately X-ray crystallography provides a lot of these terms, more than enough. This is known in the church as "God's blessing on X-ray crystallographers."

CHAPTER 5

DIFFRACTION FROM POINTS, PLANES, MOLECULES, AND CRYSTALS

Two basic questions we encounter in understanding X-ray crystallography are how do crystals diffract X rays, and how can we describe that process in mathematical terms? These questions may be addressed in a number of ways, but the approach we choose here is divide and conquer. Diffraction from a crystal can be deconstructed into two problems, each worked out separately, and finally, their solutions combined.

As discussed in Chapter 3, a crystal may be conceived as (1) the contents of a unit cell, namely the set scattering points, or atoms of the asymmetric units related by space group symmetry, distributed in space according to (2) a point lattice characterized by the unit cell vectors, or lattice parameters. That is, a crystal is the product of a unit cell and a lattice, to use some loose mathematical jargon, though it is not so loose as it seems. The product of two spatial functions, or distributions, is called a convolution. This is simply the repetition of one distribution (e.g., the contents of a single unit cell) according to the specifications of another (e.g., a lattice). A graphic example of a convolution was provided in Chapter 3. Think of filling a test tube rack with test tubes, or an egg carton with eggs.

The concept of a repeated distribution is important because it can be shown (we will forego a painful formal proof here) that the Fourier transform (or diffraction pattern) of the convolution of two spatial functions is the product of their respective Fourier transforms. This was demonstrated physically using optical diffraction in Figure 1.8 of Chapter 1. In principle, this means that if we can formulate an expression for the Fourier transform of a single unit cell, and if we can do the same for a lattice, then if we multiply them together, we will have a mathematical statement for how a crystal diffracts waves, its Fourier transform.

The reader probably has it branded indelibly upon his/her mind by now that Fourier transform and diffraction pattern are mathematical and physical correlates. Hence the point will be belabored no more, and only one or the other will be used. The questions now become specific. How does a collection of atoms diffract X-rays, and how does a point lattice diffract X-rays?

Introduction to Macromolecular Crystallography, Second Edition By Alexander McPherson

DIFFRACTION PATTERN OF AN ARBITRARY ARRAY OF POINTS IN SPACE

In Figure 5.1, if waves arriving along a direction $\bar{k}_0$ (defined according to some coordinate system) encounter scattering points, each point will scatter the waves as if it was itself the originator of a spherical wave emitted in all directions. These scattered waves, it is important to note, give rise to diffraction effects only when distances between the points are close to the wavelength of the radiation. The points we are dealing with here will always be assumed to have that property. For any direction in space we can define a vector $\bar{k}$ originating at each scattering point and traveling in that direction in space to where we wish to observe the diffracted radiation. The vectors are normal to the wavefronts of the spherical waves they represent.

Figure 5.1 shows an incident wave $\bar{k}_0$ illuminating two points P_0 and P_1 and an arbitrary direction $\bar{k}$, along which we seek the consequences of the scattered waves. That is, we want to know what is the amplitude and phase of the resultant wave at any point along the direction $\bar{k}$ created by the interference of the two scattered waves. Because $\bar{k}$ is general, we can ask this question for any direction and the analysis will be the same. For simplicity, let us assume that the points at x_0, y_0, z_0 and x_1, y_1, z_1 scatter X rays equally, meaning they have the same number of electrons. Hence the amplitudes of the two scattered waves will be the same. The crucial question in combining the two scattered waves is: What are their

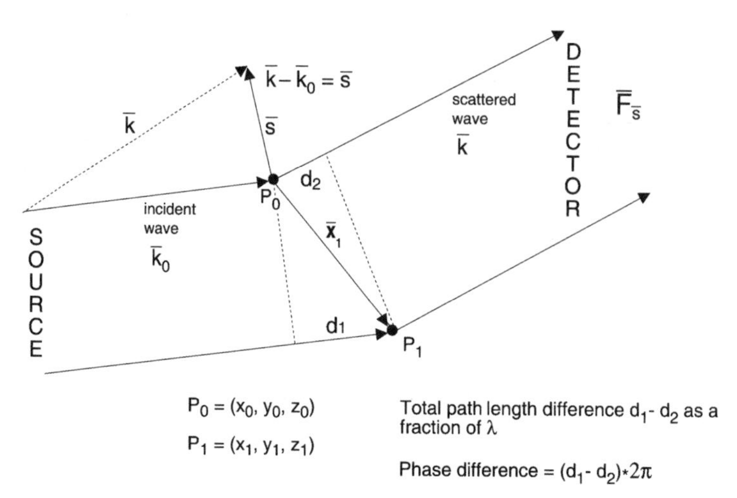

FIGURE 5.1 An incident wavefront traveling along direction $\bar{k}_0$ illuminates a point P_0 at x_0, y_0, z_0, which is chosen as the origin of the coordinate system, and another point P_1 at general position x_1, y_1, z_1. The distance from the source, and that to the detector, is very much greater than the distance between points P_0 and P_1; hence the incident rays along $\bar{k}_0$, as well as the diffracted rays along some general direction $\bar{k}$ can be considered parallel. The two points scatter the wave and the two diffracted waves interfere with one another along $\bar{k}$ to produce a resultant wave $\bar{F}_{\bar{s}}$ at the detector. The phase difference between the two scattered waves is a function of the path length difference from the source to x_0, y_0, z_0 and to x_1, y_1, z_1, and then to the detector. $\bar{x}_1$ is the vector from the point at the origin P_0 to the point P_1 at x_1, y_1, z_1. The vector $\bar{s}$, called the diffraction vector, is equal to the vector difference between the direction of diffraction and the incident wave normal, $\bar{k} - \bar{k}_0$, and it bisects the angle between them.

relative phases? As we saw earlier, we can choose the phase of the first wave, from P_0, at x_0, y_0, z_0, to be zero (remember, we get the first one free when we establish our reference system for the waves). Then what is the relative phase of the wave scattered by P_1, at x_1, y_1, z_1?

The relative phases of waves along any direction $\bar{k}$ depend on the difference in path length that the incident wave, traveling along $\bar{k}_0$, had to go to reach the points, plus the difference in path length that the scattered waves traversed in reaching the observer, or detector along $\bar{k}$. Losses and gains on the two legs, measured as a fraction of wavelength λ, determine relative phase. As was seen in Chapter 4, a total difference of λ, or $n\lambda$, corresponds to a phase difference of 2π or 360°. We will assign an amplitude of $1/\lambda$ to $\bar{k}_0$ and $\bar{k}$ so that the path length difference comes out as a fraction of the wavelength λ and we can thereafter ignore λ in our calculations.

In Figure 5.1 the point at x_1, y_1, z_1 scatters behind the point at x_0, y_0, z_0 by a distance d_1, but its scattered wave along $\bar{k}$ has a shorter distance to travel to reach the detector by d_2. In vector terms, d_1 and d_2 are the projections of the vector $\bar{x}_1$, which defines the position of the point at x_1, y_1, z_1 with respect to x_0, y_0, z_0, on $\bar{k}_0$ and $\bar{k}$, respectively. The projection of any vector on another vector is the scaler product (or dot product) of the two vectors.

Thus the total path length difference as a fraction of the wavelength λ is

$$d_2 - d_1 = (\bar{x}_1 \cdot \bar{k}) - (\bar{x}_1 \cdot \bar{k}_0) = \bar{x}_1 \cdot (\bar{k} - \bar{k}_0).$$

The phase difference of the two waves therefore will be

$$\phi_1 = 2\pi\bar{x}_1 \cdot (\bar{k} - \bar{k}_0),$$

where ϕ_1 is the phase of the wave scattered by the point at x_1, y_1, z_1, $\phi_0 = 0$ for the wave scattered by the point at x_0, y_0, z_0. The resultant wave, which we will denote $\bar{F}_{\bar{k}}$ propagating in the direction $\bar{k}$ will be the sum of the two scattered waves, which is

$$\begin{aligned} & \cos 0 + i \sin 0 \,(\text{wave from } x_0, y_0, z_0) \\ + \; & \underline{\cos\phi_1 + i \sin\phi_1 \,(\text{wave from } x_1, y_1, z_1)} \\ \bar{F}_{\bar{k}} = \; & (\cos 0 + \cos\phi_1) + i(\sin 0 + \sin\phi_1) \\ = \; & 1 + \cos\phi_1 + i \sin\phi_1. \end{aligned}$$

Or in Eulerian notation,

$$\bar{F}_{\bar{k}} = \exp i0 + \exp i\phi_1 = 1 + \exp i\phi_1.$$

The continuous distribution of waves $\bar{F}_{\bar{k}}$ in any and all directions $\bar{k}$, for an incident direction $\bar{k}_0$, is the diffraction pattern of the two points x_0, y_0, z_0 and x_1, y_1, z_1. This simple diffraction pattern, the physical distribution of resultant waves in space arising from the scattering of two points is, in mathematical terms, the Fourier transform of the two points. If more points, each designated by a subscript j, are added to the set, as in Figure 5.2, as we might have for atoms comprising molecules in a unit cell, then the formulation of the

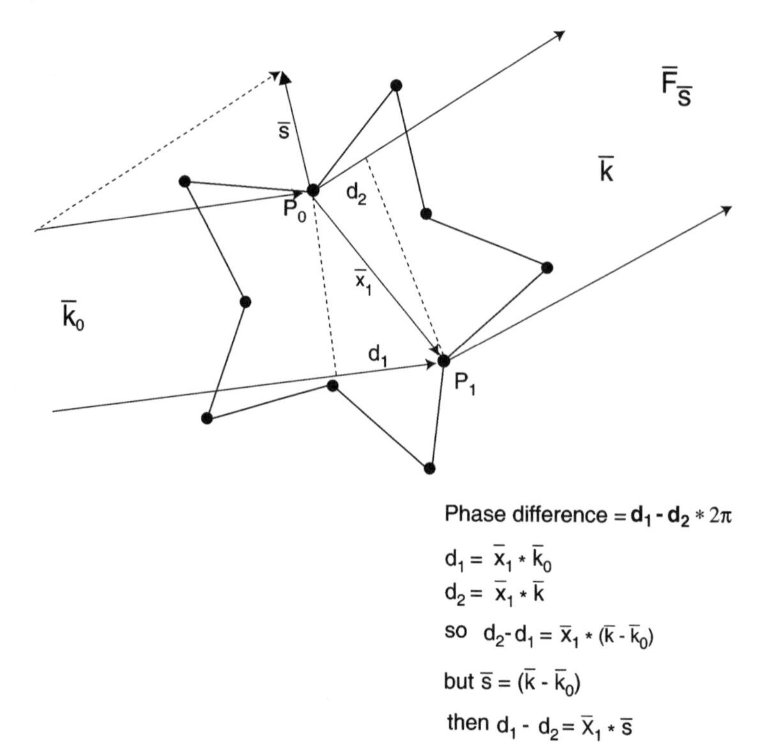

FIGURE 5.2 If the points in Figure 5.1 are but two among many, the analysis is the same except that the diffracted waves from all scattering points must be included in defining the resultant $\bar{F}_{\bar{s}}$. If the coordinates x, y, z of each scattering point are known, then that atom's contribution to the resultant wave $\bar{F}_{\bar{s}}$ can be formulated. $\bar{F}_{\bar{s}}$ is the summation of all the component waves. The distances d are in fractions of λ.

diffraction pattern is derived in exactly the same way; only there are more points and more waves that contribute to the final result. Again, we would seek to determine all path length differences in a systematic way, calculate the relative phases, and add them together. The expression for the resultant wave in any direction $\bar{k}$ will then be

$$\bar{F}_{\bar{k}} = \sum_{j=1}^{N} \exp i\phi_j = \sum_{j=1}^{N} \cos 2\pi[\bar{x}_j \cdot (\bar{k} - \bar{k}_0)] + i \sin 2\pi[\bar{x}_j \cdot (\bar{k} - \bar{k}_0)],$$

where the summations are over all N scattering points j, each located at

$$\bar{x}_j = (x_j, y_j, z_j).$$

This expression is somewhat cumbersome, and it can be made more pointed as well. In physics we never use two vectors where one would suffice, so we define the difference vector $\bar{s} = \bar{k} - \bar{k}_0$, where $\bar{s}$ is called the diffraction vector. In terms of $\bar{s}$, the expression for the Fourier transform of a set of scattering points becomes

$$\bar{F}_{\bar{s}} = \sum_{j=1}^{N} \cos 2\pi[\bar{x}_j \cdot \bar{s}] + i \sin 2\pi[\bar{x}_j \cdot \bar{s}].$$

This formulation cleverly expands the value and scope of the expression describing the Fourier transform because $\bar{s}$ is dependent on both $\bar{k}_0$ and $\bar{k}$, and defines the relationship between them. Since altering $\bar{k}_0$ alters $\bar{F}$ just as choosing a different $\bar{k}$ does, using $\bar{s}$ as the variable (the difference in direction between $\bar{k}$ and $\bar{k}_0$) encompasses all possibilities of incident and diffraction directions. Thus the expression above is the comprehensive Fourier transform of the distribution of points for all $\bar{k}_0$ and $\bar{k}$. The diffraction vector $\bar{s}$ we will see re-appear in a variety of manifestations as we proceed.

It is wise at this point to introduce one additional refinement into the expression for the diffraction pattern that further expands its application. Above, we made the tacit assumption that all points scattered with the same intensity; for instance, they all represented atoms that had the same number of electrons. This, of course, is unrealistic, for most molecules are composed of different kinds of atoms. We can easily correct for this assumption, however, by making the amplitude of the wave scattered by each point equal or proportional to its scattering power, its atomic number Z_j. In doing this, we rewrite the Fourier transform as

$$\bar{F}_{\bar{s}} = \sum_{j=1}^{N} Z_j[\cos 2\pi(\bar{x}_j \cdot \bar{s}) + i \sin 2\pi(\bar{x}_j \cdot \bar{s})].$$

Now let us ask what is needed to numerically evaluate this expression for the Fourier transform. The diffraction vector $\bar{s} = \bar{k} - \bar{k}_0$, as well as λ, are experimental variables that are chosen, and the Z_j are known for each atom as well. The only remaining variables are the $\bar{x}_j$, and these can be generated from the atomic coordinates x_j, y_j, z_j. Thus all we really need to compute the resultant waves making up the diffraction pattern, for any array of scattering points, are their relative positions in space.

We might note in passing that were the Fourier equation applied to asymmetric units related by space group symmetry in a crystallographic unit cell, the expressions for symmetry equivalent atomic positions assume considerable value. Their application can reduce the number of terms in the summation by the number of symmetry equivalent positions. We need, in practice, to consider only the atoms comprising a single asymmetric unit in the actual calculations.

What would the diffraction pattern from an array of scattering points similar to that in Figure 5.2 actually look like? Let's carry out an optical diffraction experiment on the similar, but simpler, array of points in Figure 5.3*a*. If we do that for a single incident direction $\bar{k}_0$ perpendicular to the plane of the points, then the diffraction pattern we observe, the Fourier transform of the object in Figure 5.3*a* is seen in Figure 5.3*b*. Each point of light or dark in the diffraction pattern in Figure 5.3*b* represents the diffracted intensity for a particular direction $\bar{k}$ (or even better, $\bar{s}$).

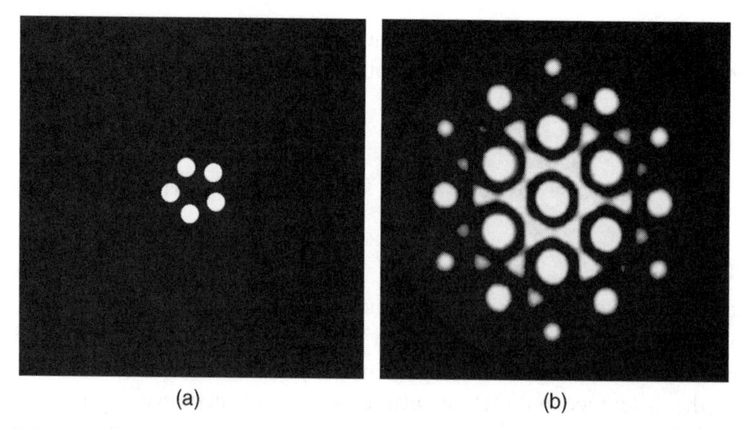

FIGURE 5.3 The object in this optical diffraction experiment is the five point array in (*a*). Its diffraction pattern is seen in (*b*). Each point in (*a*) is characterized by a position $\boldsymbol{x}$, $\boldsymbol{y}$, z, and each point in diffraction space (*b*) is characterized by a diffraction vector $\bar{s}$ and a resultant wave $\bar{F}_{\bar{s}}$, which produces an intensity. The diffraction pattern in (*b*) is simply the collection of $\bar{F}_{\bar{s}}$ for all scattering vectors $\bar{s}$.

DIFFRACTION FROM EQUALLY SPACED POINTS ALONG A LINE

By extension to larger arrays of scattering points, the procedure above serves as a means for deriving the Fourier transform of a general object, such as a single molecule, or even a group of molecules, composed of scattering points arranged arbitrarily in space. That is, the transform is for atoms within an asymmetric unit that are crystallographically independent of one another. Let us also remember, that we are proceeding on the premise that an expression for the diffraction pattern of a crystal, its Fourier transform, can be derived by multiplying such a continuous transform of the contents of a single unit cell by that of the point lattice which characterizes the crystal.

To accomplish our task, we must formulate an expression for the scattering of waves by a point lattice. Let us begin by considering the scattering of waves by a one-dimensional periodic lattice, a line of points separated by constant distances $\boldsymbol{a}$, that is, points related in a periodic manner by a vector $\bar{a}$, where $\bar{a}$ has a magnitude comparable to that of λ.

In Figure 5.4*a* we have such a line of points periodically placed according to $\bar{a}$, with an incident wave $\bar{k}_0$ having some arbitrary direction defined by θ_1, and some general scattering direction $\bar{k}$, defined by θ_2. What is $\bar{F}_{\bar{s}}$? If we examine Figure 5.1, we can see that the first two points in our line of points are no different than the two scattering points P_0 and P_1 in Figure 5.1, in which case $\bar{x}_1$, from that figure, becomes $\bar{a}$. The phase difference ϕ for the waves scattered by the first two points is

$$\phi = \bar{a} \cdot (\bar{k} - \bar{k}_0)2\pi.$$

And the phases for subsequent points in the line will be

$$2\bar{a} \cdot (\bar{k} - \bar{k}_0)2\pi,$$

$$3\bar{a} \cdot (\bar{k} - \bar{k}_0)2\pi,$$

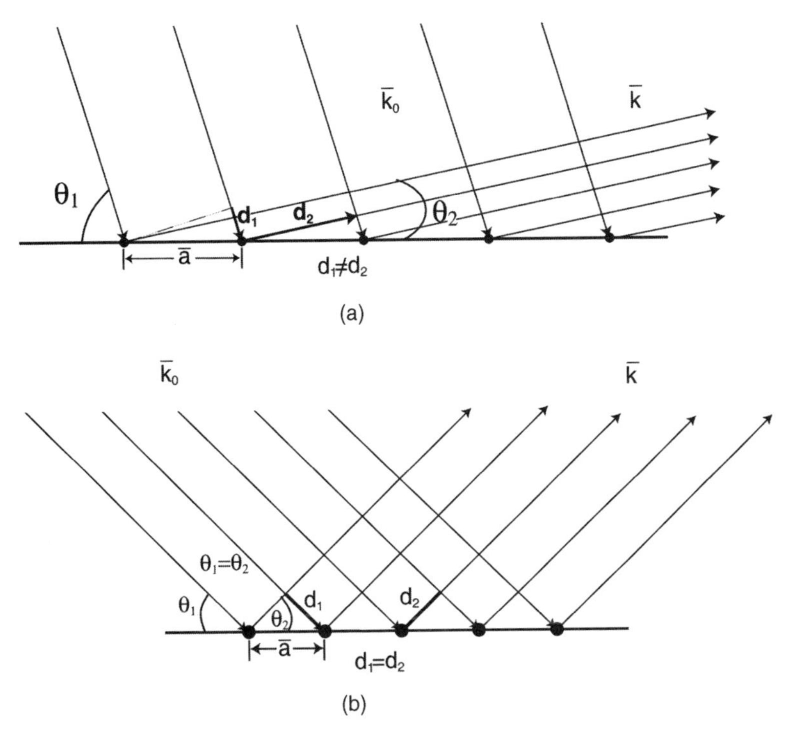

FIGURE 5.4 In (*a*) a wavefront $\bar{k}_0$ encounters a periodic, linear array of scattering points translationally related by $\bar{a}$. The wavefront makes an angle θ_1 with the line of points. $\bar{k}$ represents some completely arbitrary scattering direction, which is defined by the angle θ_2 that it makes with the line of points. If $\theta_1 \neq \theta_2$, then $d_1 \neq d_2$, and the waves scattered by the successive points along the row will be out of phase by integral multiples of some constant angle. The waves will therefore produce a resultant wave of zero amplitude. If, as in (*b*), $\theta_1 = \theta_2$, the angle of incidence equals the angle of reflection, then $d_1 = d_2$. The waves scattered by all points in the line have identically the same phase, and perfect constructive interference will prevail. The resultant wave will have amplitude N times the amplitude of any individual scattered wave, where N is the total number of points in the line.

or in terms of a diffraction vector $\bar{s} = (\bar{k} - \bar{k}_0)$, $4(\bar{a} \cdot \bar{s})2\pi$, and so on. Then for all of the points N in the line,

$$\bar{F}_{\bar{s}} = \sum_{n=1} \exp i[2n\pi(\bar{a} \cdot \bar{s})].$$

In general, $(\bar{a} \cdot \bar{s})$ will not be integral, so $2n\pi(\bar{a} \cdot \bar{s})$ will be multiples of some nonzero phase angle. $\bar{F}_{\bar{S}}$ will therefore be the summation of a long series of N waves, each out of phase with the one before and after by a constant amount.

This, as we saw in Chapter 3, is a Dirac delta function, which always sums to zero except when $(\bar{a} \cdot \bar{s})$ becomes integral, that is, when $\bar{a} \cdot \bar{s}$ is a multiple of λ. Because $\bar{a}$ is invariant, only when $\bar{k}_0$ and $\bar{k}$ articulate certain relationships that result in specific directions for $\bar{s}$ will that be true. When $\bar{s}$ takes on those unique directions, then waves scattered by all N points have relative phases of zero and thus constructively interfere. The amplitudes of all of the N scattered waves then add arithmetically.

This is a one-dimensional illustration of Bragg's law. The most important consequence of this example is to show that the diffraction pattern of a periodic distribution of points is not continuous in diffraction space, as it is for a set of scattering points arbitrarily distributed in real space, but it is discrete and periodic. The Fourier transform of a one-dimensional point lattice, that is, the locations in diffraction space where it takes on nonzero values, is itself a discontinuous one-dimensional point lattice. But then, we knew that already from the optical diffraction pattern obtained experimentally in Figure 1.6 of Chapter 1. We will presently see that the same is true not only for a one-dimensional lattice but also in three dimensions as well.

Let us return to Figure 5.4 for a moment. For a line where the points are distributed along its length arbitrarily, or a line that is a continuum of points, the results are somewhat different. In those cases the scattering points along the line will be spatially related by all variety of vectors $\bar{v}$, having arbitrary distances between them. For those cases, $\bar{v} \cdot \bar{s}$ can satisfy the diffraction condition for any and all possible $\bar{v}$ only when $\bar{v} \cdot \bar{s} = 0$. This means when $\bar{s}$ is perpendicular to the line. It can easily be shown that $\bar{s} = \bar{k} - \bar{k}_0$ is perpendicular to the line only when $\theta_1 = \theta_2$ (or, equivalently, when $d_1 = d_2$ in Figure 5.4). This means that a line of arbitrarily spaced scattering points, or a continuous line, will only allow a nonzero resultant wave to be observed when the angle of incidence and the angle of reflection are equal. Thus any continuous line, or line of arbitrarily spaced points, will act as a simple mirror that reflects waves according to the usual laws of optics.

DIFFRACTION FROM A PLANE, FAMILIES OF PLANES, AND LATTICES OF POINTS

If a continuous line of points reflects waves according to the usual laws of optics, that is, as a one-dimensional mirror, then it follows that a continuous plane will as well, since any plane can be decomposed into an infinite number of lines. Similarly a plane of points having periodic spacings along two directions (a two dimensional point lattice), like that in Figure 5.5, must behave as we saw for a line of periodically spaced points in Figure 5.4. Thus we would expect that a two-dimensional point lattice would yield a diffraction pattern that is also a two-dimensional point lattice. That is, of course, exactly what we showed using optical diffraction in Chapter 1.

It may at this point be starting to become obvious why, in Chapter 3, we bothered to gather all the points of a lattice into planes and the planes into families. Indeed we now know how all of the lattice points collected on one plane diffract X rays. Thus we can figure out how an entire lattice diffracts, its Fourier transform, by asking how all the planes in a particular family diffract.

In Figure 5.6 are the top few planes of the multitude comprising the $hk\ell$ family of planes passing through a three-dimensional point lattice. We know from the discussion above that we can consider the planes to simply reflect waves from direction $\bar{k}_0$ making an angle θ with the planes because, if the angle of reflection does not equal the angle of incidence, the waves from a two-dimensional plane of points destructively interfere and sum to zero. The only direction $\bar{k}$ in which diffraction needs to be considered is when the angle of incidence is equal to the angle of reflection. Now we have to consider how the reflected waves from the consecutive lattice planes interfere to produce diffraction from the entire family.

In Figure 5.6 both the incident and reflected waves, $\bar{k}_0$ and $\bar{k}$, respectively, make the same angle θ with every plane. A general lattice point on the top plane of the family is illuminated ahead of that on the one below by distance d, ahead of a point on the next plane down by

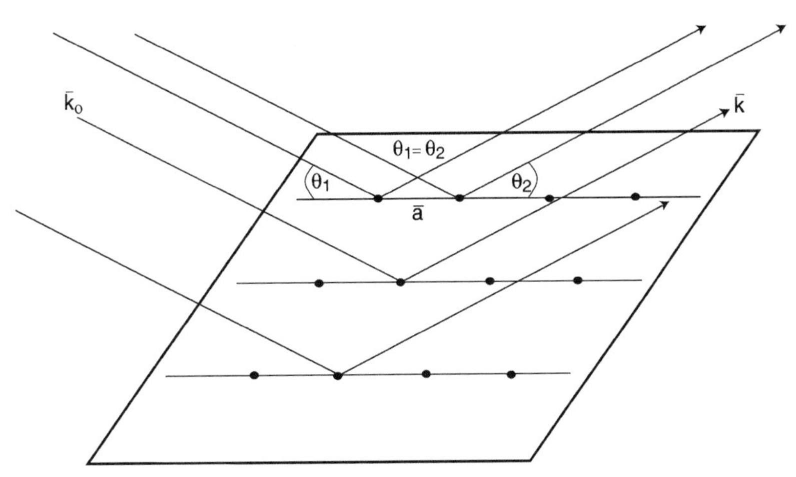

FIGURE 5.5 A wavefront along $\bar{k}_0$ encounters a two-dimensional point lattice and is scattered in the general direction $\bar{k}$. All the points in the planar lattice can be organized into lines of periodically spaced points as in Figure 5.4. Thus all points on any two-dimensional plane through a three-dimensional lattice can be considered to reflect waves according to conventional laws of optics, just as did the one-dimensional line of points.

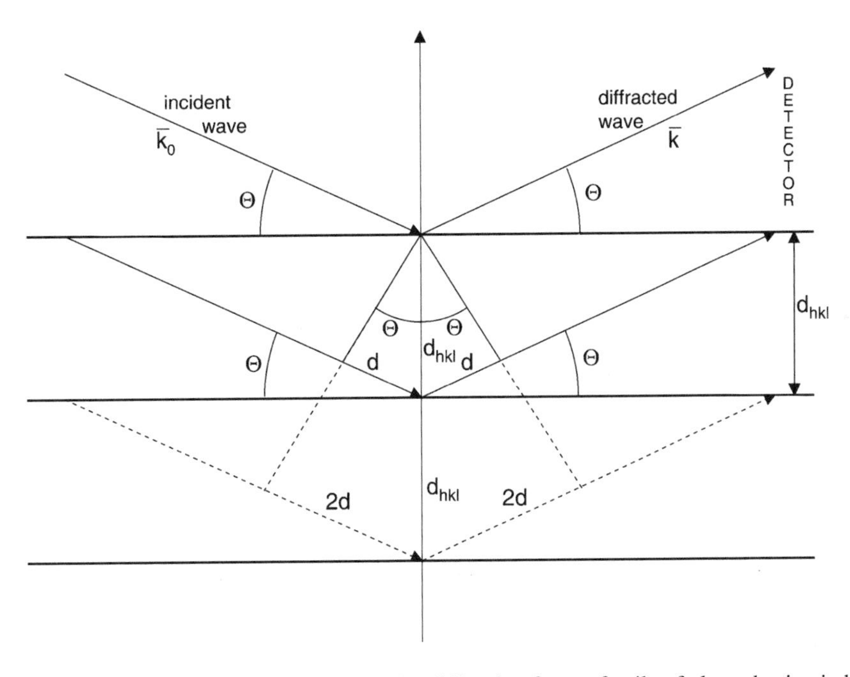

FIGURE 5.6 The derivation of Bragg's law for diffraction from a family of planes having indexes $hk\ell$ and interplanar spacing $d_{hk\ell}$. The incident wave direction is $\bar{k}_0$, and the diffraction direction is $\bar{k}$. Because the planes may be considered to reflect waves according to the conventional laws of optics, both $\bar{k}_0$ and $\bar{k}$ must make a common angle θ with the planes. To ensure that the total path length difference of the waves in reaching the detector, $2d$, is an integral number of wavelengths $n\lambda$, and that constructive interference occurs for all planes in the family, $n\lambda = 2d_{hk\ell} \sin\theta$.

$\mathbf{2}d$, and so on. The reflected rays from successive planes also reach the detector ahead of one another by d, $\mathbf{2}d$, and so on. Hence the source to detector path length differences for waves from consecutive planes are $\mathbf{2}d$, $\mathbf{4}d$, $\mathbf{6}d$, and so on.

From earlier examples, it should be intuitively clear that unless the path length difference $2d = 0$, or $2d$ is some integral multiple n of λ, we will once again have an infinite series of waves out of phase with one another by some constant increment, and hence have destructive interference (i.e., a resultant wave of zero amplitude). When, however, $2d = n\lambda$, we will have perfect constructive interference and every plane, and therefore every point in the lattice will sparkle in perfect phase. In Figure 5.6, $d_{hk\ell}$ is the interplanar spacing for the $hk\ell$ family of planes, which we can calculate for any crystal from the unit cell parameters (see Table 3.2 of Chapter 3). From Euclid's geometry we also know that the angles opposite d_1 and d_2 are both also equal to θ. Trigonometry then tells us that $d_1 = d_2 = d_{hk\ell} \sin\theta$. To obtain constructive interference, we already noted that d_1 plus d_2 must equal $n\lambda$; hence

$$n\lambda = 2d_{hk\ell} \sin\theta.$$

This is the familiar formulation of Bragg's law for a three-dimensional point lattice. It says that the Fourier transform of a point lattice is absolutely discrete and periodic in diffraction space, and that we can predict when a nonzero diffraction intensity will appear for any family of planes $hk\ell$, and what the angle of incidence and reflection θ must be in order for an intensity to appear. Bragg's law, notice, is completely independent of atoms, or molecules, or unit cell contents. The law is imposed by the periodicity of the crystal lattice, and it strictly governs where we may observe any nonzero intensity in diffraction space. It tells us when the resultant waves produced by the scattering of all of the atoms in the many individual unit cells, each represented by a single lattice point, are exactly in phase.

Bragg's law is essentially a three-dimensional case of the Dirac delta function, seen in Chapter 4, in that it specifies the diffracted radiation to be identically zero except for a discrete set of angular relationships between two variables $\bar{k}$ and $\bar{k}_0$, or more precisely, for specific values of their difference, $\bar{s}$. Cast in terms of vectors, where $\bar{d}$ is the plane normal and has length equal to $\lambda/d_{hk\ell}$ the diffracted ray may be written as

$$\bar{F}_{\bar{s}} = \sum_{n=1}^{N} \exp 2\pi i(\bar{d} \cdot \bar{s}),$$

where N is the number of planes in the family, a very large number.

Bragg's law tells us something more. For the families of planes in Figure 5.6, $\bar{k} - \bar{k}_0$, the diffraction vector $\bar{s}$, is perpendicular to the family of planes as illustrated in Figure 5.7. We also see in Figure 5.7 that when Bragg's law is satisfied, $\bar{s}$ has a length of $1/d_{hk\ell}$. In other words, for any family of planes $hk\ell$, the diffraction vector $\bar{s}$, which specifies when diffraction occurs for any distribution of points, including a lattice of points, must be the corresponding reciprocal lattice vector $hk\ell$. Thus the diffraction from a family of N planes $hk\ell$ can be written

$$\bar{F}_{\bar{s}} = \bar{F}_{\bar{h}} = \sum_{n=1}^{N} \exp 2n\pi i(\bar{h} \cdot \bar{d})$$

As was asserted earlier, the collection of all the reciprocal lattice vectors, $\bar{h}$, or their end points, constitute reciprocal space. Reciprocal space is the Fourier transform of the real crystal lattice and constitutes all the possible diffraction vectors for the crystal.

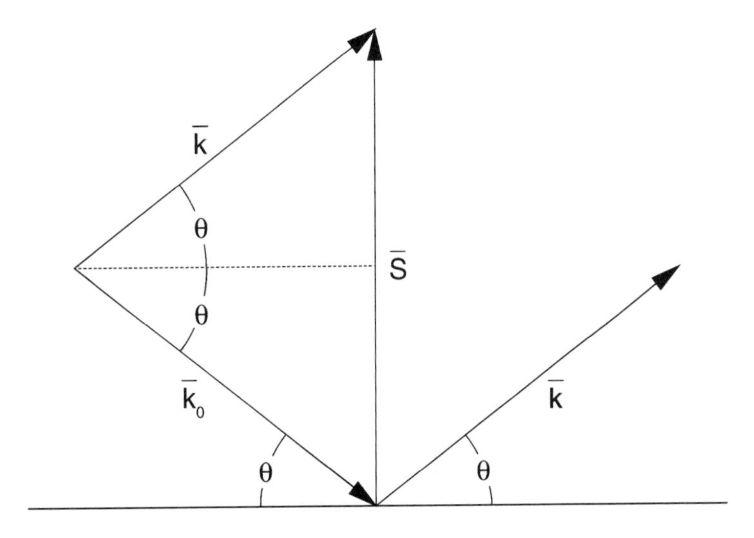

From Euclid's geometry $\bar{S}$ is perpendicular to the planes

$|\bar{S}| = |\bar{k}_0| \sin\theta + |\bar{k}| \sin\theta$

but $|\bar{k}_0| = |\bar{k}| = 1/\lambda$

then $|\bar{S}| = 2 \sin\theta/\lambda = n/d$

thus $\bar{S} = n\bar{h}$, where $\bar{h}$ is the reciprocal lattice vector for this family of planes

FIGURE 5.7 If we take the difference between the incident and reflected rays $\bar{k}$ and $\bar{k}_0$ from (*a*) to form the diffraction vector $\bar{s}$, then, by Bragg's law, it is clear that $\bar{s}$ must be identical to the reciprocal lattice vector $\bar{h}$ for any family of planes in diffracting position. This must be true no matter what the actual distribution of atoms, or scattering material, around the planes in the family may be.

In defining the families of planes in a crystal that contain all lattice points, we ignored the fact that any family of planes, of spacing $d_{hk\ell}$ could always be subdivided further. Indeed many families of planes are simply subdivisions of others (**222, 444, 888**, etc.). Hence there are an infinite number of families. When the interplanar spacing becomes small enough, however, it is no longer of consequence so far as X-ray diffraction is concerned. Bragg's law, $\lambda = 2d \sin\theta$, must be satisfied in order for a specific family of planes to diffract X rays. This we saw from above, or $\sin\theta = \lambda/2d$. The maximum value of $\sin\theta$, however, is 1. Thus the minimum d spacing that a family of planes can have and still diffract X-rays is $d = \lambda/2$.

For CuK_α radiation of **1.54 Å**, the minimum interplanar spacing capable of diffracting X rays is **0.77 Å**. Families of planes with $d < 0.77$ Å in this case are of no interest. Because of this limit to d spacing, the reciprocal lattice is also not infinite but bounded by some wavelength-dependent resolution limit.

CONTINUOUS AND DISCONTINUOUS TRANSFORMS

The Fourier transform of an object, a distribution of scattering points like those in Figures 5.2 and 5.3, whether it be continuous or discontinuous, is its diffraction pattern.

It is the pattern of radiation arising from the interference of the waves scattered by each point in the object. From Figure 5.1, for example, we deduced an equation, a scattering function, for two points within the larger object in Figure 5.2, and by summation we obtained a diffraction pattern for all of the points making up the object. We derived the Fourier transform of the entire object. If we had actually calculated the values for this Fourier transform everywhere in space around the object, that is, at all $\bar{s}$, we would have found that it was continuous, meaning measurable intensity would be present for all $\bar{s}$. Values of zero might appear from time to time, but not in any regular way. It would have had the appearance of the optical diffraction patterns seen in Figures 1.4 through 1.6 in Chapter 1.

If, on the other hand, the distribution of scattering points in our array is not random, but is discrete and periodic (i.e., takes on nonzero values only at regularly spaced points), for example, as the line of points in Figure 5.4 or the plane of points in Figure 5.5, then the diffraction pattern it produces, its Fourier transform, is also discrete and periodic. This we saw by optical diffraction in Figures 1.8 and 1.9 of Chapter 1. The waves scattered by the points in the line, or the planes comprising a family, destructively interfere and sum to zero except at certain fixed angles dependent on the spacing between the points or planes, and the wavelength of the radiation.

Thus there are two different kinds of scattering distributions: continuous, nonperiodic and discrete, periodic. Their corresponding Fourier transforms have the same characteristics. We may now ask, What happens when we combine the two? What are the characteristics of the diffraction pattern produced by their combination, their product? This was illustrated by optical diffraction in Figure 1.8 of Chapter 1, and here in Figure 5.8. If we take the object (it could be a molecule) in Figure 5.8*a*, which has the continuous transform seen in Figure 5.8*b*, and repeat it according to the point lattice of Figure 5.8*c*, which has the transform shown in Figure 5.8*d*—that is, we take their product, we convolute them to produce the scattering distribution in Figure 5.8*e*—then we may ask what the Fourier transform will be. Because discrete, periodic scattering distributions give rise to Fourier transforms having these same properties, we should expect the diffraction pattern of the convoluted distributions to be discrete as well. On the other hand, there is a continuous scattering distribution, an object, at each allowed point in the array, not just a single point. What we in fact observe is that while the diffraction pattern of the distribution of points in Figure 5.8*e* is discrete and periodic, the intensities in the diffraction pattern vary from point to point. They are not all of identical intensity as they were in Figure 5.8*d*.

Because the observed diffraction pattern is a product of the diffraction patterns from the two distributions, what is observed at each nonzero point in the combined transform, or diffraction pattern determined by the periodic point lattice in Figure 5.8*c*, is the value of the Fourier transform at that point from the continuous distribution in Figure 5.8*a*. That is, the combined diffraction pattern samples the continuous Fourier transform of the object making up the array, that of Figure 5.8*b*, but only at those discrete points permitted by the array's periodic, discrete transform seen in Figure 5.8*d*.

Extended to three dimensions, the diffraction pattern of a crystal is the Fourier transform of the molecules in the unit cells of the crystal, but visible only at discrete reciprocal lattice points permitted by the underlying crystal lattice. This is the price we pay to simultaneously observe the cooperative diffraction from all of the unit cells in the crystal, to obtain enough intensity to measure. While a single unit cell would have given a continuous transform, it would not yield experimentally measurable intensity. A crystal yields sufficient intensity, but only grudgingly, at discrete points.

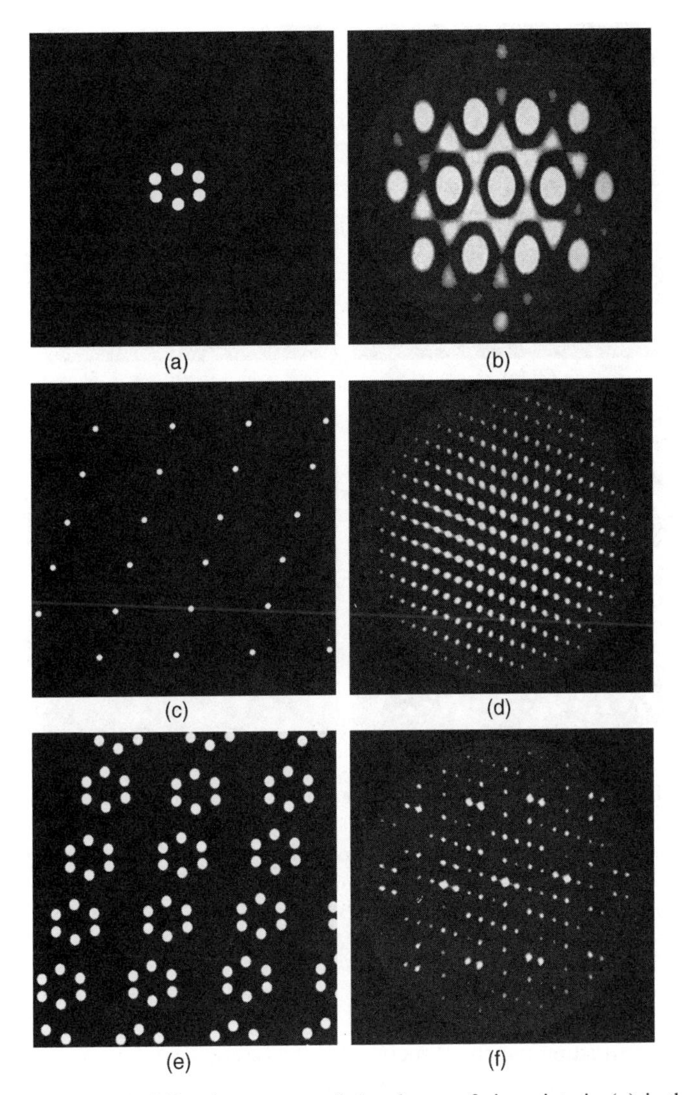

FIGURE 5.8 The optical diffraction pattern of the cluster of six points in (*a*) is the continuous Fourier transform shown in (*b*). The points in (*a*) might well represent the atoms in a molecule. In (*c*) is a point lattice, and in (*d*) its diffraction pattern, or Fourier transform. Note that the diffraction intensities decline in proceeding from the center of the pattern toward the edge (from low to high resolution, or low to high $\sin\theta$), but at any specific resolution their values are all the same. There is no evident variation in intensities over the pattern as we see in X-ray diffraction patterns from real crystals. In (*e*) the assembly of scattering points in (*a*) has been set down at every point in the lattice in (*b*); that is, a convolution has been performed. In (*f*) is the diffraction pattern of (*e*). The intensity at any point in (*f*) could be obtained by superimposing the pattern in (*d*) upon the pattern in (*b*) and multiplying the value at every point above by every point below. That is, the continuous transform in (*b*) would be sampled at those discrete points specified by (*d*). Thus the diffraction pattern of the convolution of two functions is the product of their individual diffraction patterns, the product of their separate Fourier transforms.

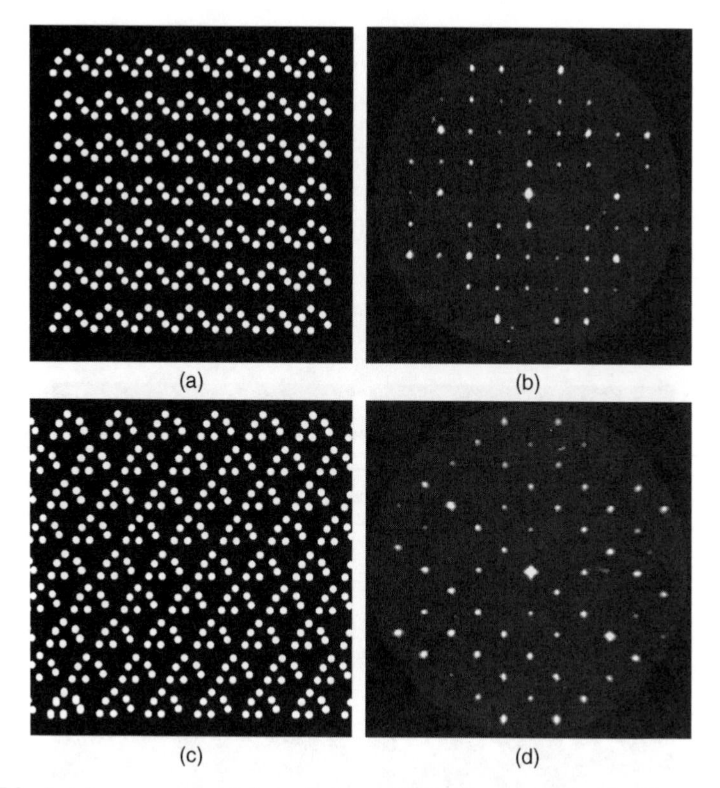

FIGURE 5.9 In (*a*) an assembly of six scattering points has been convoluted with an orthorhombic point lattice, and in (*c*) the same arrangement with a monoclinic lattice. The corresponding diffraction patterns are, respectively, orthorhombic and monoclinic as well. The diffraction patterns, however, are very different in their intensities because the point lattices sample the continuous transform of the six scattering points, the asymmetric unit of these two-dimensional crystals, at different places.

Because the diffraction pattern of a crystal is the periodic superposition (or product, or convolution) of the continuous transform of the unit cell contents with the lattice transform, other interesting consequences follow. For example, the locations of reflections in the diffraction pattern of a crystal, the net or lattice on which they fall, is entirely determined by the lattice properties of the crystal, namely the unit cell vectors. They in no way depend on the structure or properties of the molecules that fill the unit cells. On the other hand, the intensity we measure at each point in the diffraction pattern, and its associated phase, is entirely determined by the distribution of electrons, the positions of atoms x_j, y_j, z_j, within the unit cells.

Consider another optical diffraction experiment, that in Figure 5.9. Here the identical pattern of six arbitrarily related points (the atoms in an asymmetric unit?) is repeated according to two different point lattices in Figures 5.9*a* and 5.9*c*. Because the repeating object (the six-point motif) is the same for both arrays, the same continuous transform must be sampled in both diffraction patterns in Figures 5.9*b* and 5.9*d*. But the two diffraction patterns appear very different, why? Because the continuous transform of the six points is sampled by the two point lattices at different places.

The alert reader may note that the point lattice shown in Figure 5.9*a* is a primitive orthogonal lattice, and so is the diffraction lattice in Figure 5.9*b*. The point lattice shown in Figure 5.9*c*, on the other hand, is a monoclinic lattice, and so is its diffraction lattice in Figure 5.9*d*. Further inspection would also show that the distances between reflections along vertical and horizontal axes in Figures 5.9*b* and 5.9*d* are reciprocally related to the corresponding axes in Figures 5.9*a* and 5.9*c*.

Recall that most unit cells, triclinic being the exception, contain not one molecule but small integral sets of molecules related by space group symmetry. In addition, the symmetry elements that relate the asymmetric units inside the cells, when combined with the lattice translations, become global and extend throughout the crystal. Thus both the crystal lattice and the unit cell contents share the same symmetry properties. As you might then expect, the lattice of the diffraction pattern, the reciprocal lattice, will possess the symmetry properties of the real lattice. A tetragonal crystal lattice will have a tetragonal reciprocal lattice. Because the unit cell contents have this same symmetry, the distribution of intensities in the diffraction pattern will as well. The symmetry that we observe in the diffraction pattern will therefore correspond with the symmetry of the asymmetric units and the lattice of the real crystal that gave rise to it. This will be explored in more detail in Chapter 6 where numerous examples will be offered as evidence.

What is the relationship between the lattice of the crystal and the lattice on which the diffraction intensities fall, the lattice of the transform? The relationship is that between the real space lattice of the crystal and the reciprocal lattice. The point where a wave diffracted by a particular family of planes $\boldsymbol{hk\ell}$ appears in the diffraction pattern is related to the origin of the diffraction pattern by the reciprocal lattice vector $\bar{\boldsymbol{h}} = \boldsymbol{hk\ell}$. The direction of the reciprocal lattice vector $\bar{\boldsymbol{h}}$ is normal to the family of planes, and its length is $\mathbf{1}/\boldsymbol{d_{hk\ell}}$. The end point of the reciprocal lattice vector defines a permissible point in diffraction space where a diffraction wave may be observed. That wave, having both amplitude and phase, is the Fourier transform of that particular family of planes $\boldsymbol{hk\ell}$.

The inverse relationship between the crystal's real space lattice and its reciprocal lattice defines the distances between adjacent reflections along reciprocal lattice rows and columns in the diffraction pattern. Conversely, measurement of the reciprocal lattice spacings yields the unit cell parameters. Angles between the axes of the reciprocal lattice can similarly be used to determine unit cell axial angles.

DIFFRACTION FROM A CRYSTAL

Our strategy was to derive an expression for diffraction by an arbitrary set of scattering points, such as the atoms comprising the contents of a crystallographic unit cell, and then for a three-dimensional point lattice. With these in hand we could then form their product and arrive at an expression for the Fourier transform, or diffraction pattern of the crystal. This process was illustrated using optical diffraction in Figures 5.8 and 5.9. We can demonstrate that process of combining, or convoluting the unit cell and lattice transforms using several approaches, some rather intuitive, some mathematical, and others somewhere between.

Let us look at this idea of convoluting the unit cell and lattice transforms in multiple ways. This is informative because each approach offers a somewhat different perspective, and each may illuminate subtleties that another obscures.

Convoluting Molecules with a Lattice

Convolution of the molecules in the unit cell with the crystal lattice yields a Fourier transform, or diffraction pattern that is the product of their separate transforms. The Fourier transform, or scattered radiation from the atoms in a set of molecules, for any diffraction vector $\bar{s}$, is a single resultant wave. While this resultant wave is produced by the combined scattering of all of the points in the object, we could just as well consider the resultant scattering of rays by all the atoms to be represented by the single resultant wave emanating from a single point in the object as in Figure 5.10. This is similar to our use of the center of mass, or center of gravity for an object, where we consider the totality of some physical property of the object as being concentrated, or originating, at a single point.

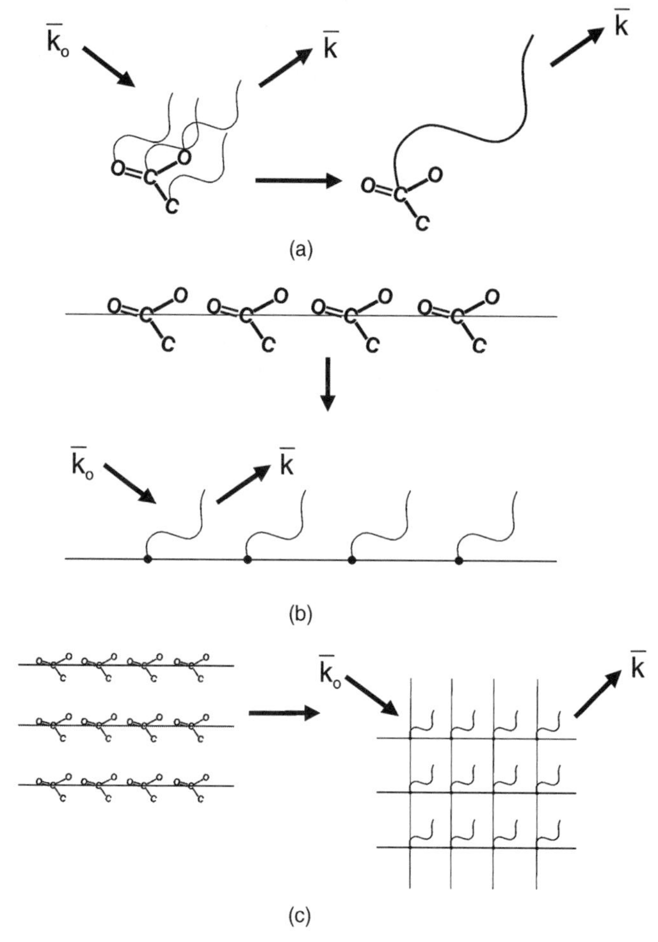

FIGURE 5.10 The waves scattered by each of the four atoms of the molecule in (*a*) will combine to produce a single resultant wave that we can consider arising entirely from a single point in the molecule, here arbitrarily chosen to be a carbon atom. In (*b*), the scattering from all of the atoms in the row of molecules can be substituted by the single resultant wave from each molecule placed at appropriate, periodic points in a row. In (*c*), the scattering from all of the molecules in a large, periodic array is simply the same resultant wave produced by one molecule but arising from each point of the appropriate point lattice describing the periodic arrangement. The diffraction from such an array, a lattice of points, we know is governed by Bragg's law.

If we do this, then the product of the transform of the object and the lattice becomes, as in Figure 5.4, simply the line of points, where each point serves as an identical source of a common wave corresponding to the scattering of the entire continuous object for some diffraction vector $\bar{s}$. Although the lattice points produce a wave for any and all diffraction vectors $\bar{s} = (\bar{k} - \bar{k}_0)$, because the waves arise from points in a lattice, the waves cancel, or sum to zero except when all the points belong to a family of planes $\boldsymbol{hk\ell}$ for which Bragg's law is satisfied, that is, when $\bar{s} = \bar{h}$. When this condition is met, the waves emitted from each point constructively interfere and sum in an arithmetic manner. The lattice then multiplies the resultant wave from the object, the atoms within the unit cell, by the total number of unit cells in the crystal and allows us to observe it, but only for specified values of $\bar{s}$, namely only at those points in diffraction (Fourier transform) space where $\bar{s} = \bar{h}$.

What is seen for one dimension is quite the same for the two- or three-dimensional cases as well. Just as the resultant wave created by the interference of the scattered waves from all of the atoms in the molecules could be considered as arising from discrete lattice points, the same is true for a real crystal. We can consider the resultant waves produced by the scattering of all of the atoms in the unit cells to simply be emerging from a single lattice point common to each cell, as in Figure 5.10. Because the contents of the unit cell are continuous and nonperiodic, their transform, or resultant waves $\bar{F}_{\bar{s}}$ will be nonzero for all $\bar{s}$. Because the lattice points in a crystal are discrete and periodic, however, the waves from all lattice points will constructively interfere and be observable only in certain directions according to Bragg's law, that is, when $\bar{s} = \bar{h}$.

Unit Cells and Bragg Planes

Figure 5.11 illustrates a two-dimensional array of four unit cells in a crystal and a family of planes having integral indexes. By virtue of their integral indexes, the family of planes intersects the cell edges and samples the contents of all unit cells in exactly the same way. Thus, whatever is true for one unit cell is true for every unit cell in the crystal. Although we have reduced this to a two-dimensional case, the same argument holds for three dimensions,

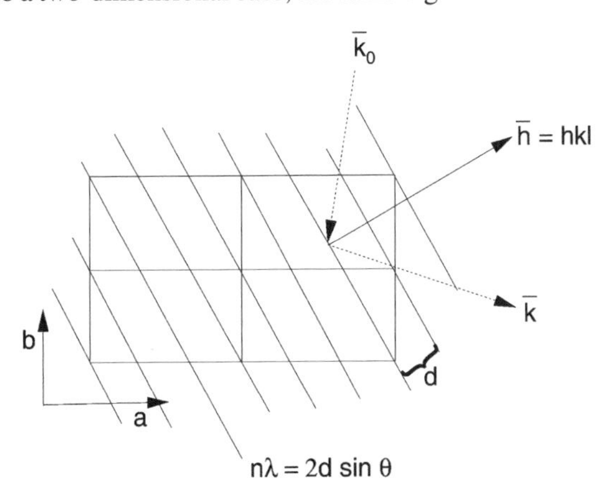

FIGURE 5.11 When Bragg's law is satisfied, then the diffraction vector, $\bar{s} = \bar{k} - \bar{k}_0$, must be equal to the reciprocal lattice vector $\bar{h} = \boldsymbol{hk\ell}$. When this condition is satisfied the diffraction of waves for all planes in the family $\boldsymbol{hk\ell}$ constructively interfere. Notice here that while an individual plane may sample a series of contiguous unit cells at different places, the family of planes samples every unit cell exactly the same. Thus the diffraction from the family of planes is identical for every unit cell in a crystal.

and therefore a real crystal. For any family of parallel planes with a common interplanar spacing $d_{hk\ell}$, X rays scattered by successive planes will constructively interfere and give rise to a diffracted ray only when the incident radiation, or normal to the wavefront, strikes the planes at such an angle that Bragg's law is satisfied. Only then will the waves reflected by all of the planes in the set have exactly the same phase and constructively interfere. The scattering points or electron density could be disposed anywhere in the cell, and this criterion would still have to be met.

THE STRUCTURE FACTOR FOR A CRYSTAL

Let us now look at a diffraction experiment from the perspective of an individual unit cell in Figure 5.12*a* that contains some distribution of atoms, scattering points, organized by chemical bonds into a molecule. For simplicity the unit cell contains only a single asymmetric unit, that is, the space group is $P1$, although interaxial angles are taken as $90°$ (a very rare occurrence, but nonetheless observed in some crystals). Subdividing this unit cell (and all others in the crystal as well) is a family of planes of interplanar spacing $d_{hk\ell}$ having integral indexes $hk\ell$.

If Bragg's law is not satisfied for this family of planes, it doesn't matter what the distribution of electron density in the unit cell is, since no diffraction from the crystal occurs. Scattered waves from any one unit cell will be out of phase with those from all others, producing destructive interference.

When Bragg's law is satisfied, the resultant wave produced by this unit cell will be duplicated by every unit cell in the crystal, and they will sum together, or constructively interfere, to yield an observable resultant wave. But what will the wave from an individual unit cell be? What will be its amplitude and phase?

The wave produced by all of the atoms in a unit cell, when Bragg's law is satisfied, will simply be the sum of the waves scattered by the individual atoms within the cell. To obtain the diffracted wave produced by the entire crystal (which is what we measure as a diffraction intensity), we need only sum the waves scattered by each atom in one unit cell and then multiply by the total number of unit cells, N, in the crystal. What are the amplitudes and relative phases of these individual waves with respect to one another, or to some other reference wave?

Consider the same unit cell, under the same conditions, in Figure 5.12*b*. If an atom lay exactly on an $hk\ell$ plane, then the phase of the wave scattered by that atom would be zero with respect to the plane. Remember that all points on a plane scatter with identical phases (see above). Because there is one wavelength difference in the path length for waves scattered by successive planes in the family when Bragg's law is satisfied, atoms lying exactly on other planes would also scatter with a phase of $\phi = 0 + 2\pi = 0 + n\pi = 0$.

Consider further an atom that lies exactly between two planes. If the difference in phase between two successive planes is 2π, then this atom would scatter with a phase of $2\pi/2 = \pi$. In fact any atom in the cell lying an arbitrary distance D from the nearest plane would scatter with a phase of $(D/d_{hk\ell})2\pi$. In other words, the phase angle of the wave scattered by any atom in the unit cell is simply a function of its distance from a plane of that particular family. Can we determine how far any atom is from an $hk\ell$ plane and, therefore, its phase? We can indeed, so long as we know the positions, the x, y, z coordinates, of the atoms in the unit cell.

The origin of the unit cell may be arbitrarily chosen. To simplify our analysis, in Figure 5.12*b* we take it at the intersection of the cell edges so that any point p_j in the

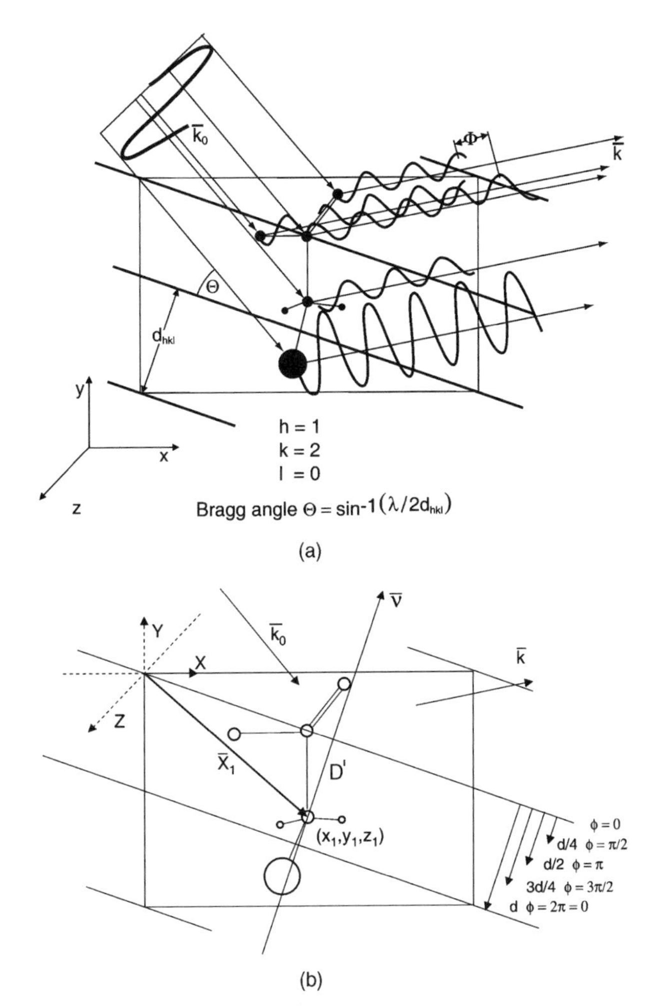

FIGURE 5.12 In (*a*) is an experiment designed to yield a resultant diffraction wave from all of the atoms in the unit cell of a crystal. The diffracted wave is called a structure factor. The incident wave is from direction $\bar{k}_0$, and the diffracted wave is along $\bar{k}$. Here the Bragg condition is satisfied for an arbitrary $hk\ell$ family of planes of spacing $d_{hk\ell}$, passing through an orthorhombic unit cell containing a single molecule. Every atom scatters a wave in the direction $\bar{k}$ with an intensity proportional to its atomic number, and with a phase ϕ_j dependent on its position x_j, y_j, z_j. In (*b*) the vector $\bar{x}_1$ defines the position in the unit cell of an atom at x_1, y_1, z_1 with respect to the crystallographic origin. A working vector of unit length $\bar{v}$ is drawn normal to the family of planes $hk\ell$. The distance of the atom at x_1, y_1, z_1 from the plane above is D. The phase of the wave scattered by the atom at x_1, y_1, z_1, or any atom at x_j, y_j, z_j, will be a function of this distance as indicated at far right. The structure factor for the entire crystal $\bar{F}_{hk\ell}$ will be the sum of the waves scattered by each of the atoms in the unit cell multiplied by the number of unit cells in the crystal.

unit cell is identified by $\bar{x} = (x_j, y_j)$ where x_j and y_j are fractional coordinates. The set of planes shown in Figure 5.12*b* have indexes $h = 1, k = 2$ and interplanar distances of

$$\frac{1}{d} = \sqrt{\frac{h^2}{a^2} + \frac{k^2}{b^2}}$$

or

$$d = \sqrt{\frac{a^2}{h^2} + \frac{b^2}{k^2}}$$

The working vector $\bar{v}$ having unit length and which is normal to the planes can be drawn. It passes not only through this unit cell but identically through all unit cells in the crystal. We use this unitary vector simply to define a direction. The phase of a wave scattered by an atom with respect to the set of planes will be $\phi = 2\pi(D/d_{hk\ell})$, where D is the atom's distance from the nearest plane. But D is the projection of $\bar{x}_1$ onto $\bar{v}$, that is, $D = \bar{x}_1 \cdot \bar{v}$, so that $\phi = 2\pi(\bar{x}_1 \cdot \bar{v}/d_{hk\ell})$.

The vector $\bar{v}$ is unit length and normal to the planes, $d_{hk\ell}$is the distance between the planes, and therefore $\bar{v}/d_{hk\ell}$ is the reciprocal lattice vector $\bar{h}$. Hence $\phi = 2\pi(\bar{x}_1 \cdot \bar{h})$. Only those points in the unit cell where there is an atom need to be considered, since the scattering from any other point will be zero. We can replace $\bar{x}_1$ with $\bar{x}_j$, where $\bar{x}_j$ represents the coordinates x_j, y_j, z_j of the jth atom. The resultant diffracted ray from the entire unit cell, then, is the sum of the waves scattered by all N of the atoms j in the unit cell:

$$\bar{F}_{\bar{s}} = \bar{F}_{\bar{h}} = \sum_{j=1}^{N} Z_j \exp 2\pi i(\bar{x}_j \cdot \bar{h}) = \sum_{j=1}^{N} Z_j[\cos 2\pi(\bar{x}_j \cdot \bar{h}) + i \sin 2\pi(\bar{x}_j \cdot \bar{h})].$$

To take into account the disparity in electrons, and hence scatting power, for the various atoms in the unit cell, the atomic number Z_j has been introduced as a means of defining amplitudes for the component waves. The total diffracted wave for the entire crystal will be the product of the equation above with the total number of unit cells in the crystal.[1] For a single unit cell, like that in Figures 5.12*a* and 5.12*b*, an atom's contribution to the total structure factor of one unit cell can also be illustrated in vector terms as in Figure 5.13. To put the expression in the correct units, it is necessary to multiply the summation by a constant, the volume of the unit cell. Here, that is simply $V = a \times b \times c$. Thus

$$\bar{F}_{\bar{h}} = V \sum_{j=1}^{N} Z_j \exp 2\pi i(\bar{x}_j \cdot \bar{h}).$$

A further, rather minor alteration is necessary to this equation if it is to better reflect physical reality. The electrons surrounding atomic nuclei occupy a volume defined by the

[1] In fact, we do not know the number of unit cells in the crystal, but that doesn't matter. We don't really have to multiply by that number because all we are interested in is the relative amplitudes of the diffracted waves and their phases. Multiplying all scattered waves by a constant will have no effect other than scaling.

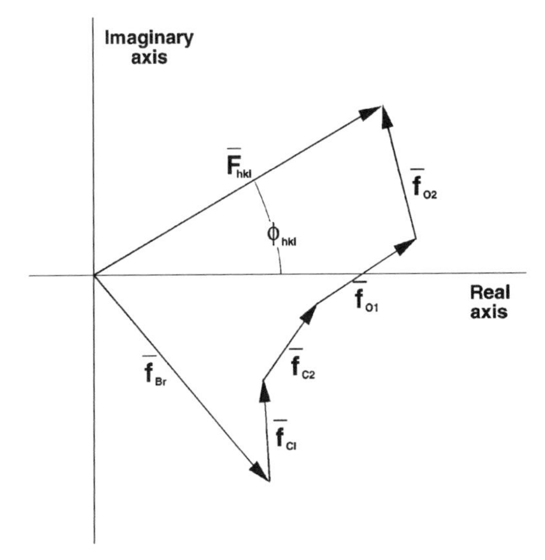

FIGURE 5.13 The structure factor $\bar{F}_{hk\ell}$ is a wave, and it therefore has an amplitude and phase. It can also be described as a vector in the complex plane, as was seen in Chapter 4. The individual wave contributions to $\bar{F}_{hk\ell}$ by each atom in the molecule can also be described by vectors. The sum of these vectors for all atoms yields $\bar{F}_{hk\ell}$. The vectors added here correspond to contributions from the five atoms of the molecule filling the unit cell in Figure 5.12.

atomic radius. When those electrons scatter X rays, there is some internal destructive interference among waves scattered from different elements of that volume. This is increasingly severe as $\sin\theta$ increases. To accommodate this decline in atomic scattering intensity with Bragg angle, we replace Z_j with a parameter f_j, which we call the scattering factor for atom j. These f_j are known for all kinds of atoms and can be retrieved from tables as a function of the Bragg angle, which of course is directly calculable from $hk\ell$ and the unit cell dimensions. Some f_j and their variation with scattering angle are shown in Figure 5.14. The equation, in trigonometric form, for the scattered wave from the crystal in terms of the three components of $\bar{x}_j$ and $\bar{h}$ then becomes, using the expanded form for $\bar{h} \cdot \bar{x}_j$

$$\bar{F}_{\bar{h}} = V \sum_{j=1}^{N} f_j[\cos 2\pi(hx_j + ky_j + lz_j) + i \sin 2\pi(hx_j + ky_j + lz_j)].$$

The function $\bar{F}_{\bar{h}}$ is called the structure factor for the $hk\ell$ family of planes in the crystal. Because it is a wave, or vector, it has both a magnitude and a phase. The set of all $\bar{F}_{\bar{h}}$ for all values of $\bar{h}$ comprise the set of diffracted rays resulting from all possible families of planes in the crystal and thereby constitutes the diffraction spectrum of the crystal. The magnitude of $\bar{F}_{\bar{h}}$ is readily measured as the square root of the observed diffracted intensity, that is, $\sqrt{I_{\bar{h}}} = F_{\bar{h}}$, but there is no experimental means presently available to directly measure its phase $\phi_{\bar{h}}$.

To reiterate, the ensemble of $\bar{F}_{\bar{h}}$ is a discrete, finite set and, due to Bragg's law, not a continuum. $\bar{F}_{\bar{h}}$ is nonzero only at specific reciprocal lattice points $hk\ell$ in diffraction space determined by the parameters of the crystal lattice. The intensity value at those points

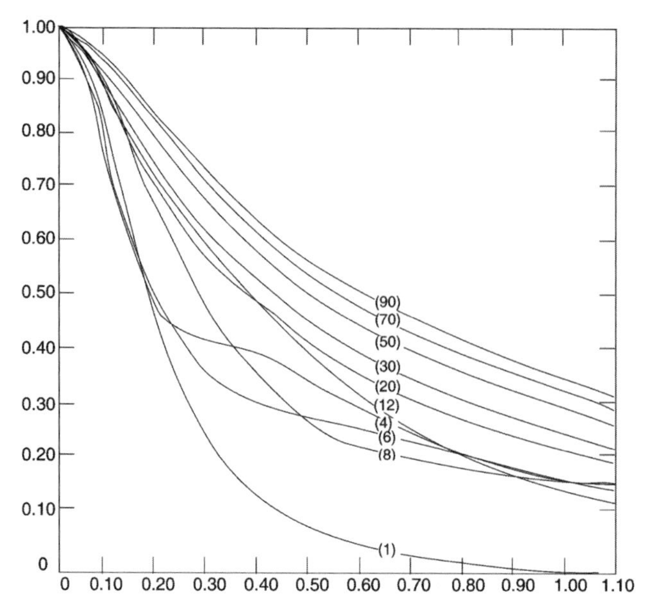

FIGURE 5.14 The curves represent the scattering power of various atoms, ranging from atomic number 1 to atomic number 90, as a function of $\sin^2\theta/\lambda^2$, that is, as a function of resolution. Values for the scattering factor can be read directly from the graphs (or from tables) and used in the structure factor equation to calculate $\bar{F}_{hk\ell}$.

is directly a function of the distribution of atoms in the unit cell. Furthermore, the set of diffraction intensities is identified by correspondence with the set of reciprocal lattice points, which is the Fourier transform of the crystal lattice and forms itself a lattice in diffraction space.

THE STRUCTURE FACTOR AS A PRODUCT OF TRANSFORMS

The last approach is for the more mathematically inclined. The expression for $\bar{F}_{\bar{h}}$ was formulated above with the implicit assumption that Bragg's law was obeyed. It can be derived in another manner that explicitly contains Bragg's law.

It was shown that the scattering function for a general scattering distribution such as the N atoms in a unit cell is

$$\bar{F}_{\bar{s}} = \sum_{j=1}^{N} f_j \exp i2\pi(\bar{x}_j \cdot \bar{s}),$$

where $\bar{x}_j$ is the vector specifying the position of the atom j and f_j is a measure of its scattering power.

For a discrete set of planes in a lattice the scattering function is

$$\bar{F}_{\bar{s}} = \sum_{n=1}^{NP} \exp i2n\pi(\bar{d}_{hk\ell} \cdot \bar{s}),$$

where $\boldsymbol{n}$ is over all planes $\boldsymbol{NP}$ in the set, that is, the entire crystal. The combined Fourier transform of the two functions, their joint scattering distribution, is the product of their separate transforms, and for a crystal this is $\bar{F}_{\bar{s}}$ (lattice) $\times$ $\bar{F}_{\bar{s}}$ (unit cell). Then

$$\bar{F}_{\bar{s}}(crystal) = \sum_{j=1}^{N} f_j \exp i2\pi(\bar{x}_j \cdot \bar{s}) \sum_{n=1}^{NP} \exp i2n\pi(\bar{d}_{hk\ell} \cdot \bar{s}).$$

The summation of exponential terms on the right is a Dirac delta function, a discrete function, which is everywhere zero except when the argument is zero or integral. The summation on the left is a continuous function, which determines the value of the entire transform at those nonzero points. Now $\bar{d}_{hk\ell}$ is normal to the set of planes of a particular family, and $|\bar{d}_{hk\ell}|$ is the interplanar spacing. In order for $\bar{d}_{hk\ell} \cdot \bar{s} = \pm 1$, $\bar{s}$ must be parallel with $\bar{d}_{hk\ell}$ and have magnitude $1/|\bar{d}_{hk\ell}|$; that is, $\bar{s} \equiv \bar{h}$ the reciprocal lattice vector. If $\bar{s} \neq \bar{h}$, then there is destructive interference of the waves diffracted by different unit cells, and the resultant wave from the crystal is zero. The elements of the diffraction spectra, the structure factors, for the crystal can therefore be written as

$$\bar{F}_{\bar{s}}(crystal) = \bar{F}_{\bar{h}} = \sum_{n=1}^{NP} \sum_{j=1} f_j \exp i2\pi(\bar{x}_j \cdot \bar{h}),$$

$$\bar{F}_{\bar{h}} = NP \sum_{j=1} f_j \exp i2\pi(\bar{x}_j \cdot \bar{h}),$$

where $\boldsymbol{NP}$ is the total number of planes in the crystal. Again, to assure mathematical propriety and have the equation in the correct units, we must multiply by the volume of the unit cell. [2] Hence

$$\bar{F}_{\bar{h}} = V \sum_{j=1}^{N} f_j \exp i2\pi(\bar{x}_j \cdot \bar{h})$$

We have now shown, by three different approaches, that if one knows the atomic coordinates x_j, y_j, z_j of all of the atoms j in a unit cell, and their scattering factors f_j, then one can precisely predict the amplitude and phase of the resultant wave scattered by a specific family of planes $hk\ell$. We can calculate this for any and all families of planes in the crystal, hence the amplitude and phase can be calculated for every structure factor in the X-ray diffraction pattern. Given the structure of a crystal, namely the coordinates of the atoms in the unit cell, we can predict the entire diffraction pattern, the entire Fourier transform of the crystal. This is an enormously powerful statement. It means that if, by some means, we can deduce the positions of the atoms in a crystal structure, then we can immediately check the correctness of that deduction by seeing how well we can predict the relative values of the intensities in the diffraction pattern.

[2] To save paper, we are here compressing notation to the vector forms for position $\bar{x} = x, y, z$ and indexes $\bar{h} = h, k\,\ell$, and using Euler's form of the complex numbers. We will expand them back to their full forms later. Again, it is unnecessary to actually multiply by NP, which we do not even know, since it is a constant and doesn't change the relative intensities or phases of the structure factors.

This is impressive you might agree, but the real objective in X-ray crystallography is to record the diffraction pattern and to figure out from the diffraction intensities the structure of the crystal, and hence the $x,\ y,\ z$ coordinates of the atoms in the molecules that make it up. Seldom, it would seem, that we would want to predict diffraction patterns if we already know the crystal structure. Here, however, the symmetrical nature of the Fourier transform proves invaluable. If a scattering distribution, a collection of atoms, is Fourier transformed to create a diffraction pattern, then the Fourier transform of the diffraction pattern must reproduce the original scattering function, the crystal structure. This being so, it then follows that if we record the diffraction pattern of a crystal, then the Fourier equation will permit us to calculate its transform and obtain the atomic coordinates defining the crystal structure. We might anticipate that equation as probably looking a lot like the structure factor equation but going in the opposite direction, from reciprocal space back into real space. Before we examine that Fourier equation, however, it is prudent to first address some other important points regarding structure factors.

TEMPERATURE FACTORS

An additional refinement of the structure factor equation, again involving the scattering term, is necessary to make it more precise. Remember, the observed diffraction intensities, those actually produced by the scattering of the atoms in the unit cells, represent reality. Hence we must try to take into account any physical property or effect which significantly alters the way atoms scatter X rays. An atom is not, in general, completely stationary; it vibrates because of thermal motion about some mean position (x_j, y_j, z_j), which we take to be its atomic center. As a consequence, its electron complement is spread out over a larger volume in space than the atomic radius and the scattering factor f_j above would suggest. In addition, different atoms in a molecule (or asymmetric unit) may show different amounts of thermal motion, or statistical disorder, which has a similar effect. The spreading of the electrons of an atom over a larger volume essentially serves to attenuate the scattering power of the atom, and because it arises from internal interference effects, this attenuation grows exponentially more severe as a function of scattering angle θ.

Therefore, to better approximate the real scattering efficiency of each atom, the scattering factor f_j in the structure factor equation is multiplied by an exponential term that effectively reduces the f_j as a function of $\sin\theta$, where θ is the Bragg angle. The new term has the form

$$\exp\left(\frac{-B_j \sin^2\theta}{\lambda^2}\right),$$

where B_j is a constant, generally different for each atom in the asymmetric unit, that is proportional to its thermal motion. B_j is called the atom's temperature factor. In well-executed structure analyses it is possible to estimate the temperature factors quite accurately by their influence on the observed diffraction intensities. When the temperature factor term is included, the structure factor equation then assumes the form

$$\bar{F}_{hk\ell} = V \sum_{j=1}^{N} f_j \left[\exp \frac{-B_j \sin^2\theta}{\lambda^2}\right] \big(\cos 2\pi(hx + ky + lz) + i \sin 2\pi(hx + ky + lz)\big).$$

It should be noted that more complicated forms of the temperature factor term can be employed when the crystal structure analysis is particularly precise and the resolution high. These expressions take into account the possible anisotropy of the thermal motion or statistical disorder. In the most sophisticated cases, six parameters are used to define the three-dimensional ellipsoids of thermal motion, which serve to describe anisotropic temperature factors. These should not be a source of concern to the reader at this time.

CENTERS OF SYMMETRY

If a crystal contains a center of symmetry among its space group elements, then for every atom at point $\bar{x}_j = (x_j, y_j, z_j)$, there is a corresponding atom at $-\bar{x}_j = (-x_j, -y_j, -z_j)$. The structure factor equation for $\bar{F}_{\bar{h}}$ will therefore contain a term

$$f_j \exp i2\pi(\bar{h} \cdot -\bar{x}_j) = f_j[\cos 2\pi(\bar{h} \cdot -\bar{x}_j) + i \sin 2\pi(\bar{h} \cdot -\bar{x}_j)]$$

corresponding to every term

$$f_j \exp i2\pi(\bar{h} \cdot \bar{x}_j) = f_j[\cos 2\pi(\bar{h} \cdot \bar{x}_j) + i \sin 2\pi(\bar{h} \cdot \bar{x}_j)].$$

Now, for any angle α, $\cos\alpha = \cos -\alpha$ but $\sin -\alpha = -\sin\alpha$. Thus imaginary parts of the two terms will always have opposite signs, and those terms will always subtract to zero in pairs. The scattering contribution for the centrosymmetrically related pairs of atoms will then reduce to

$$\bar{F}_{\bar{h}} = V \sum_{j=1}^{N} f_j \cos 2\pi(\bar{h} \cdot \bar{x}_j).$$

Note that this is a strictly real function; that is, $\bar{F}_h$ has no imaginary component. Since the phase of any reflection is $\phi = tan^{-1}(\sin\phi/\cos\phi)$ but $\sin\phi = 0$, then $\phi = tan^{-1}0$. Hence $\phi = 0$ or π. Thus the phase angle of $\bar{F}_{\bar{h}}$ must be either 0 or π. Since $\cos 0 = +1$ and $\cos\pi = -1$, then $\bar{F}_{\bar{h}} = \pm F_{\bar{h}}$. This means that for crystals whose space group contains a center of symmetry, only the signs of structure amplitudes $F_{\bar{h}}$ comprising the diffraction pattern need to be sought, not all possible phase angles from 0 to 2π. This is a very significant simplification, and it makes the solution of the structures of centrosymmetric crystals considerably less difficult. There are only two choices for the phase of each structure factor.

This might seem of scant use in protein crystallography, since we have no centric space groups. Crystals of biological macromolecules, as previously pointed out, cannot possess inversion symmetry. Sets of centric reflections frequently do occur in the diffraction patterns of macromolecular crystals, however, because certain projections of most unit cells contain a center of symmetry. The correlate of a centric projection, or centric plane in real space, is a plane of centric reflections in reciprocal space. A simple example is a monoclinic unit cell of space group $P2$. The two asymmetric units have the same hand, as they are related by pure rotation, and for every atom in one at x_j, y_j, z_j there is an equivalent atom in the other at $-x_j, y_j, -z_j$. If we project the contents of the unit cell on to a plane perpendicular to the y axis, namely the xz plane, by setting $y = 0$ for all atoms, however, then in that projection for every atom at x_j, z_j there is an equivalent atom at $-x_j, -z_j$ The projection is therefore centrosymmetric.

Consider then the plane of reflections in reciprocal space having indexes $h0l$. We can write their structure factors as

$$\bar{F}_{h0\ell} = \frac{1}{V}\sum_j f_j[\cos 2\pi(hx_j + \ell z_j) + i\sin 2\pi(hx_j + lz_j)].$$

Once again we see that for every $\sin 2\pi(hx_j + lz_j)$ there will be a term $\sin 2\pi(-hx_j - lz_j)$, and the terms will cancel in pairs. Hence the phases can only be 0 and π for this class of reflections. They are centrosymmetric, even though the three-dimensional structure is noncentric.

FRIEDEL'S LAW

Friedel's law arises from certain properties of the Fourier transform, and it has a significant consequence in diffraction space. It places an important restriction on the relative amplitudes and phases of the structure factors for the $hk\ell$ and $-h-k-\ell$ families of planes. The $hk\ell$ and $-h-k-\ell$ families are not really two separate families but one family. We can think of the structure factor $\bar{F}_{hk\ell}$ as arising from scattering from the tops of the planes in the family, and $\bar{F}_{-h-k-\ell}$ as arising from scattering from the bottoms of the planes in that same family. Intuitively we might expect the scattering to be the same from a family of planes, whether we use the tops or the bottoms, as they are, after all, imaginary planes. The amplitudes of the scattered waves are in fact the same, but the phases are not identical; they have opposite sign.

Figure 5.15 illustrates the Friedel relationship of $\bar{F}_{\bar{h}}$ and $\bar{F}_{-\bar{h}}$ in physical terms. The heavy planes in Figure 5.15 have indexes $hk\ell$ and the light planes are any arbitrary planes

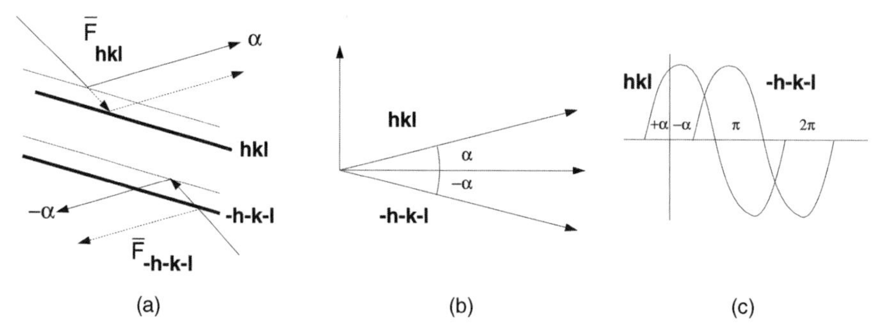

FIGURE 5.15 Illustration of Friedel's law. The heavy planes are members of the family $hk\ell$, and the light planes represent any arbitrary planes of scattering material lying between. The radiation scattered by the electron density on consecutive planes of the family will have phase differences of $2n\pi$ when Bragg's law is satisfied, for both $\bar{F}_{hk\ell}$ and $\bar{F}_{-h-k-l}$. The electron density on the light planes, however, will scatter with a phase $+\alpha$ ahead for $\bar{F}_{hk\ell}$, but $-\alpha$ behind for $\bar{F}_{-h-k-l}$. Since this will be true for all of the scattering material between the planes of the family $hk\ell$, the phase of $\bar{F}_{hk\ell}$ and that of $\bar{F}_{-h-k-\ell}$ must be equal but of opposite sign. The magnitudes of $\bar{F}_{hk\ell}$ and $\bar{F}_{-h-k-l}$ must, however, be the same.

lying between. In the case of $\bar{F}_{\bar{h}}$ the light plane scatters with a phase angle $+\alpha$ ahead of $\boldsymbol{hk\ell}$, and for $\bar{F}_{-\bar{h}}$ the phase lags by $-\alpha$. Since these relationships will be true for all the scattering matter lying on all planes between the $\boldsymbol{hk\ell}$ planes, the phases for $\bar{F}_{\bar{h}}$ and $\bar{F}_{-\bar{h}}$, as shown in Figure 5.15*b* and *c*, will always be of opposite sign. The amplitudes of $\bar{F}_{\bar{h}}$ and $\bar{F}_{-\bar{h}}$, however, will be the same. Thus the diffraction pattern, the intensity weighted reciprocal lattice, contains a center of symmetry regardless of the crystal that produced it, and in particular, the intensities of reflections related by inversion through the origin will always be identical in magnitude but have phases of opposite sign.

ANOMALOUS DISPERSION EFFECTS

We must here consider one last point regarding the scattering of X rays by a single atom. We have already seen that we must replace Z_j by f_j, the scattering factor, and introduce a temperature factor B_j to make things better comport with physical reality. Anomalous dispersion is a very minor correction, or alteration in the scattering of X rays by an atom, and in previous times we might have passed it by with no more than a brief nod to its existence. Although the measurable effect on structure amplitudes $F_{hk\ell}$ is generally small, we must now, however, give it more serious attention. This is because the anomalous dispersion effect is the basis for what has become one of the most powerful methods for solving crystal structures. We will treat this approach to structure solution in a later chapter, but let us deal briefly with the effect here.

To this point we have assumed that an atom, be it heavy or otherwise, scatters as a point source of scattering power f_j having phase ϕ_j. Although the detailed physical explanation is outside the scope of this book and involves quantum mechanical properties, it must be pointed out that this is not entirely true. An atom scatters X rays in a somewhat more complex fashion, in that its scattered radiation is composed of two components. The major component, which arises from normal Thompson scattering, and is by far the largest component, has phase ϕ dependent on the atom's position as we have assumed. But there is also a minor component of the scattering that has phase $\phi + \pi/2$. This is because the electrons of the atom also absorb a small amount of radiation due to electron resonance phenomena and re-emit it with a phase change. This second component is called the anomalous dispersion, and to be entirely correct, we should properly describe the radiation scattered by an atom as a complex number,

$$f = f_0 + f' + i\Delta f''.$$

The magnitude of the imaginary anomalous component, $\pi/2$ out of phase with respect to the real scattering component, depends both on the atomic number of the atom and its degree of absorption of the incident radiation wavelength. That is, its magnitude is wavelength dependent. In general, the anomalous component is very small, usually no more than 1% or less of the real scattering factor f. In some cases, however, it can be significant, and it can produce pronounced effects in the diffraction pattern. For example, with $\mathbf{CuK_\alpha}$ radiation, generally used by protein crystallographers, the lanthanide series of elements and uranium may scatter with an anomalous dispersion that is 10% or more of the real component.

Because part of the anomalous dispersion component is $\pi/2$ out of phase with the isomorphous, real component, the net observable effect is a breakdown of Friedel's law regarding the perfect equality of the magnitudes of $F_{hk\ell}$, and $F_{-h-k-\ell}$. That is, the two need not be absolutely equivalent but can demonstrate some slight difference $\Delta F_{anom} = F_{hk\ell} - F_{-h-k-l}$. This difference will normally be imperceptible and within the expected statistical error of most X-ray diffraction intensity measurements, but with care in data collection, and judicious choice of X-ray wavelength, it can be measured and used to obtain phase information in conjunction with isomorphous replacement phase determination, or even independently, as described in Chapter 8.

THE ELECTRON DENSITY EQUATION

To this point we have focused single-mindedly on understanding and writing an expression that describes the diffraction from a crystal, its Fourier transform, if we know the atomic coordinates x_j, y_j, z_j in the crystallographic unit cell. The reader may be impatient by now as the real life objective is to do the opposite, to define the x_j, y_j, z_j coordinates of the atoms when we can measure the $\bar{F}_{\bar{h}}$ making up the diffraction pattern. It is now time to go the other way, but understanding the meaning of the $\bar{F}_{\bar{h}}$ makes that task easier.

In deriving the structure of a molecule, or distribution of atoms using X-ray crystallography, we do not directly obtain the x, y, z coordinates of the atoms. That is, we don't solve some system of linear equations whose solution is the set of numerical values for x, y, z. We employ a Fourier transform equation that incorporates the diffraction data, the structure factors, and yields the value of the electron density $\rho(x, y, z)$ at any and all points x, y, z within the crystallographic unit cell. From the peaks and features of this continuous electron density distribution in the unit cell we then infer the locations of the atoms, and hence their coordinates. This will be described as it is done in practice, in Chapter 10. Following this, the coordinates are improved by applying refinement procedures, as also outlined in Chapter 10. Here, however, our objective is to understand this Fourier transform equation, namely the electron density equation.

The unit cell of a crystal contains a set of atoms or scattering material occupying locations distributed throughout the unit cell. The electron density therefore can be considered a continuous function of position $\rho(\bar{x})$ or $\rho(x, y, z)$ throughout a single unit cell, where $\bar{x} = x, y$, and z are fractional coordinates. This was illustrated in Figure 3.1 of Chapter 3. The function takes on high values near atomic centers and tends to zero elsewhere. Because $\rho(\bar{x})$, the electron density, is identical from unit cell to unit cell, it is a three-dimensional periodic function in space exactly analogous to the one-dimensional resultant wave of Figure 4.9 in Chapter 4. The period of $\rho(x, y, z)$ is $|\bar{a}|$ along $\bar{a}$, $|\bar{b}|$ along $\bar{b}$, and $|\bar{c}|$ along $\bar{c}$. Because it is a periodic wave function, it can be analyzed into a spectrum of component density waves of successively higher frequency, and conversely, it can be synthesized from those spectral components, also as illustrated in Figures 4.9 through 4.11 of Chapter 4.

In the case of a periodic, three-dimensional function of x, y, z, that is, a crystal, the spectral components are the families of two-dimensional planes, each identifiable by its Miller indexes $hk\ell$. Their transforms correspond to lattice points in reciprocal space. In a sense, the planes define electron density waves in the crystal that travel in the directions of their plane normals, with frequencies inversely related to their interplanar spacings.

(Remember from Chapter 4 that the periods and frequencies of waves are reciprocally related.) Exactly those properties are expressed by their reciprocal lattice vectors $\bar{\boldsymbol{h}}$. The amplitudes of these electron density waves vary according to the distribution of atoms about the planes. Although the electron density waves in the crystal cannot be observed directly, radiation diffracted by the planes (the Fourier transforms of the electron density waves) can. Thus, while we cannot recombine directly the spectral components of the electron density in real space, the Bragg planes, we can Fourier transform the scattering functions of the planes, the $\bar{F}_{hk\ell}$, and simultaneously combine them in such a way that the end result is the same, the electron density in the unit cell. In other words, each $\bar{F}_{hk\ell}$ in reciprocal, or diffraction space is the Fourier transform of one family of planes, $\boldsymbol{hk\ell}$. With the electron density equation, we both add these individual Fourier transforms together in reciprocal space, and simultaneously Fourier transform the result of that summation back into real space to create the electron density.

We have seen that the diffracted waves $\bar{F}_{hk\ell}$, from a particular family of planes $\boldsymbol{hk\ell}$, when Bragg's law is satisfied, depends only on the perpendicular distances of all of the atoms from those $\boldsymbol{hk\ell}$ planes, which are $\bar{h} \cdot \bar{x}_j$ for all atoms j. Therefore each $\bar{F}_{hk\ell}$ carries information regarding atomic positions with respect to a particular family $\boldsymbol{hk\ell}$, and the collection of $\bar{F}_{hk\ell}$ for all families of planes $\boldsymbol{hk\ell}$ constitutes the diffraction pattern, or Fourier transform of the crystal. If we calculate the Fourier transform of the diffraction pattern (each of whose components $\bar{F}_{hk\ell}$ contain information about the spatial distribution of the atoms), we should see an image of the atomic structure (spatial distribution of electron density in the crystal). What, then, is the mathematical expression that we must use to sum and transform the diffraction pattern (reciprocal space) back into the electron density in the crystal (real space)?

We are instructed in how to formulate the electron density equation by the structure factor equation. The expression for $\rho(x, y, z)$, the electron density, must have the same form and be symmetrical with that for $\bar{F}_{hk\ell}$. Therefore variables and known, or measured, entities in real and reciprocal space (in the crystal and in the diffraction pattern respectively) must be substituted for one another. In the trigonometric term of the structure factor equation, called the "kernel" of the transform, $\bar{h} \cdot \bar{x} = hx + ky + lz$ is the same since $\bar{h} \cdot \bar{x} = \bar{x} \cdot \bar{h}$. That doesn't change at all. In the equation for $\bar{F}_{hk\ell}$, however, the summation was over all x_j, y_j, z_j in the unit cell. The x_j, y_j, z_j served as the running variables along with the coefficients f_j (which are equivalent to $\rho(x_j, y_j, z_j)$). The $\boldsymbol{hk\ell}$ at the same time were the running indexes of the entities being calculated, the $\bar{F}_{hk\ell}$. In the equation for $\rho(x\, y\, z)$, the roles are reversed. For given values of x, y, z in real space, summation is over $\boldsymbol{hk\ell}$ in reciprocal space, which provides running indexes of the coefficients in the series.

In the structure factor equation the coefficients in the summation were all nonzero electron densities f_j occurring at x_j, y_j, z_j, which is really $\rho(x_j, y_j, z_j)$, and $\bar{F}_{hk\ell}$ were the entities being calculated. Hence the coefficients in the electron density equation yielding $\rho(x, y, z)$ must be the reciprocal space entities $\bar{F}_{hk\ell}$. Finally, to keep units consistent, and the mathematics consistent with Monsieur Fourier, the sign of the imaginary term must be changed to minus, and the constant V must be inverted to $1/V$, the volume of the reciprocal unit cell. Thus the electron density equation assumes the form.

$$\rho(x, y, z) = \frac{1}{V} \sum_{\bar{h}=-\infty}^{+\infty} \bar{F}_{\bar{h}}[\cos 2\pi(\bar{h} \cdot \bar{x}) - i \sin 2\pi(\bar{h} \cdot \bar{x})].$$

The electron density equation, as presented here, is formally correct, but it does lend itself to some simplification, which ultimately makes it both more realistic and more computationally efficient. Before embarking on this bit of refinement, it is useful to once again recall two important corollaries:

1. From fundamental trigonometry,

$$\cos -\alpha = \cos \alpha$$

for any angle α, but

$$\sin -\alpha = -\sin \alpha.$$

2. From Friedel's law (anomalous dispersion aside),

$$F_{-h-k-\ell} \equiv F_{hk\ell}$$

and

$$\phi_{-h-k-l} = -\phi_{hk\ell}.$$

Now [3]

$$\rho(\bar{x}) = \frac{1}{V} \sum_{\bar{h}=-\infty}^{+\infty} \bar{F}_{\bar{h}} \exp i2\pi(\bar{h} \cdot \bar{x}),$$

but

$$\bar{F}_{\bar{h}} = F_{\bar{h}} \exp i\phi_{\bar{h}},$$

where $\phi_{\bar{h}}$ is the phase angle in radians of $\bar{F}_{\bar{h}}$. Then

$$\rho(\bar{x}) = \frac{1}{V} \sum_{\bar{h}=-\infty}^{+\infty} F_{\bar{h}} \exp\ i2\pi(\bar{h} \cdot \bar{x}) \exp\ i\phi_{\bar{h}}.$$

To form the product of two exponential functions, the exponents are added,

$$\rho(\bar{x}) = \frac{1}{V} \sum_{\bar{h}=-\infty}^{+\infty} F_{\bar{h}} \exp\ i[2\pi(\bar{h} \cdot \bar{x}) + \phi_{\bar{h}}].$$

[3] To save paper, we are here compressing notation to the vector forms for position $\bar{x} = x,\ y,\ z$ and indexes $\bar{h} = h, k\ \ell$, and using Euler's form of the complex numbers. We will expand them back to their full forms later.

Now reverting to complex number notation, we obtain

$$\rho(\bar{x}) = \frac{1}{V} \sum_{\bar{h}=-\infty}^{+\infty} F_{\bar{h}}[\cos(2\pi(\bar{h} \cdot \bar{x}) + \phi_{\bar{h}}) - i \sin(2\pi(\bar{h} \cdot \bar{x}) + \phi_{\bar{h}})].$$

By Friedel's law, $\phi_{-\bar{h}} = -\phi_{\bar{h}}$ and $\sin[2\pi(-\bar{h} \cdot \bar{x}) + \phi_{-\bar{h}}] = -\sin[2\pi(\bar{h} \cdot \bar{x}) + \phi_{\bar{h}}]$. Thus all **sin** terms, the imaginary terms, cancel in pairs, while all **cos** terms are added.

Therefore we can write the electron density equation as

$$\rho(\bar{x}) = \frac{1}{V} \sum_{\bar{h}=-\infty}^{+\infty} F_{\bar{h}} \cos[2\pi(\bar{h} \cdot \bar{x}) + \phi_{\bar{h}}],$$

or, expanding our notation, as

$$\rho(x, y, z) = \frac{1}{V} \sum_{\bar{h}=-\infty}^{+\infty} F_{hk\ell} \cos[2\pi(hx + ky + lz) + \phi_{hk\ell}].$$

We should note in passing that the original expression for $\rho(x, y, z)$ was a product of two complex numbers. This implied that $\rho(x, y, z)$ was also complex. We know, of course, that electron density has no imaginary component and that $\rho(x, y, z)$ must always be a strictly real function. We see here that because the Fourier transform is defined as a series summed from $-\infty$ to $+\infty$, and because of Friedel's law, $\rho(x, y, z)$ does indeed conform to reality. It's somehow reassuring to know that mathematics and reality are not in conflict, isn't it?

What this equation states is that you can calculate directly the value of the electron density $\rho(x, y, z)$ at any point x, y, z in the unit cell by summing over all of the structure factors $\bar{F}_{hk\ell}$, each multiplied by a trigonometric term, the kernal. Thus all structure factors contribute to the calculation of the electron density at each point x, y, z in the unit cell, just as each structure factor $\bar{F}_{hk\ell}$ was created from waves contributed by all of the atoms in the crystal. By systematically going through the entire unit cell on a fine grid of points x_j, y_j, z_j, and calculating $\rho(x_j y_j, z_j)$ at each grid point, a map, or image of the electron density throughout the entire cell can be obtained. While this may appear to be a computationally daunting task, with today's computers it is trivial, requiring no more than a few seconds time. It is discussed further in Chapter 10.

THE PHASE PROBLEM

At this point the most problematic feature of the process emerges. Inspection of the electron density equation as it was initially stated shows that the coefficient of each term in the summation for $\rho(x, y, z)$ at any value of x, y, z is $\bar{F}_{hk\ell}$. The structure factor $\bar{F}_{hk\ell}$ is, as we have seen, a wave. It is a complex number; it has an amplitude and a phase. In the final form of the equation we see that this feature persists in the form of the phase angle for each structure factor that must be included in the kernal. To calculate $\rho(x, y, z)$, then, it is necessary to know both the amplitude $\boldsymbol{F_{hk\ell}}$, and the phase $\boldsymbol{\phi_{hk\ell}}$ of each structure factor.

Here we have encountered the crucial, essential difficulty in X-ray diffraction analysis. It is not experimentally possible to directly measure the phase angles $\boldsymbol{\phi_{hk\ell}}$ of the structure factors. The best that our sophisticated detectors can provide are the amplitudes of the structure factors $\boldsymbol{F_{hk\ell}}$ but not their phases. Thus we cannot proceed directly from the measured diffraction pattern, the measured intensities, through the Fourier equation to the crystal structure. We must first find the phases of the structure factors. This central obstacle in structure analysis has the now infamous name, ***The Phase Problem***. Virtually all of X-ray diffraction analysis, not only macromolecular but for all crystals, is focused on overcoming this problem and by some means recovering the missing phase information required to calculate the electron density.

CHAPTER 6

INTERPRETATION OF DIFFRACTION PATTERNS

When you have mounted a protein crystal, or any crystal, for data collection, you may know with certainty very little about the orientation of planes within the crystal, unless you have made an exhaustive survey of the morphology of the crystal using special techniques and instruments, including polarized light. Even if you have made such an optical investigation, you probably will still have only limited information, particularly for macromolecular crystals because they have such weak optical properties. Besides, protein crystallographers are seldom, if ever, trained to do such things anymore, and virtually none of them any longer know how (if you would like to be different, see the book by Wood, 1977). Let us assume then that when you enter into data collection from a crystal, you don't know the orientation of any of the families of planes, how to bring them into diffraction position (i.e., have them satisfy Bragg's law), or how to manipulate the crystal to get from one set of planes to another.

Given these unknowns, it might appear that X-ray data collection would be a very difficult process indeed. It is not, in fact. X-ray crystallographers only rarely think about planes in the crystal, or their orientation. They use instead the diffraction pattern to guide them when they orient and manipulate a crystal in the X-ray beam. Recall that the net, or lattice, on which the X-ray diffraction reflections fall is the reciprocal lattice, and that every reciprocal lattice point, or diffraction intensity, arises from a specific family of planes having unique Miller indexes $\boldsymbol{hk\ell}$.

DIFFRACTION PATTERNS, PLANES, AND RECIPROCAL SPACE

If we expose a crystal to a collimated (parallel) beam of X rays, we will observe on our film (or area detector, or image plate) a set of reflections corresponding to some families of planes, which by chance, happen to be in diffracting position. An example is shown in Figure 6.1. If we reorient the crystal in the beam through some rotation, other planes

Introduction to Macromolecular Crystallography, Second Edition By Alexander McPherson

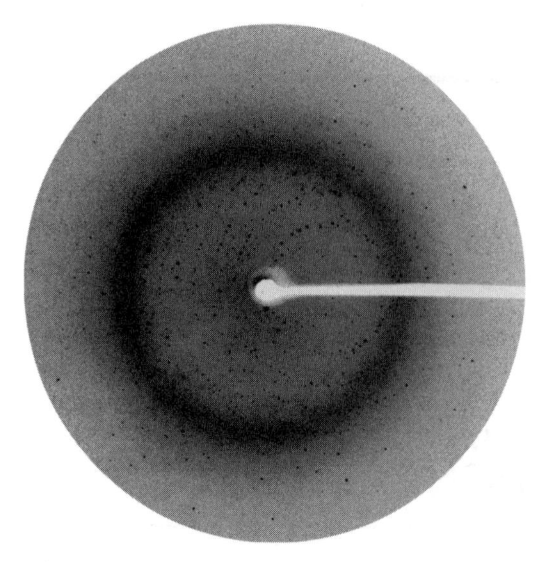

FIGURE 6.1 If a completely stationary crystal in a random orientation is subjected to X rays, some families of planes will, by chance, be in diffracting position; that is, they will satisfy Bragg's law. Those families of planes will therefore produce diffraction intensities. A crystal of cytochrome P450 in an arbitrary orientation gives rise to the reflections seen here.

are brought into diffracting position. So we can then observe their diffraction intensities, their reciprocal lattice points, and the distribution of those reciprocal lattice points with respect to our experimental, laboratory coordinate system. We cannot discern the families of planes in the crystal, we cannot see or detect them (they're imaginary anyway), but we can readily observe their reciprocal lattice vectors, or points. Remember, the orientation of the reciprocal lattice is locked to that of the real lattice, hence to the families of planes. When the crystal lattice rotates, its reciprocal lattice rotates accordingly, and so does its image on our film or detector. If we continuously reorient the crystal in a systematic way, then we can observe the appearance of entire planes of diffraction intensities as in Figures 6.2 and 6.3.

Remember further that each reciprocal lattice point represents a vector, which is normal to the particular family of planes $\boldsymbol{hk\ell}$ (and of length $\mathbf{1/}\boldsymbol{d_{hk\ell}}$) drawn from the origin of reciprocal space. If we can identify the position in diffraction space of a reciprocal lattice point with respect to our laboratory coordinate system, then we have a defined relationship to its family of planes, and the reciprocal lattice point tells us the orientation of that family. In practice, we usually ignore families of planes during data collection and use the reciprocal lattice to orient, impart motion to, and record the three-dimensional diffraction pattern from a crystal. Note also that if we identify the positions of only three reciprocal lattice points, that is, we can assign $\boldsymbol{hk\ell}$ indexes to three reflections in diffraction space, then we have defined exactly the orientation of both the reciprocal lattice, and the real space crystal lattice.

EWALD'S SPHERE

You may by now be convinced that the diffraction pattern and its underlying reciprocal net reveal the dispositions of families of planes, but it is probably not at all clear how this relationship can be used in the laboratory. Exactly how are the reciprocal lattice vectors

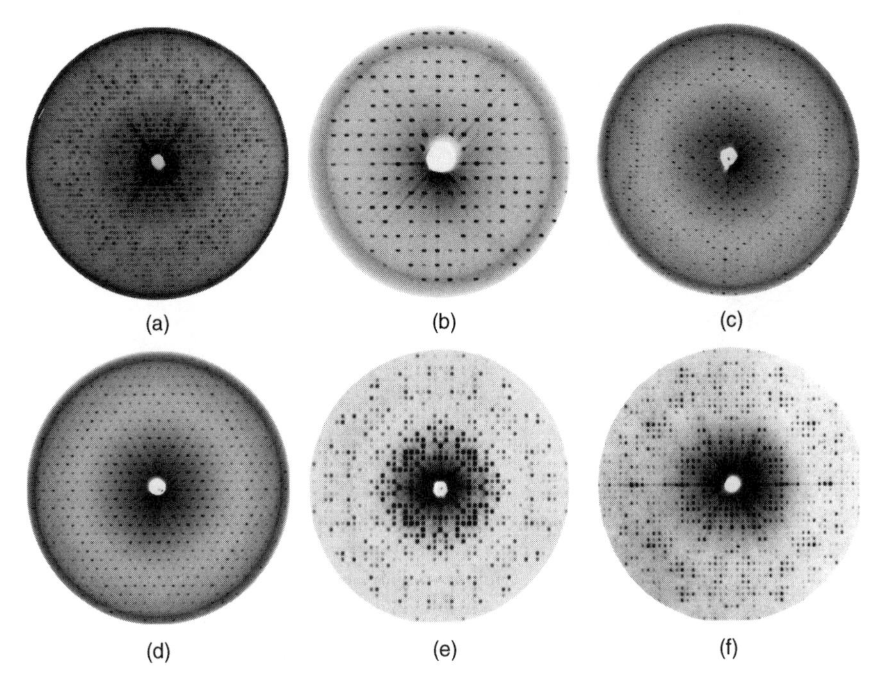

FIGURE 6.2 Diffraction photographs from a variety of protein crystals exhibiting a diversity of reciprocal lattice symmetries. With two images from a crystal, usually orthogonal to one another, the entire symmetry of the three-dimensional lattice can generally be deduced. The protein crystals and the symmetry of the diffraction patterns are (*a*) $\boldsymbol{P2_12_12_1}$ porcine α-amylase, the symmetry is ***mm*** (mirror-mirror); (*b*) triclinic (***P*1**) glycerol-3PO_4-dehydrogenase, the symmetry is $\bar{1}$; (*c*) centered orthorhombic (***C*222**) fructose 1,6 bisphosphatase, the symmetry is ***mm***; (*d*) cubic crystal of phaeseolin viewed along body diagonal, symmetry is **6*mm***; (*e*) tetragonal ***I*422** lactate dehydrogenase, the symmetry is **4*mm***, (*f*) tetragonal crystal of lactate dehydrogenase viewed along a twofold axis, symmetry is ***mm***.

oriented with respect to the physical crystal? How do we get from one family of planes to another? How do we bring a particular family into diffracting position so that we can observe its reflection? To assist us in this matter, Ewald (1921) developed the construction shown in Figure 6.4.[1] This diagram illustrates a family of Bragg planes, its corresponding reciprocal lattice, the recording device (film), the X-ray beam, and an imaginary globe called the sphere of reflection (now more commonly called Ewald's sphere). The objective of this construction is to illustrate the relationship between the components of the diffraction experiment, and to serve as a guide to the crystallographer as to how he may systematically alter the orientation of the crystal (real space) to record desired portions of the diffraction pattern (reciprocal space).

In Figure 6.4 the plane passing through $\boldsymbol{O}$ is a member of a family of planes having Miller indexes $\boldsymbol{hk\ell}$ of periodic spacing $\boldsymbol{d}$, which makes an angle $\boldsymbol{\theta}$ with the X-ray beam, where $\boldsymbol{\theta}$ is the proper Bragg angle for the family of planes. A sphere (Ewald's sphere) is constructed of radius $\mathbf{1/\lambda}$ centered at $\boldsymbol{O}$. The origin $\boldsymbol{O}$ is chosen as the center of the crystal,

[1] Actually, he had no intention of helping us, because he invented his construction before the first X-ray diffraction data was ever collected. He must have invented it for other uses, or he must have been a man of almost supernatural vision.

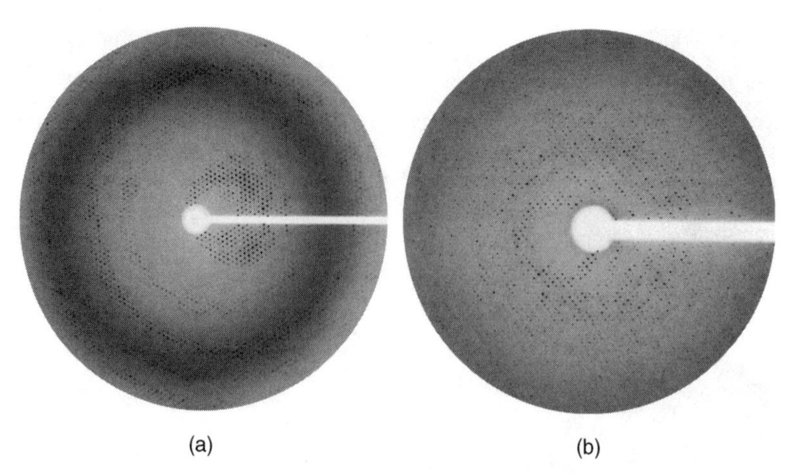

FIGURE 6.3 There are numerous experimental geometries for, and approaches to recording diffraction intensities from crystals. Those seen in Figure 6.2 utilize the precession method. Here are two images from crystals of (*a*) Bence–Jones protein and (*b*) satellite tobacco mosaic virus recorded using the rotation method, which is the method currently used for virtually all rapid data collection. Rotation angles are generally **0.5**° to **2**°, depending principally on the unit cell dimensions and the mosaicity of the crystal. A complete, three-dimensional data set representing all possible orientations of the crystal may be comprised of more than a hundred diffraction images, depending primarily on the symmetry of the particular crystal and its unit cell dimensions.

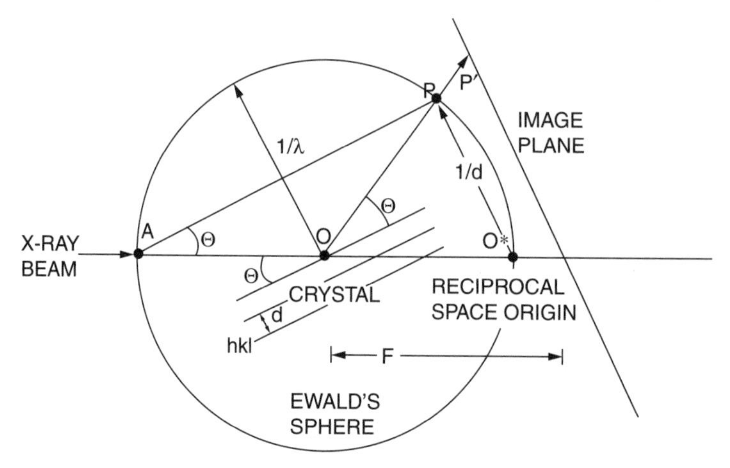

FIGURE 6.4 Ewald's sphere, a construction relating Bragg's law as it applies in real space with the reciprocal space requirement for constructive interference, and the expected location of diffraction intensities. The origin of the crystal is at point O, the center of a sphere of radius $1/\lambda$. The origin of reciprocal space is at O^*. The X-ray beam is along the horizontal and passes through O and O^*. The chord O^*P is the reciprocal lattice vector corresponding to the set of planes $hk\ell$. When a family of planes $hk\ell$ is in reflecting position, its corresponding reciprocal lattice point P lies on Ewald's sphere. The diffracted ray is emitted along OP and will strike a film plane, placed a distance F behind the crystal, at the point P'.

and the point O^* may be arbitrarily assigned as the origin of reciprocal space. Remember that while the reciprocal lattice is locked in terms of orientation to the real crystal, it bears no translational relationship; hence we can put the origin of reciprocal space anywhere we like. Ewald liked O^*. The chord **AP** is drawn (by simple geometry, it must also be parallel with the $hk\ell$ planes) forming the triangle APO^*. Because APO^* is a triangle within a hemisphere having one side a diameter, it must be a right triangle (Euclid, 300 BC) and O^*P must be perpendicular to AP and therefore to the $hk\ell$ planes. Bragg's law states that a set of planes is in reflecting position only when $n\lambda = 2d \sin\theta$. Therefore, when the $hk\ell$ family of planes is in diffracting position, $\sin\theta = \lambda/2d = (O^*P)(\lambda/2)$, and therefore the length of $O^*P = 1/d$. Now O^*P is normal to the $hk\ell$ planes and has length $1/d$; hence O^*P is the reciprocal lattice vector corresponding to the set of planes, and P is the corresponding reciprocal lattice point $hk\ell$.

Ewald's sphere demonstrates that when a family of planes is oriented with respect to the X-ray beam so that Bragg's law is satisfied, its reciprocal lattice point lies exactly on (not inside) the sphere. The converse is also true: when a reciprocal lattice point passes through the sphere, a diffracted ray is produced. Ewald's sphere provides an alternative conception of diffraction geometry based on bringing reciprocal lattice points through the sphere of reflection, rather than bringing particular sets of crystal planes into diffracting position. Since the reciprocal points fall on a regular lattice array that is clearly defined, the motions and orientations required to observe particular reflections are more readily visualized. We can more readily identify reciprocal lattice points in diffraction space than we can planes in a crystal. The only necessary condition is that we be able to define exactly the relationship between the reciprocal lattice axes a^* b^* c^* and the axes of our laboratory coordinate system, and that we can do.

Further examination of the diagram permits one to predict precisely where the diffracted ray corresponding to a particular reciprocal lattice point, or family of planes, will appear with respect to our detector. The diffracted ray may be thought of as arising from the origin of the crystal at O. It makes an angle 2θ with the X-ray beam and intersects the sphere at the position of the reciprocal lattice point P. If a film is placed at some distance F behind the sphere so that it is parallel with O^*P, the diffracted ray will intercept the film at P'. The point P' on the film is the projection of the reciprocal lattice point P. A film containing a distribution of reflections is always some kind of projection of a portion of the reciprocal lattice onto a plane. It is common, though not strictly correct, to refer to a reflection on a film as a reciprocal lattice point.

While the absolute distance between points in reciprocal space is a function of λ and the reciprocal lattice parameters a^*, b^*, c^*, α^*, β^*, and γ^*, the absolute distances between maxima observed on the film will be expanded in proportion to F, which acts as a constant magnification factor.

The axes of the reciprocal lattice, remember, maintain a fixed orientation with respect to the real axes of the crystal by definition, regardless of the crystal's orientation. That is, if the crystal is rotated, the reciprocal lattice is rotated as well. If the crystal is continuously reoriented in a specific manner about its center by some constant motion, all of the points on a single reciprocal lattice plane, or region of reciprocal space, can be made to systematically pass through the sphere of reflection. If the film is maintained constantly parallel with a reciprocal lattice plane by mechanical linkage to the crystal, a magnified but otherwise undistorted replica of the reciprocal lattice plane will be recorded on the film. This principle, proposed by de Jong and Bouman (1938), was the basis for some of the more widely used film based data collection techniques, including the precession method (Buerger, 1942, 1944) used to solve the first protein crystal structures.

CRYSTAL SYMMETRY AND THE SYMMETRY OF THE DIFFRACTION PATTERN

An important property of Fourier transforms that we did not emphasize in the previous chapter is that spatial relationships in one space are maintained in the corresponding transform space. That is, specific relationships between the orientations in real space of the members of a set of objects are carried across into reciprocal space. This is particularly important in terms of crystallographic symmetry, and we will encounter it again when we consider the process known as molecular replacement (see Chapter 8).

A consequence of this property of the Fourier transform is that symmetry elements, namely space group symmetry, that characterize the arrangement of asymmetric units in the crystal also apply to the distribution of reflections in reciprocal space. We mean by this, not only that the diffraction pattern (the reciprocal lttice) exhibits the same Bravai lattice (see Chapter 3) as the real crystal, but also that the distribution of the intensities reflects the symmetry of the asymmetric units in the crystallographic unit cell, that is, the space group symmetry. Recall that spacings between reflections and the net on which they fall in the diffraction pattern tell us the kind of unit cell we have, and the unit cell dimensions (as their reciprocals). Thus the distribution of reflections and their intensities in diffraction space tell us everything we need to know about the crystal. Well, almost everything.

What has been said here is true but obscures another fundamental property of the Fourier transform, one that complicates matters a bit but not hopelessly so. The Fourier transform fails to directly carry translational relationships from one space to another, in particular, from real space into reciprocal space. This means that the transform does not discriminate between asymmetric units based on the distances between them. The immediate relevance of this is that a set of asymmetric units related by a screw axis symmetry operator (which has translational components) in real space is transformed into diffraction space as though it simply contained a pure rotation axis. The translational components are lost. If our crystal has a $\mathbf{6_1}$ axis, we will see sixfold symmetry in the diffraction pattern. If we have $\mathbf{2_12_12_1}$ symmetry in real space, the diffraction pattern will exhibit **222** (or more properly, ***mmm***) symmetry.

"Now you tell us!" You might be thinking, but do not despair. It is true that the translational components of symmetry elements are lost during transformation into reciprocal space and simply appear as the corresponding rotational symmetry element, that's the bad news. The good news is that the translational components of any symmetry element do leave cryptic evidence in the diffraction pattern of their presence in the crystal. Furthermore we know exactly the nature and form of that evidence, it is distinct and clear to the eye of the crystallographer, and we know exactly where to look for it.

The evidence for the existence of screw axis symmetry is manifested in certain subclasses of reflections that are "systematically absent." These "systematic absences," we will see, fall along axial lines (***h*00, 0*k*0, 00*l***) in reciprocal space and clearly signal not only whether an axis in real space is a screw axis or a pure rotation axis, but what kind of a screw axis it is, for instance, $\mathbf{4_1}$ or $\mathbf{4_2}$, $\mathbf{6_1}$ or $\mathbf{6_3}$. Thus the inherent symmetry of the diffraction pattern (which we call the Laue group), plus the systematic absences, allow us to unambiguously identify (except for a few odd cases) the space group of any crystal.

SYMMETRY AND SYSTEMATIC ABSENCES

Let us look at this question of diffraction pattern symmetry in a slightly different, but no less correct way in the hope that we may gain some insight into the sources of those "systematic absences." Reflections that appear along a reciprocal lattice line that passes through the origin of reciprocal space, are identical to those that would appear if the entire electron density of the crystal were projected onto the corresponding line in real space, and then transformed, using the structure factor equation. For example, if all of the electron density in the unit cell was projected onto the ***a*** axis in real space, and the one-dimensional crystal's diffraction pattern subsequently generated, then it would be identical to the ***h*00** line of reflections (along ***a****) in the three-dimensional diffraction pattern. This is true of all corresponding lines in real and reciprocal space that pass through the origin. Transform of the ***h*00** line of reflections using the electron density equation, the F_{h00} reflections alone, would in turn yield the distribution of electron density of the unit cell projected onto the real ***a*** axis of the crystal.

Similarly, if all of the electron density in the unit cell were projected onto a single plane, let us say the ***ab*** plane in real space, then the diffraction pattern of this two-dimensional crystal would be the ***hk*0** zone of reflections in the three-dimensional diffraction pattern. As with line projections, this theorem holds true for any and every plane passing through the origin.

This idea has some useful consequences in terms of interpreting diffraction patterns. For example, consider the case of a twofold axis in real space, along ***b*** and perpendicular to the ***ac*** plane, as we would have in a monoclinic crystal of space group ***P*2** or ***C*2**. The projection of all of the electron density in the unit cell having this dyad symmetry onto the ***ac*** plane, would of course also have twofold symmetry. Because this projection has dyad symmetry, the corresponding diffraction pattern, which is the ***hk*0** plane of reflections in reciprocal space, would also have twofold symmetry, namely reflections $F_{hk0} = F_{-h-k0}$.

If, instead of a monoclinic crystal, we considered a tetragonal crystal having a fourfold axis along ***c*** and therefore perpendicular to the ***ab*** plane, then the plane projection would also have fourfold symmetry. So too would the corresponding reflections on the ***hk*0** zone of reciprocal space have a fourfold distribution:

$$F_{hk0} = F_{-hk\ell} = F_{kh\ell} = F_{-kh\ell}.$$

Similar kinds of relationships would arise among equivalent reflections for threefold or sixfold axes in the crystal as well. This tells us that symmetry elements in real space, the crystal, may be identified by searching the appropriate zones, or planes of reciprocal space for symmetrical patterns of diffraction intensities. If we see fourfold or threefold or sixfold distributions of reflections in the diffraction patterns, then they must imply corresponding symmetry relationships in the crystal.

Consider now a $\mathbf{2_1}$ screw axis along ***b*** in real space, perpendicular to the ***ac*** plane. When the contents of the unit cell are projected onto this plane, the translational component of the $\mathbf{2_1}$ operator is lost, and the projection is identical to that which would have been obtained simply from a twofold axis. Thus we might expect the corresponding plane of reflections in reciprocal space, the ***h*0*ℓ*** zone of reflections, to exhibit a twofold symmetrical distribution, whether a twofold axis or a twofold screw axis is present, and indeed that is the case. The same logic would apply for any rotation/screw axis choice. Because their projections in real space are identical, the Fourier transforms of those projections must be as well, and we

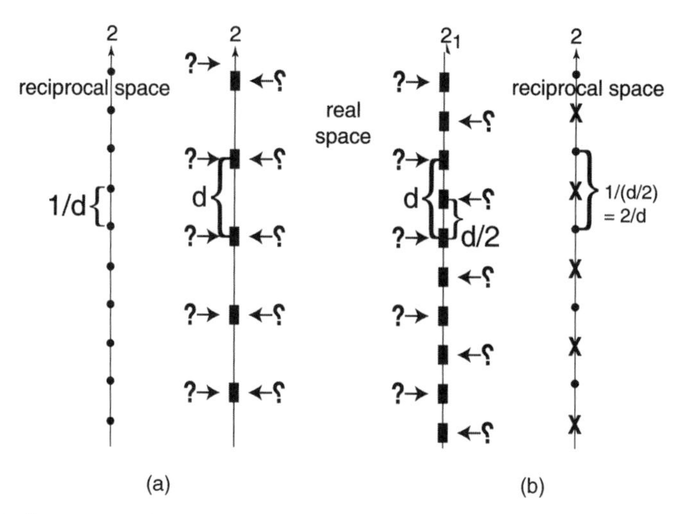

FIGURE 6.5 An asymmetric unit in real space is designated by the symbol for a question mark (**?**). Twofold axis (**2**) related asymmetric units in (*a*) when projected on the vertical dyad axis produce densities repeated through the crystal, along the axis, with periodicity ***d***, the unit cell dimension. This gives rise in reciprocal space, to the left, of diffraction spots of periodicity **1/*d***, the actual reciprocal unit cell dimension. The $\mathbf{2_1}$ screw axis related asymmetric units on the right, when projected onto the axis, yield a periodic spacing of ***d*/2** , half the true unit cell dimension. In reciprocal space the spacing of points is therefore **2/*d***, twice the true reciprocal space dimension. Thus, if a $\mathbf{2_1}$ axis is present, the one-dimensional diffraction pattern appears to have systematically absent reflections along the corresponding reciprocal axis.

cannot discriminate them. Thus the presence of rotational symmetry on a plane of reflections in the diffraction pattern can tell us that an operator is present in the crystal, but not whether it is a pure rotational operator or a screw operator; for example we cannot tell the difference between a **2** and $\mathbf{2_1}$ axis, or a $\mathbf{4_1}$ or $\mathbf{4_2}$ or **4** fold axis, and so on.

In the case of symmetry operators containing translational components, that is, screw operators or centering operations (and glide planes in crystals of conventional molecules), there is a saving grace. The evidence of a translation operator or component is not, in fact, lost upon transformation into reciprocal space. If a symmetry axis is present along a certain direction in real space, say along ***c***, we may then consider the projection of the entire contents of the unit cell onto the ***c*** axis as described above. As illustrated in Figure 6.5, if the axis is a pure rotational operator, then the projection on ***c*** will have an arbitrary, nonrepetitive distribution along its entire length from 0 to 1. If the axis contains a translational component, however, this will not be true. There will be a repeat within the projected density between 0 and 1. It will be a double repeat for a $\mathbf{2_1}$ axis, triple repeat for a $\mathbf{3_1}$ axis, quadruple repeat for a $\mathbf{4_1}$ axis, and so on.

The subperiodicity of the axial projection within the unit cell in real space of course has an effect in diffraction space. Recall the reciprocal relationship between the two. If a distance is halved in real space, it is doubled in reciprocal space; if a periodicity is quartered in real space, it is multiplied fourfold in reciprocal space, and so on. Thus, along the corresponding reciprocal space axis $\mathbf{c}^*$, or the **00*l*** line of diffraction intensities, reflection spacing must correspond to the increased periodicity, or smaller repeat distance in real space. In reciprocal space, reflections will appear less frequently, at greater intervals. For a $\mathbf{2_1}$ axis, the appearance of reflections will have double the normal periodic interval. For a $\mathbf{2_1}$ axis, reflections will occur along the **00*l*** line of reciprocal lattice points only for $l = \mathbf{2, 4, 6, 8 \ldots}$, that is, only for reflections where l is an even integer. For a $\mathbf{4_1}$ screw axis, the interval will be quadrupled, and only reflections $l = \mathbf{4, 8, 12, 16...}$ will be nonzero; that is, the only allowed reflections will be $\mathbf{00}l = \mathbf{4}n$.

Another way of looking at this outcome is that in certain one-dimensional projections, screw symmetry operators produce an apparent but precise sub periodicity, or sub cell, within the actual unit cell. Internal destructive interference of the waves produced by the subperiodic structures with one another causes reflections of certain classes always to sum to zero. In the case of the $\mathbf{2_1}$ axis above, the class of reflections becoming zero were those for which $\mathbf{00}l$ was odd. These classes of missing reflections (systematic absences), are used, in inspecting the diffraction pattern of a crystal, to discriminate between rotational and screw symmetry operators, between twofold and $\mathbf{2_1}$, or threefold and $\mathbf{3_1}$ axes in real space. Figure 6.6 illustrates the appearance of screw axis produced systematic absences for some real cases.

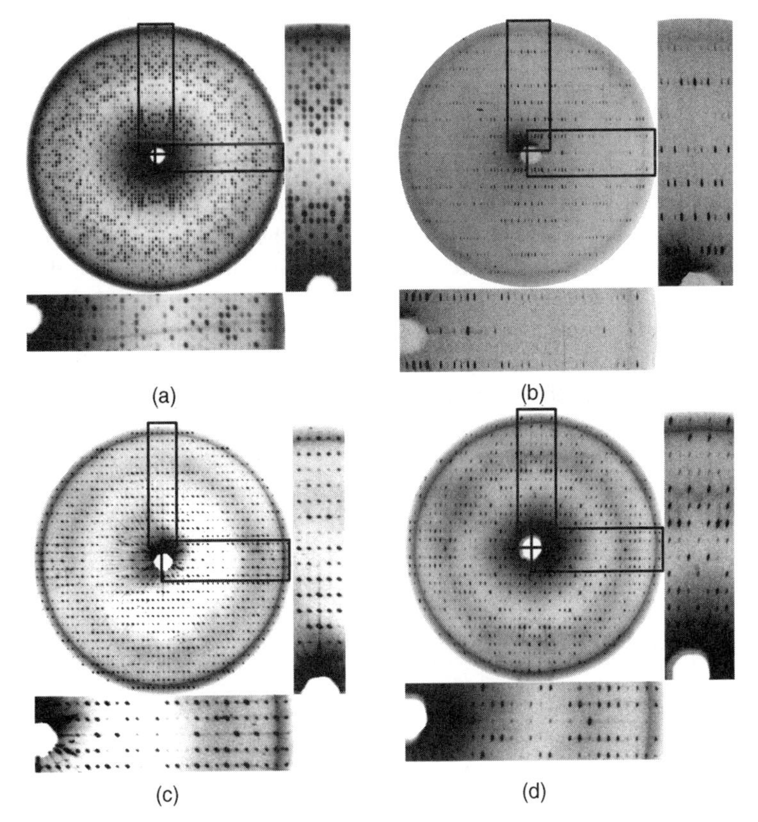

FIGURE 6.6 Shown here are planes of diffraction intensities for four different orthorhombic crystals characterized by twofold and $\mathbf{2_1}$ screw axes. Associated with each are details of the relevant reciprocal lattice axes where the presence of systematic absences, namely reflections with odd indexes absent, might be expected to occur. In (*a*) is the $\boldsymbol{h0\ell}$ plane from $\boldsymbol{P2_12_12_1}$ α-amylase, which has $\mathbf{2_1}$ systematic absences along both the $\boldsymbol{h00}$ line (vertical axis) and $\mathbf{00}l$ line (horizontal axis). In (*b*) is the $\boldsymbol{h0\ell}$ plane from $\boldsymbol{P2_122_1}$ yeast phenylalanine tRNA. Absences consistent with a $\mathbf{2_1}$ axis are apparent along the horizontal $\mathbf{00}l$ line but not along the vertical $\boldsymbol{0k0}$ line, which corresponds to a twofold axis. In (*c*) is $\boldsymbol{P222_1}$ RNase B, the $\boldsymbol{0k\ell}$ zone, and in (*d*), the $\boldsymbol{h0\ell}$ zone from a crystal of $\boldsymbol{P2_12_12_1}$ RNase A plus $\boldsymbol{d(pA)_4}$ complex. In (*c*), alternate reflections (all having odd indexes) are absent only along the $\mathbf{00}l$, horizontal line, while in (*d*), they are absent along both axial lines, the $\boldsymbol{h00}$ and the $\boldsymbol{00\ell}$.

We arbitrarily chose, in the example above, to project the unit cell contents onto the ***c*** axis, but we might just as well have chosen the $\mathbf{2_1}$ axis to lie along ***a*** or ***b***, the corresponding outcome would have been similar. Thus, if a space group contains multiple $\mathbf{2_1}$ axes, such as $\boldsymbol{P2_12_12_1}$, with screw operators along all three directions, then we would expect to find systematic absences along each of the three reciprocal lattice rows $\boldsymbol{h00}$, $\boldsymbol{0k0}$, and $\boldsymbol{00l}$. The presence of one screw axis is independent of others in this regard. We would say, then, that space group $\boldsymbol{P2_12_12_1}$ is characterized by systematic absences such that the only permitted reflections are $\boldsymbol{h00 = 2n}$, $\boldsymbol{0k0 = 2n}$, and $\boldsymbol{00l = 2n}$. Look in the *International Tables*, Volume I, and see if this is correct.

Similarly we might have considered a $\mathbf{3_1}$, a $\mathbf{4_3}$, a $\mathbf{6_1}$, or any other kind of screw axis as well as the $\mathbf{2_1}$. Note, however, that the translational components for these screw axes are not $\frac{1}{2}$, but $\frac{1}{3}$, $\frac{1}{4}$ and $\frac{1}{6}$, respectively. Thus we might expect that not every other reflection will be present, but only every third ($\boldsymbol{00l = 3n}$), fourth ($\boldsymbol{00l = 4n}$), or sixth ($\boldsymbol{00l = 6n}$), respectively. A pure rotation axis, having no translational component, gives no systematic absences in the diffraction pattern. Absences produced by higher symmetry screw axes are shown in Figure 6.7.

The same idea pertains when there are pure translational operators within the crystal, as when centering operations (see Chapter 3) relate equivalent sets of asymmetric units by translations of half of the unit cell lengths (e.g., ***C*** centering: $\mathbf{0, 0, 0}$ and $\frac{1}{2}, \frac{1}{2}, \mathbf{0}$; etc.). Centering operations also produce subperiodicities within the unit cell, and as above, these subperiodicites in real space produce systematic absences in diffraction space.

Because the subperiodicities produced in projections by centering operations apply to two or three directions in real space, and are not simply confined to a line in space, the systematic absences they produce are more extensive. As we might expect, if entire additional sets of asymmetric units are produced by centering operations, as they are for ***C*** or ***I*** or ***F*** centering, then half or more of the expected reflections in reciprocal space are systematically absent. This gives rise to diffraction patterns that look like three-dimensional chessboards, with alternate reflections absent along all rows and columns. This idea is illustrated in Figure 6.8. Because of these very characteristic checkered patterns, centered crystal lattices can be immediately recognized from their diffraction patterns. Some examples are shown in Figures 6.9 and 6.10. As for the screw symmetry operators, systematic absences for all kinds of centering operations, and for all space groups are contained in the *International Tables*, Volume I.

ANALYSIS OF DIFFRACTION PATTERNS

Modern computer systems armed with current crystallographic programs can analyze X-ray diffraction data collected on two-dimensional electronic detectors, such as image plates or CCD arrays, and immediately provide plausible alternatives for the likely crystallographic unit cell. Frequently, however, it falls to the crystallographer to refine the choices and select among them in order to arrive at the correct unit cell, unit cell dimensions, space group of the crystal, and other properties essential for the full three-dimensional structure determination. The machine usually cannot make all decisions independently, and indeed should not be allowed to do so. It must be remembered that if the unit cell dimensions are very much in error, or wrong, or the space group symmetry is incorrect, then there is no way that the structure can ever be solved. Getting these things correct from the start is imperative.

Despite the convenience, speed, and general precision offered by data analysis programs, it is important for the crystallographer to know what is being sought, and what consider-

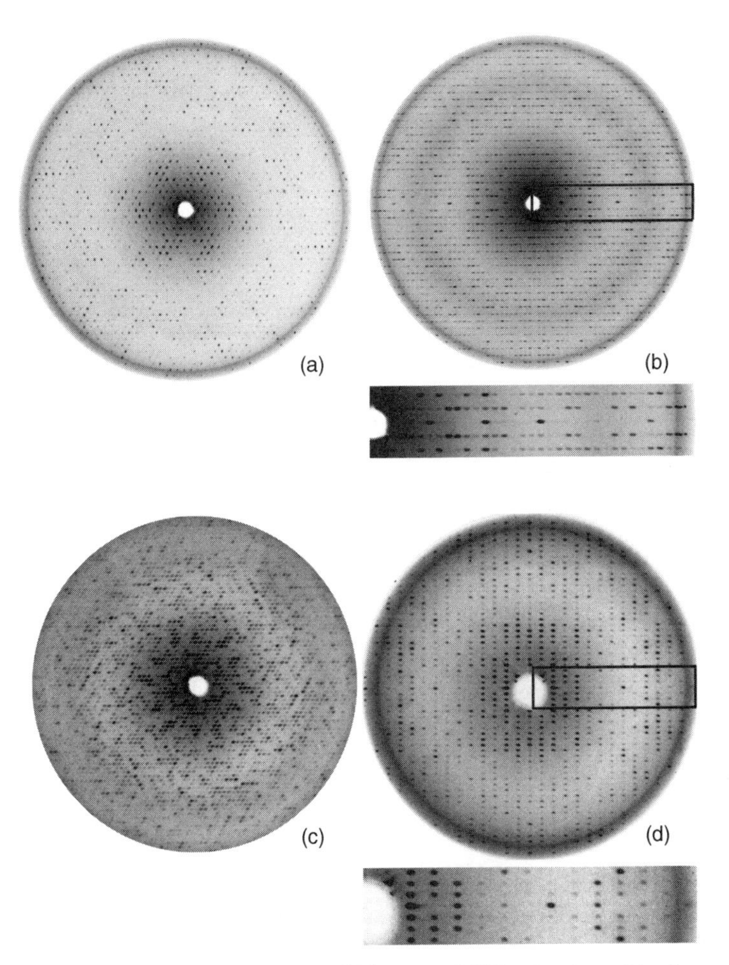

FIGURE 6.7 In (*a*) is the ***hk*0**, and in (*b*) the ***h*0*ℓ*** zones of diffraction intensities from a crystal of concanavalin B having space group $P6_1$. The sixfold symmetry produced by the 6_1 axis is clearly evident in the ***hk*0** zone, but the presence of only reflections for which $l = 6n$ along the horizontal **00*l*** axis shows the order of the screw axis. In (*c*) and (*d*) are the ***hk*0** and ***h*0*ℓ*** zones, respectively, of reflections from $P6_3$ canavalin. Again, the sixfold symmetry is apparent in the ***hk*0** zone, but only alternate even reflections ($00l = 2, 4, 6, 8 \ldots$) are present along the horizontal **00*l*** line. Note the difference in the systematic absence pattern for $P6_1$ concanavalin B and $P6_3$ canavalin crystals that allows discrimination of the two space groups.

ations are being made by the various computer algorithms. That is, what features of the diffraction pattern are being investigated, how, and for what purpose? To address this question, let us assume that we have the necessary diffraction patterns in hand, or at least certain representative zones, or planes, or sectors of reciprocal space that contain sufficient data for accurate judgments, particularly of symmetry relationships. This may be a complete three-dimensional data set from which certain subsets can be selected and examined, or it may simply be a set of films, from a precession camera for example (the classical instrument for conducting a preliminary X-ray diffraction analysis), of the major zones (***hk*0, *h*0*l*, 0*kl***) of the diffraction pattern. We can ask, If no computer were available to analyze the diffraction pattern, how would we go about doing this ourselves, and what might we expect to learn?

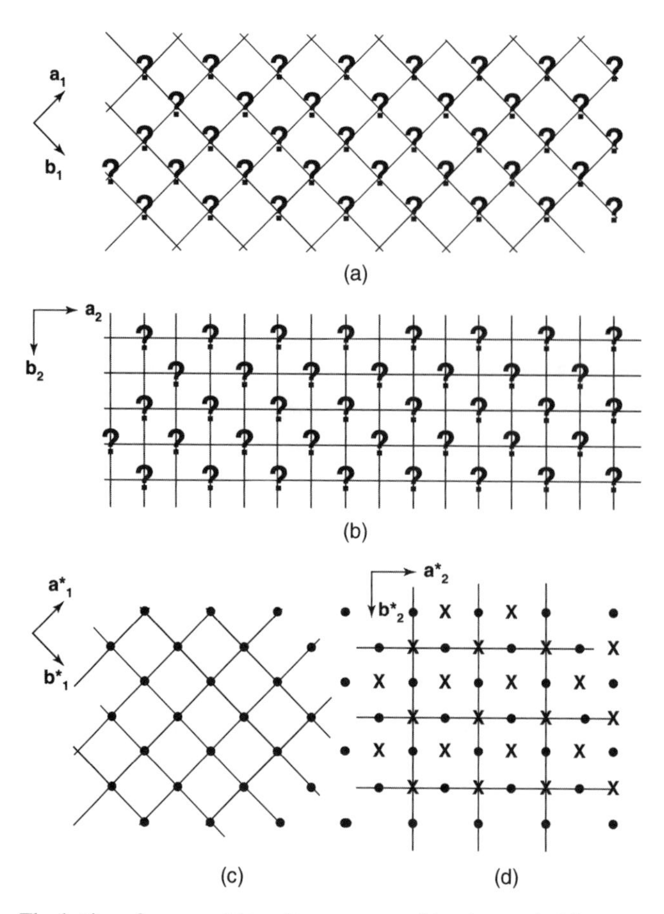

FIGURE 6.8 The lattice of asymmetric units, represented by the symbol for a question (**?**), is the same in both (*a*) and (*b*). The corresponding reciprocal lattice is accordingly the same for both (*c*) and (*d*). If the unit cell is chosen as in (*a*) to be a primitive unit cell, then every lattice point in (*c*) is occupied. If, however, the real unit cell is chosen to be centered as in (*b*), then half of the reciprocal lattice points, indexed according to this centered cell, are systematically absent, and a checkerboard pattern of diffraction intensities is observed.

We approach this problem systematically by asking a series of questions, and from their answers, either fixing certain properties or eliminating others as impossible. The questions we seek to answer, and the order of inquisition, is as follows:

1. What is the crystal class (triclinic, monoclinic, orthorhombic, trigonal, tetragonal, hexagonal or cubic)?
2. Is the crystal lattice primitive or centered? That is, is it primitive ***P***, ***C*** face centered, body centered ***I***, or face centered ***F***?

From these two questions we determine the Bravais lattice

3. What are the unit cell dimensions? $\boldsymbol{a}$, $\boldsymbol{b}$, and $\boldsymbol{c}$ and the unit cell angles $\boldsymbol{\alpha}$, $\boldsymbol{\beta}$, and $\boldsymbol{\gamma}$ in angstroms and degrees?

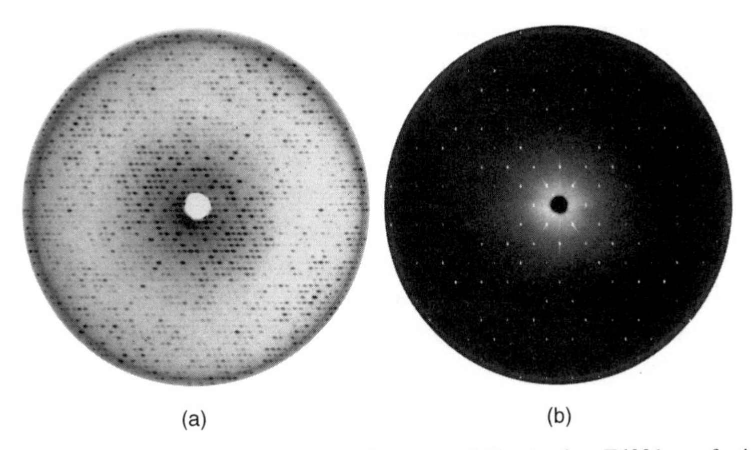

FIGURE 6.9 When the X-ray beam is directed along a twofold axis of an ***F*****422** horse ferritin crystal the diffraction pattern in (*a*) is obtained. The systematic absences due to the *F* centering produce an immediately recognizable checkerboard pattern of reflections. In (*b*) is the **0*kl*** diffraction plane from a monoclinic crystal of Gene 5 DNA Unwinding Protein having space group ***C*****2**. Along both columns and rows, alternate reflections are systematically absent. The indexes of the reflections that are present correspond exclusively to those for which $\boldsymbol{k + l = 2n}$.

4. What is the symmetry of the reciprocal lattice? That is, what are the symmetry operators that relate sets of identical intensities? The symmetry group that we observe for a crystal in reciprocal space, namely the diffraction pattern symmetry, is called the Laue symmetry, or Laue group.
5. What systematic absences are present?

In responding to all five questions, we obtain the crystal class, unit cell dimensions, Bravais lattice, and space group. All are essential crystal properties, but the diffraction pattern contains other useful information as well.

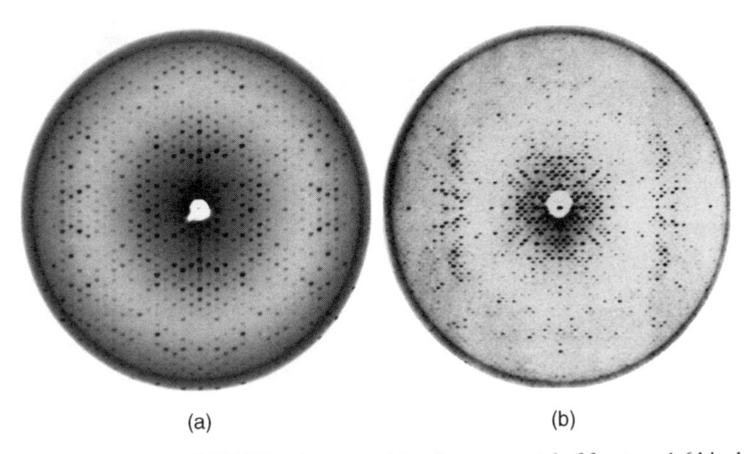

FIGURE 6.10 In (*a*) are the ***h*0*l*** diffraction intensities from a crystal of fructose 1,6 bisphosphatase from rabbit liver having space group ***I*****222**. In (*b*) is the ***hk*0** plane of reflections from $\boldsymbol{C222_1}$ canavalin. In both images the checkerboard patterns produced by systematic absences resulting from the *I* and *C* centering operations are evident.

6. What is the resolution of the diffraction pattern, namely what is the family of Bragg planes in the crystal of smallest interplanar spacing that is represented by a measurable intensity in the diffraction pattern? Definition of "measurable intensity" may here become a contentious question.
7. What is the mosaic spread, or angular spread of the diffraction intensities? Basically this means, "How large is the spot on the image," and it, along with the resolution limit, serve as measures of the order of the crystal.
8. How many molecules or chemical units are there in the crystallographic asymmetric unit? This is a question that is usually easily answerable, but in ambiguous cases must be viewed with caution, as it may significantly affect other phases of the analysis.

The first determination that must be made is the net, or lattice, upon which all of the reflections in the diffraction pattern fall, namely the reciprocal lattice. That is, we must draw three axes through the origin of reciprocal space and choose three spacings so that lines drawn parallel to the axes and separated by the assigned spacings include every reflection in the diffraction pattern. These will become the $\boldsymbol{a}^*$, $\boldsymbol{b}^*$, and $\boldsymbol{c}^*$ axes in reciprocal space. Don't worry about getting them mixed up in the initial assignment, as you can always switch their designations later. Generally, this choice of reciprocal axes is fairly evident for diffraction patterns of high symmetry, but it can be challenging for low symmetry triclinic, and to some extent, monoclinic crystals. Sometimes the conclusions drawn at this stage must be reevaluated later after consideration of symmetry properties, which in the end supercede all others. Consider the examples presented in Figures 6.11 and 6.12.

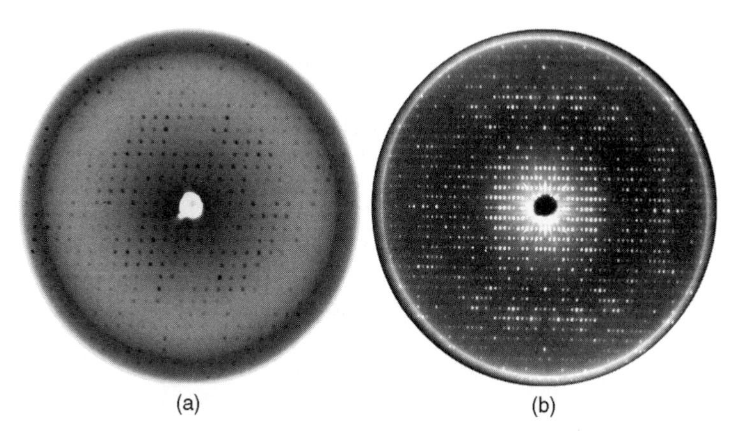

FIGURE 6.11 In the diffraction pattern from triclinic, ***P*1** crystals of glycerol-3PO_4 dehydrogenase in (a), there is no symmetry other than the center of symmetry produced by Friedel's law. The axes for reciprocal space may therefore be arbitrarily chosen. The axes yielding the largest reciprocal unit cell are chosen, since this corresponds to the smallest crystallographic unit cell. In (*b*) the photograph is from an orthorhombic crystal of RNase A+$d(pA)$4. The reflections fall on an orthogonal net whose axes are along the horizontal and vertical directions. More important, choice of these two axes as $\boldsymbol{a}^*$ and $\boldsymbol{b}^*$ preserves the inherent ***mm*** symmetry of the diffraction intensity distribution. Axes and unit cells are always chosen to express the highest possible symmetry of the diffraction pattern. It is more important to preserve the highest symmetry than to have the smallest possible crystallographic unit cell in terms of volume, and this may require adopting a centered unit cell.

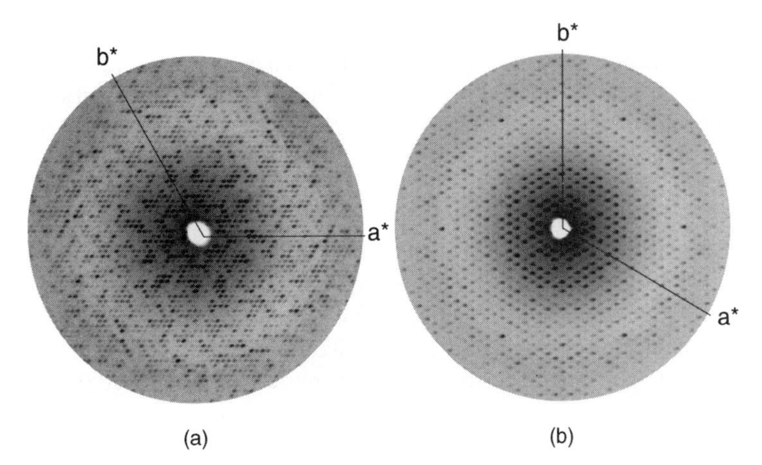

FIGURE 6.12 In (*a*) is the ***hk*0** diffraction plane of a $P6_3$ canavalin crystal having sixfold symmetry, and in (*b*) the view along the unique axis of a rhombohedral canavalin crystal that exhibits ***6mm*** symmetry. In neither case would one choose orthogonal axes on which to index the reflections. In (*a*), the natural choice would be the ***a**** and ***b**** axes indicated. In (*b*), two choices of hexagonal axes are reasonable, but those indicated are chosen as ***a**** and ***b**** because they correspond with the real unit cell of smallest volume.

Once the axes have been assigned, and knowing that their intersection at the exact center of the diffraction pattern defines the origin of reciprocal space, then every reflection in the entire diffraction pattern has a unique set of coordinates. The coordinates are always integral numbers of spacings along the three reciprocal lattice axes, and the coordinates correspond to $hk\ell$. Thus we can index the diffraction pattern and assign a name, a set of indexes to every spot. As we have seen, the reflection, or diffraction intensity found at any reciprocal lattice point $hk\ell$ represents the diffracted X rays from the corresponding $hk\ell$ family of planes in the crystal.

Consider now the angles between the axes we have chosen, and the spacings between reflections along lattice lines. If the three axes are at right angles to one another, then $\alpha = \beta = \gamma = 90°$, and we must have an orthogonal system. In some cases the best axes to choose for indexing the reflections have one angle **120°**, and then we likely have a trigonal or hexagonal system. If axes are chosen so that two interaxial angles are **90°**, but the third is not, then we must have a monoclinic system. If the axes cannot be chosen so that any angle is **90°**, then the unit cell must be triclinic. If the spacings between reflections along perpendicular rows and columns are the same, then the cell may be tetragonal or cubic. If all three spacings are different but all angles are **90°**, the system is probably orthorhombic.

Often one has several options for choosing the net, or reciprocal lattice axes. The rule is that an axial system is always chosen that preserves the highest symmetry of the diffraction pattern. That is, the axes are chosen to be consistent with the real unit cell of highest symmetry. Sometimes, before the diffraction symmetry is fully clear, incorrect axes may be chosen. The axial system, or reciprocal lattice net, can, however, always be reassigned and the $hk\ell$ indexes of the reflections reindexed at a later time. The choice of axes determines the crystal class.

The next question is whether the reciprocal lattice arises from a primitive real lattice (***P***) or from a centered lattice, and if the latter, what kind of centering (***C***, ***I***, or ***F***)? This, in practice, is equivalent to asking whether the diffraction pattern as a whole, or some planes of the pattern have the appearance of a three-dimensional chessboard, like those in Figures 6.9 and 6.10. That is, are half or more of the reflections in the diffraction pattern systematically absent throughout the entire reciprocal lattice or in one of the major zones. If not, then the lattice is primitive and that is the end of that.

If it is a chessboard pattern, then some additional investigation may be required. It may be necessary to record two or more zero levels (reciprocal lattice planes that pass through the origin of reciprocal space) of the diffraction pattern, and sometimes upper levels of the pattern (planes of reciprocal lattice points $\boldsymbol{hk\ell}$ none of whose indexes are always zero) as well. Remember, triclinic, trigonal, and hexagonal unit cells cannot be centered. Monoclinic unit cells can only be primitive or ***C*** centered (centered on the ***C*** face, or ***ab*** face), and tetragonal unit cells can only be primitive and ***I*** centered (body centered). Thus only orthorhombic and cubic unit cells remain in question, and the latter can only be ***I*** or ***F*** centered. The reader is referred again to Figure 3.15 of Chapter 3, where these cases are illustrated. The specific variety of centering, if it is present, can ultimately be determined by the pattern of systematically absent reflections, the particular subsets of $\boldsymbol{hk\ell}$ indexes with no measurable intensity. All these are detailed in the *International Tables*, Volume I, for every possible type of unit cell. From these considerations, the Bravais lattice is chosen.

Once the reciprocal lattice net has been established, it is straightforward to determine the unit cell dimensions. One simply measures the distance between rows and columns of reflections along the three axial directions, takes the reciprocals, and multiplies them by the appropriate instrumental constants to get directly the unit cell dimensions. This is so because the distances between reflections in diffraction space are exactly related to distances in real space by the reciprocal relationships $\boldsymbol{a^* = 1/a}$, $\boldsymbol{b^* = 1/b}$, and $\boldsymbol{c^* = 1/c}$ for orthogonal axes. For nonorthogonal crystal systems the calculation is somewhat more complicated because it includes trigonometric terms specific to the crystal class; even so they are relatively simple to determine. Note the striking differences in the distances between reflections for crystals having large and small unit cells in Figures 6.13 and 6.14.

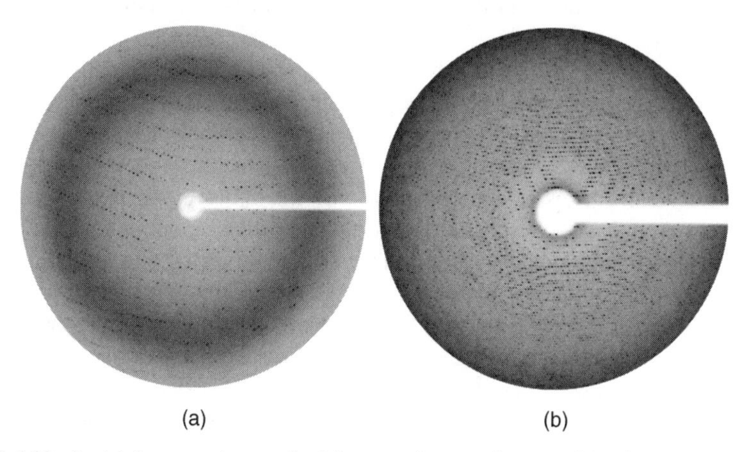

FIGURE 6.13 In (*a*) is a rotation method image of a protein crystal having moderate unit cell dimensions, a Bence–Jones protein crystal (space group $\boldsymbol{P3_121}$, $\boldsymbol{a = 153}$ **Å**, $\boldsymbol{b = 153}$ **Å**, $\boldsymbol{c = 94}$ **Å**). In (*b*) is a rotation method image from a crystal of brome mosaic virus (space group $\boldsymbol{R3}$, $\boldsymbol{a = 271}$ **Å**, $\boldsymbol{b = 271}$ **Å**, $\boldsymbol{c = 646}$ **Å**). Note the relative spacings of the reflections along the reciprocal lattice lines for the crystal having small unit cell dimensions in contrast to the crystal having very large unit cell parameters.

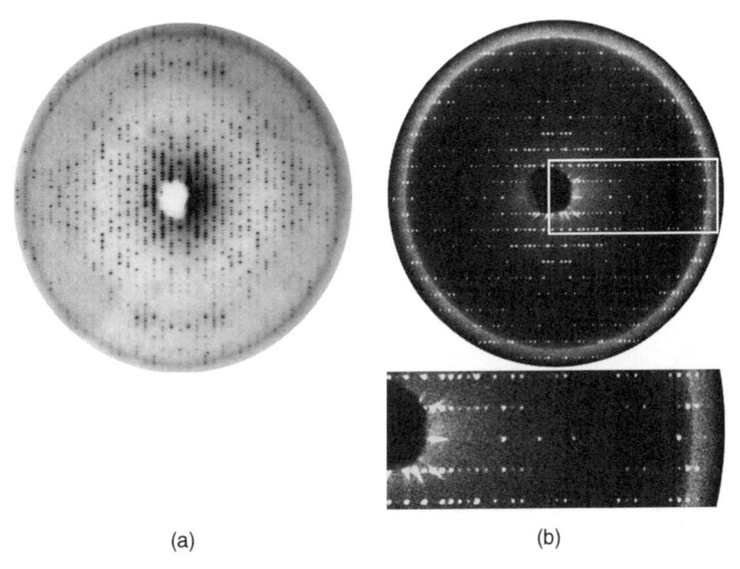

FIGURE 6.14 In the **0*kl*** diffraction image from a crystal of yeast phenylalanine tRNA (**$P2_122_1$, a = 33 Å, b = 56 Å, c = 161 Å**) shown in (*a*), note the very close spacing of reflections along the vertical ***c**** direction corresponding to the long ***c*** axis in the crystallographic unit cell, and the relatively wide spacing of reflections along the horizontal ***b**** direction, corresponding to the shorter unit cell dimension. Shown in (*b*) is the ***h*0*ℓ*** zone of diffraction intensities from a tetragonal crystal of the complex between RNase B+$d(pA)_4$ (**$P4_12_12$, a = 44.5 Å, b = 44.5 Å, c = 156.5 Å**). Again, note the difference in the distances between reciprocal lattice points corresponding to the long real ***c*** axis (horizontal) and the relatively shorter ***a*** axis (vertical).

The observed diffraction pattern on a film, or image plate, or charge coupled device (CCD) detector is a magnified image of the reciprocal lattice and this magnification factor has to be taken into account. The magnification is usually just the distance between the crystal and the film. This is evident in the Ewald sphere construction in Figure 6.4. There is a simple formula that allows this to be done (Buerger, 1944), and it is:

$$d \text{ (real space)(Å)} = \frac{\lambda(\text{Å}) \times F(mm)}{d_{image}(mm)},$$

where λ (in Å) is the wavelength of X rays used in obtaining the diffraction pattern, and F is the distance in millimeters between the crystal and the detector. Thus, by simply measuring the distance in millimeters on the detector between intensities along the three reciprocal lattice axes, a^*, b^*, and c^*, and applying the formula above, a, b, and c are obtained directly in Å. If any angle between the reciprocal axes α^*, β^*, or λ^* is not **90°** as is the β^* angle of a monoclinic lattice, then the real space angle β is simply the supplement, or **180°** minus the reciprocal space angle (which is always taken to be acute).

SYMMETRY IN DIFFRACTION SPACE

The next question, with the objective of determining the space group of the crystal, is: What is the symmetry of the entire, three-dimensional diffraction pattern? This is the most demanding aspect of the analysis and deserves some care. It is greatly simplified for

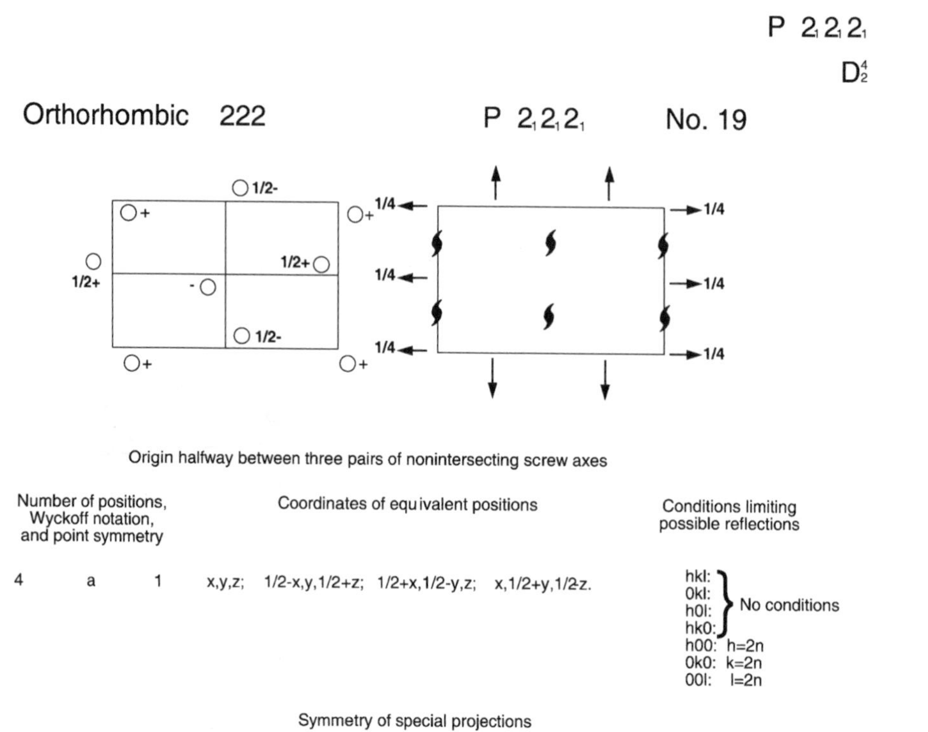

FIGURE 6.15 The diagram for space group $P2_12_12_1$ from Volume I of the *International Tables for X-ray Crystallography*. Notations and symbols are explained at the beginning of the volume.

macromolecular crystallographers because there are only 65 permitted space groups rather than the full 230. Only those space groups lacking inversion symmetry need to be considered. In addition, detailed descriptions of all symmetry groups, their equivalent positions, associated systematic absences, and all other useful properties are contained and described, as in Figures 6.15 and 6.16, in the *International Tables for X-ray Crystallography*, Volume I. The process of deducing the exact symmetry of the crystal from the symmetry of the diffraction pattern is complicated somewhat by the following:

1. Because of Friedel's law, diffraction patterns always contain a center of symmetry at the origin of reciprocal space. This means that any plane of reciprocal space that passes through the origin, in particular, the $h0l$, $0kl$, or $hk0$ zones, will display $\bar{1}$ (centric) symmetry. When this is integrated with the true symmetry of the crystal, it generally produces reciprocal lattice symmetry arrangements (called Laue groups) having higher symmetry than is really present in the crystal. One consequence of this is that a zero level (a plane of reciprocal space which passes through the origin of reciprocal space) photograph of the diffraction pattern will always exhibit an apparent twofold axis perpendicular to that reciprocal lattice plane. This is so because projection along a twofold axis appears the same as would a projection of a center

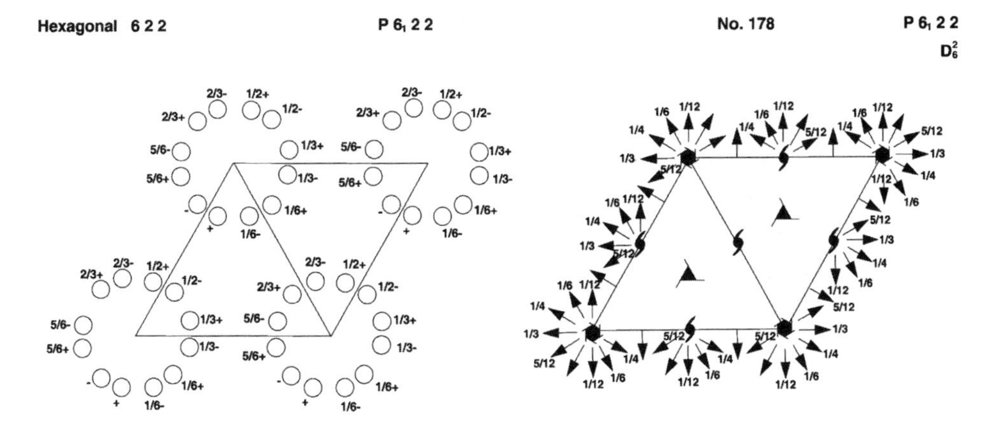

Origin at $6_1 21$ [2-axis normal to $(\bar{2}110)$]

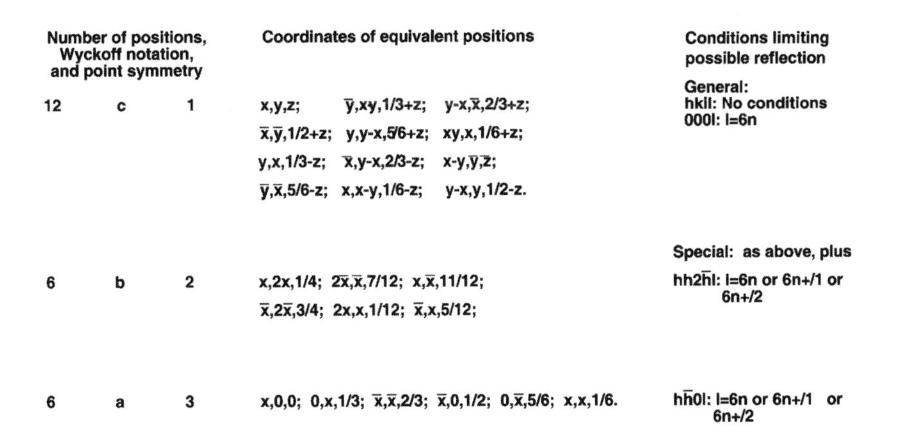

Number of positions, Wyckoff notation, and point symmetry			Coordinates of equivalent positions	Conditions limiting possible reflection
				General:
12	c	1	x,y,z; $\bar{y}$,xy,1/3+z; y-x,$\bar{x}$,2/3+z; $\bar{x}$,$\bar{y}$,1/2+z; y,y-x,5/6+z; xy,x,1/6+z; y,x,1/3-z; $\bar{x}$,y-x,2/3-z; x-y,$\bar{y}$,$\bar{z}$; $\bar{y}$,$\bar{x}$,5/6-z; x,x-y,1/6-z; y-x,y,1/2-z.	hkil: No conditions 000l: l=6n
				Special: as above, plus
6	b	2	x,2x,1/4; 2$\bar{x}$,$\bar{x}$,7/12; x,$\bar{x}$,11/12; $\bar{x}$,2$\bar{x}$,3/4; 2x,x,1/12; $\bar{x}$,x,5/12;	hh2$\bar{h}$l: l=6n or 6n+/1 or 6n+/2
6	a	3	x,0,0; 0,x,1/3; $\bar{x}$,$\bar{x}$,2/3; $\bar{x}$,0,1/2; 0,$\bar{x}$,5/6; x,x,1/6.	h$\bar{h}$0l: l=6n or 6n+/1 or 6n+/2

FIGURE 6.16 The diagram for space group ***P*6₁22** from Volume I of the *International Tables for X-ray Crystallography*. The diagram of symmetry relationships also contains, below, the equivalent positions and the expected systematic absences.

of symmetry-related arrangement. You cannot know, without additional information, whether a twofold axis is really present or if it is simply a manifestation of Freidel's law. Additional photographs will be necessary. This knot can, however, be unraveled.

2. As noted already, symmetry elements containing translational components, such as screw axes, appear in reciprocal space as the pure rotational element. The translational component, if it exists, must be deduced from systematic absences. These, however, are all explicitly described in the *International Tables*, Volume I, for each space group.
3. At least two, and occasionally more, two-dimensional planes through the diffraction pattern must be recorded and investigated in order to fix the symmetry of the three-dimensional reciprocal lattice. The nearer these are to orthogonal planes, the better. It is essential to know the spatial relationships between symmetry elements as they are identified in different images, and therefore it is necessary to know precisely the angular relationships between the images. The question to ask is: What unit cell, with what space group symmetry is consistent with the observed diffraction images? As one investigates features of the diffraction pattern in trying to find the correct unit cell

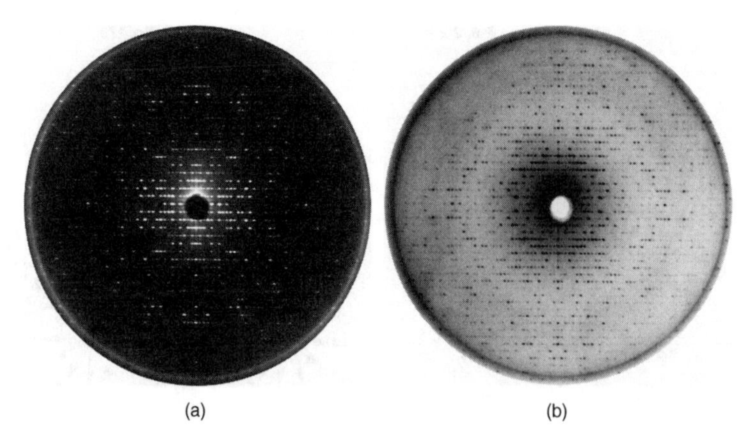

FIGURE 6.17 In (*a*) the $\boldsymbol{h0\ell}$, and in (*b*) the $\boldsymbol{hk0}$ planes of diffraction intensities from an orthorhombic crystal of canavalin (space group $\boldsymbol{C222_1}$, $\boldsymbol{a = 136.5}$ **Å**, $\boldsymbol{b = 150.3}$ **Å**, $\boldsymbol{c = 133.4}$ **Å**). The two photographs represent orthogonal views of the reciprocal lattice; that is, the crystal was rotated by **90°** about the horizontal axis between the acquisition of (*a*) and (*b*).

and the correct space group, choices that are inconsistent with the evidence are being eliminated.

4. In a few cases, the space group cannot be determined uniquely from the symmetry of the reciprocal lattice and the systematic absences. In these cases, however, only a choice between two specific possibilities remains. The two choices are usually enantiomorphic space groups, such as $\boldsymbol{P6_1}$ and $\boldsymbol{P6_5}$.

In Figures 6.17, 6.18, and 6.19 are presented pairs of diffraction intensity planes for three different protein crystals. From two photographs such as these, which are usually (e.g., Figure 6.19) orthogonal to one another, the symmetry of the entire three-dimensional diffraction pattern may be deduced. In some cases, however, additional diffraction images

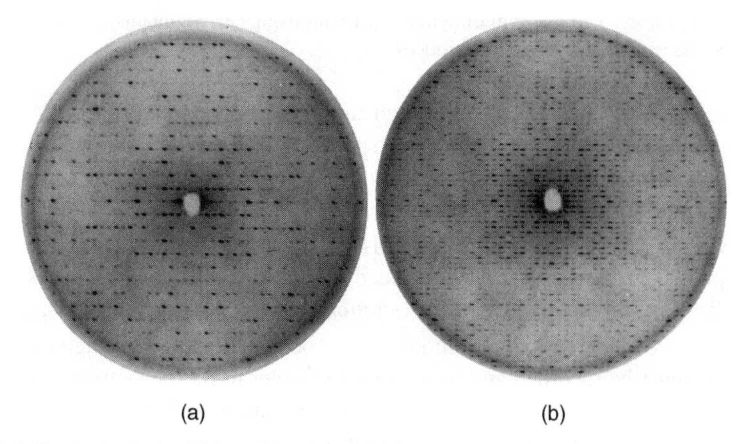

FIGURE 6.18 In (*a*) is the $\boldsymbol{hk0}$, and in (*b*) the $\boldsymbol{0kl}$ zones of reciprocal space for an orthorhombic crystal of rabbit muscle creatine kinase (space group $\boldsymbol{P2_12_12_1}$, $\boldsymbol{a = 47}$ **Å**, $\boldsymbol{b = 86}$ **Å**, $\boldsymbol{c = 125}$ **Å**). Both images exhibit ***mm*** symmetry and systematic absences characteristic of $\boldsymbol{2_1}$ axes along both the vertical and horizontal axes. The rotation angle of the crystal between (*a*) and (*b*) is **90°** about the horizontal axis.

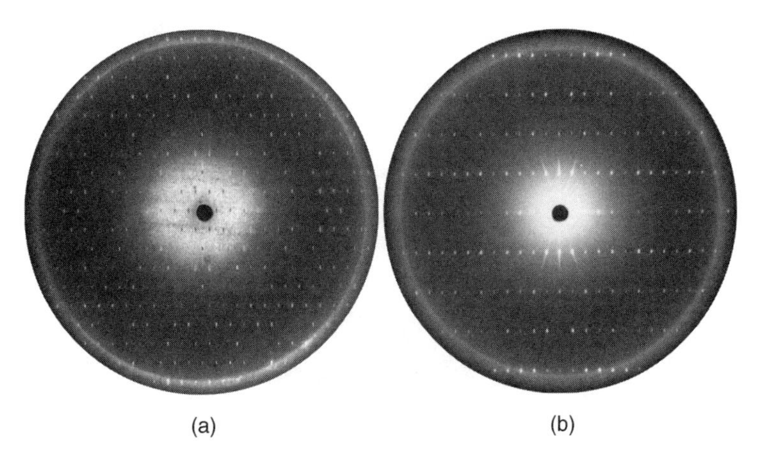

FIGURE 6.19 In (*a*) is the ***hk*0**, and in (*b*) the **0*kl*** planes of diffraction intensities from a monoclinic crystal of the Gene 5 DNA Unwinding Protein from fd bacteriophage (space group **$C2$, $a = 76.5$ Å, $b = 28.0$ Å, $c = 42.5$ Å, $\beta = 103°$**). The angle of rotation of the crystal between the image in (*a*) and in (*b*) was **103°**, the β angle, about the horizontal axis.

may be required for unambiguous symmetry assignment, or they may prove useful for confirmation of the assigned symmetry.

MORE THOUGHTS ON SPACE GROUPS

In identifying symmetry elements present in reciprocal space, we are seeking to establish symmetry relationships between intensities in various parts of the three-dimensional diffraction pattern. In doing so, it is necessary to remember that a symmetry relationship observed for a single plane of the diffraction pattern, because of Freidel's law, may not pertain to the entire pattern, and this can only be ascertained by examining additional planes through reciprocal space.

In deducing the space group of a crystal, it cannot be overemphasized that one works by the process of elimination. If, for example, the Bravais lattice is orthogonal with $a \neq b \neq c$, then the real lattice must be orthorhombic. Thus space groups corresponding to other crystal classes can be immediately eliminated. Examination of the *International Tables* reveals that there are only nine possible space groups for macromolecules in the orthorhombic system, and of these, four are primitive (***P***) and five are centered (***C***, ***I***, or ***F***). If the Bravais lattice is monoclinic, then the *International Tables* show only three possible space groups for macromolecular crystals, two are in primitive cells, $P2$ and $P2_1$, and there is only one centered space group, $C2$.

If threefold or sixfold symmetry is observed in the diffraction pattern, then a trigonal or hexagonal space group is likely. Remember, however, that cubic crystal systems also display threefold symmetry when viewed along their body diagonal (along the **111** direction). An example is shown in Figure 6.20. A threefold axis when combined with the center of symmetry produced by Friedel's law, however, appears as sixfold symmetry when only the zero level of reciprocal space perpendicular to the axis is examined. Upper level images (where $hk\ell$ and $-h-k-\ell$ reflections do not both fall on the same plane of reciprocal space) must be recorded to determine if the sixfold symmetry persists or whether it is in fact a threefold axis. This is done for a crystal of rhombohedral canavalin in Figure 6.21,

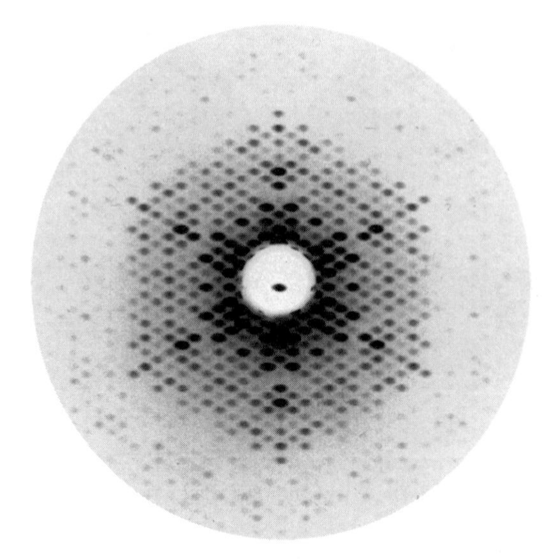

FIGURE 6.20 The diffraction image seen here exhibits ***6mm*** symmetry, which might initially suggest a hexagonal or trigonal Bravais lattice. Further investigation, however, would show that it corresponds to a view along the body diagonal, the threefold axis of the unit cell of a cubic tRNA crystal. The ***6mm*** symmetry is a consequence of combining the true threefold crystallographic symmetry operator with Friedel symmetry.

where a zero-level photograph exhibits a sixfold axis, but the second upper layer bears only threefold symmetry.

For macromolecular crystals, the symmetry of the diffraction pattern (the Laue symmetry) must be generated by Friedel's law, plus the rotational components of symmetry axes present in the crystal. Once the rotational elements have been identified, it is necessary to deduce whether they are pure rotational operators or some sort of screw axes. For dyads,

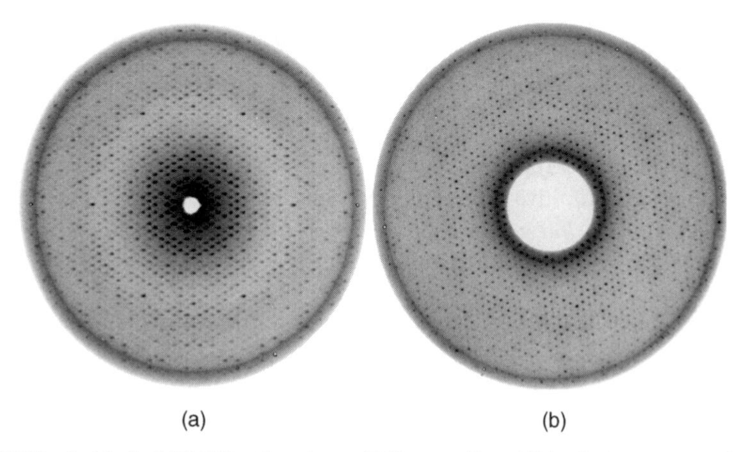

FIGURE 6.21 In (a), the ***hk*0** diffraction plane of ***R*3** canavalin exhibits ***6mm*** symmetry, but because of Friedel's law it could arise as a consequence of either a true sixfold axis or a threefold axis plus the Friedel center of symmetry. In (*b*), the ***hk*2** image, which is along the same direction but does not contain Friedel related reflections, exhibits only threefold symmetry. This demonstrates that the crystal does in fact belong to the trigonal system and not the hexagonal system.

the question is whether a twofold rotation axis exists, or whether it is a $\mathbf{2_1}$ screw axis. For three-, four-, and sixfold axes, there are more extensive choices. For example, for a fourfold screw it may be $\mathbf{4_1}$, $\mathbf{4_2}$, or $\mathbf{4_3}$.

To discriminate between the pure rotational operators and the various screw axis possibilities, the diffraction pattern must be examined for systematic absences. These are not hard to search for, as they always fall along the reciprocal lattice axial lines ($\boldsymbol{h00}$, $\boldsymbol{0k0}$, $\boldsymbol{00l}$ lines) that correspond to the directions of the putative screw axes in real space. Allowed reflections have forms such as $\boldsymbol{h00 = 2n}$ (a $\mathbf{2_1}$ axis along $\bar{a}$), $\boldsymbol{00l = 4n}$ (a $\mathbf{4_1}$ or $\mathbf{4_3}$ screw axis along $\bar{c}$), $\boldsymbol{00l = 6n}$ (a $\mathbf{6_1}$ or $\mathbf{6_5}$ screw axis along $\bar{c}$), and so on. The only serious ambiguity arises in choosing between enantiomorphic screw operators such as $\mathbf{4_1}$ and $\mathbf{4_3}$, or $\mathbf{6_2}$ and $\mathbf{6_4}$. Because handedness is lost in the transformation from real space into reciprocal space (again, because of Freidel's law), and both yield the same systematic absences, they cannot be discriminated one from the other. In those cases the space group does remain somewhat ambiguous, and the crystallographer must bear in mind that the space group could be either $\boldsymbol{P6_1}$ or $\boldsymbol{P6_5}$, $\boldsymbol{P4_122}$ or $\boldsymbol{P4_322}$, and so on.

OTHER INFORMATION IN DIFFRACTION PATTERNS

A feature that is generally accessible to the X-ray crystallographer from a preliminary analysis is the number of protein molecules, or protein subunits, in the asymmetric unit. This is very important in the actual structure analysis, as you might expect, but it may also be useful for deducing symmetry properties of crystalline oligomeric proteins. This deduction is possible because molecular symmetry elements are frequently coincident with crystallographic space group operators. That is, the crystal uses the molecular symmetry in forming symmetrical arrangements within unit cells.

The classic example is monoclinic **C2** horse hemoglobin studied by Max Perutz. In this crystal form it was known from density measurements that the **C2** unit cell contained two entire molecules of hemoglobin. There are four asymmetric units in a **C2** unit cell, however, as the *International Tables* show. Thus one-half of a hemoglobin, to satisfy space group symmetry, had to be related to the other half by a twofold axis. By this means horse hemoglobin was shown to possess a perfect twofold axis of symmetry well before it was demonstrated by any other means.

If one can measure, usually on some sort of mixed fluid gradient, the density of a protein crystal, then the number of molecules in the unit cell can be calculated from the unit cell volume (see McPherson, 1982, 1999). The amount of protein (or number of subunits) in the asymmetric unit can then be obtained from dividing by the number of asymmetric units in the cell (known from the space group).

A more convenient method that does not require direct measurement of the crystal density was introduced by Matthews (1968), though one must be cautious in its application. It has occasionally proved misleading, particularly for crystals of unusually high solvent content. Matthews pointed out (and more recent data seem to confirm it) that the ratio between the volume of the asymmetric unit (the volume of the unit cell divided by the number of asymmetric units in the cell) and the protein mass of the asymmetric unit is about $\mathbf{2.5\text{Å}^3}$/ Dalton for most crystals.

For oligomeric proteins the mass of protein in the asymmetric unit is essentially quantized; that is, it is always some integral multiple of the subunit molecular weight. Given

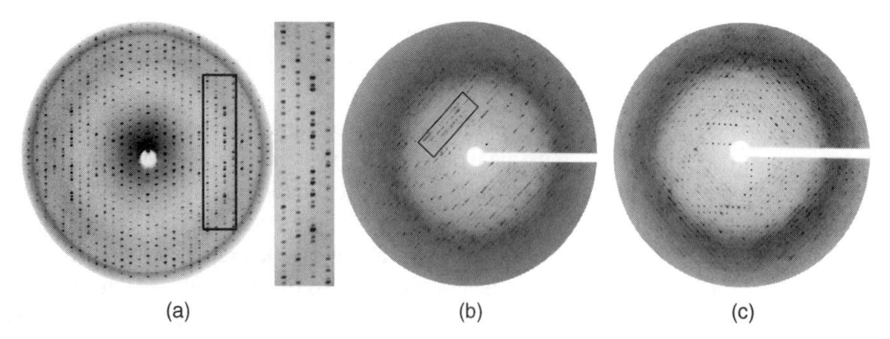

FIGURE 6.22 Diffraction images frequently reveal problems with particular crystals that are sometimes blatant but occasionally subtle. These include disorder, multiple crystals, or twinned crystals. In (*a*), the pattern initially appears very ordered and proper, but close inspection of the row of reflections indicated provides evidence that this monoclinic thaumatin crystal is in fact twinned. In (*b*), the reflections from a Bence–Jones protein crystal fall not on a single reciprocal lattice but multiple, interwoven lattices indicative of twinned or multiple crystals. In (*c*), a tetragonal crystal of Bence–Jones protein is seriously disordered as evidenced by the smeared, highly mosaic reflections and high background scatter.

the calculable asymmetric unit volume in $Å^3$, the expected ratio of **2.5** $Å^3$/Dalton, and the subunit mass M_r, then the number n by which M_r must be multiplied is straightforward.

It must be emphasized again, however, that the ratio, called V_m, of **2.5** $Å^3$/Dalton may be misleading for heavily hydrated crystals containing **70%** to **90%** solvent, and for which V_m may be **3.0** to **5.0** $Å^3$/Dalton. If a true density measure for the crystal can be obtained, it provides reassurance, or inspires retrospection.

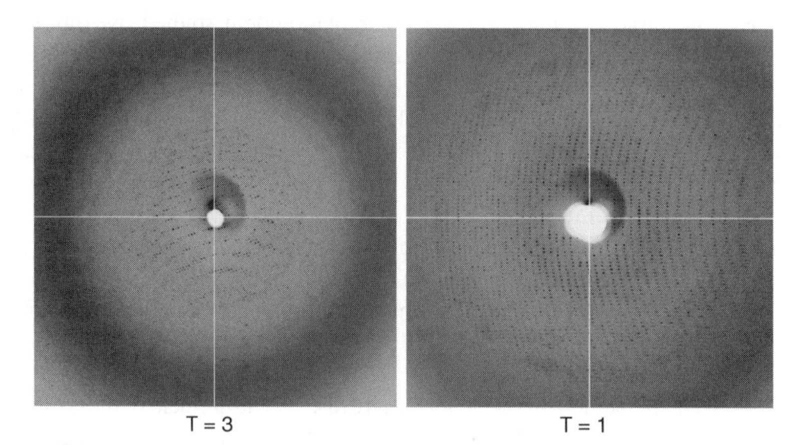

FIGURE 6.23 In (*a*) is a diffraction image, obtained by the rotation method, of a rhombohedral crystal of brome mosaic virus (**BMV**) of space group ***R*3**, recorded with synchrotron radiation. In (*b*) is a corresponding image from a tetragonal crystal of reassembled $T = 1$ **BMV** particles of space group $P4_122$. Note that the maximum distance from the primary beam of the diffraction pattern in (*b*) extends much further than does the pattern in (*a*). The most distant (highest Bragg angle) reflections in (*a*) arise from families of planes having spacings of no less than **3.4 Å**. The better ordered crystals of the $T = 1$ particles diffract to higher resolution, and the most distant reflections arise from Bragg planes of spacings that are about **2.5 Å**.

There are many insights, tricks, and unexpected relationships in the symmetries of diffraction patterns, and they can only be appreciated by practice and experience. Navigation in diffraction space is the high art of X-ray crystallography, and learning it takes time and patience. For the novice as well as the expert, however, there is always the *International Tables*, Volume I, to serve as guide and reference, and it should be consulted freely.

Diffraction patterns can also yield other kinds of information that reveal the quality and physical perfection of a specific crystal. The patterns, when examined closely, can even give warnings of subtle problems such as disorder (of many kinds) or twinning. Some examples are illustrated in Figure 6.22. When these signs appear, it is best to seek the safety of another crystal, or possibly suffer a world of grief.

The resolution of a crystal, how far it diffracts into reciprocal space, is a good measure of crystal order. It tells us immediately to what limit of precision we can expect to structurally characterize the molecules that make up the crystal. This is highly variable between protein crystals, and often between different crystal forms of the same macromolecule. Consider the diffraction patterns in Figure 6.23.

CHAPTER 7

DATA COLLECTION

When we collect X-ray diffraction data our objective is to faithfully record, ultimately in numerical form, the diffracted intensities from every family of planes in the crystal that can possibly satisfy Bragg's law (i.e., those of spacing $d > \lambda/2$). We need to do this as efficiently as possible, as there are many diffraction intensities to collect, and they must be measured as carefully and precisely as possible. It is wise to remember that once the X-ray data have been recorded, that more or less ends the diffraction experiment. The rest is just manipulating that data in real or reciprocal space. If the data contain systematic and random errors, you will have to live with those until completion of the structure analysis. The quality of the final product, the refined structure, can only be as good as the data used in solving it. Stated another way, it is possible to make bad wine from great grapes, but impossble to make great wine from bad grapes.

WHAT IS INVOLVED

The diffraction pattern from a crystal is a three-dimensional array of discrete points, reciprocal lattice points, intensity weighted by the scattering of the electrons (around atoms) distributed about the planes in the crystal. All of diffraction space, reciprocal space, is not illuminated simultaneously, however, because not all families of planes can simultaneously satisfy their specific Bragg condition. We know from the Ewald construction (see Chapter 6) that the reflections from particular families of planes, having indexes $hk\ell$, arise only when their corresponding reciprocal lattice points pass through the sphere of reflection. In data collection, then, our goal is, over time, to manipulate the orientation of a crystal, while it is being exposed to X rays, so as to systematically bring every family of planes in the crystal into diffracting position, and to record the intensities of the reflections to which they give rise.

We can see the diffraction pattern with our own eyes when we collect X-ray data because we obtain the image, the pattern of diffraction spots, on the face of our detector or film. We can't directly see the families of planes in the actual crystal, but we know, through the Ewald construction, how the diffraction pattern is related to the crystal orientation, and hence to the dispositions of the planes that pass through it. We also know from Ewald how to move the crystal about its center, once we know its orientation with respect to our laboratory coordinate system, in order to illuminate various parts of reciprocal space. In data collection we watch the diffraction pattern, not the crystal, and let the pattern of intensities guide us.

For data collection the crystal is mounted on what is called a goniostat. This device allows us to rotate the crystal about some axis passing through its center, or it may be made to systematically move the crystal about its center in some more complicated motion (as with a precession camera, or a three-circle goniostat) so as to sequentially bring all of the Bragg planes into diffracting position. Strategies for where to begin and which way to move the crystal for optimal collection efficiency can be found using completely automated procedures, given an initial orientation, and these are routinely used.

The crucial aspects of data collection are (1) producing a finely collimated beam of monochromatic X rays, (2) getting the crystal out of the mother liquor from which it was grown and centered in the X-ray beam, (3) reorienting it systematically in the X-ray beam in a continuous fashion, and (4) recording the intensities of the emitted X-ray reflections. After that, assuming you have done these things well, you can retire to your computer.

There are, of course, many practical details to consider throughout the data collection process. These include issues such as what angular increments are to be used when the crystal is rotated and an image of diffraction space is recorded? How fast should the rotation motion be? How long are you going to leave each family of planes in diffracting position in order to acquire sufficient photons to give a good measure of intensity? The intensity will be directly proportional to the time spent by the reciprocal lattice point on the sphere of reflection. How many times do you want to record each reflection, or its symmetry equivalents to ensure accuracy (redundancy)? How much radiation decay are you willing to allow before the crystal is discarded? When anomalous dispersion data is to be used for phase determination (see Chapter 8), then the wavelength of radiation chosen is crucial to the experiment, and the strategy by which the reflections are collected is as well. Fortunately many of these decisions are made for you by helpful software at the synchrotron facilities, or by your professor who, in theory, knows how to address these questions.

Once the data have been recorded, there are a host of statistical measures that provide criteria for evaluating the quality of the data. These serve as important guides to possible weaknesses in the data set, give estimates of the useful resolution of the data, and yield quantitative measures of precision. These are normally provided as a matter of course by the data-processing programs that correct, scale, and merge the measured intensities into a comprehensive set.

X-RAY SOURCES AND THE PRODUCTION OF X RAYS

The first sources for laboratory X rays were evacuated glass tubes, like that shown schematically in Figure 7.1, containing an electron source (cathode) and a target for the electrons (anode) which were maintained at some high potential with respect to one another. The anode is positive with respect to the cathode. Electrons are emitted from the hot cathode, or filament, and accelerated through an electric field to high velocity, and directed onto a

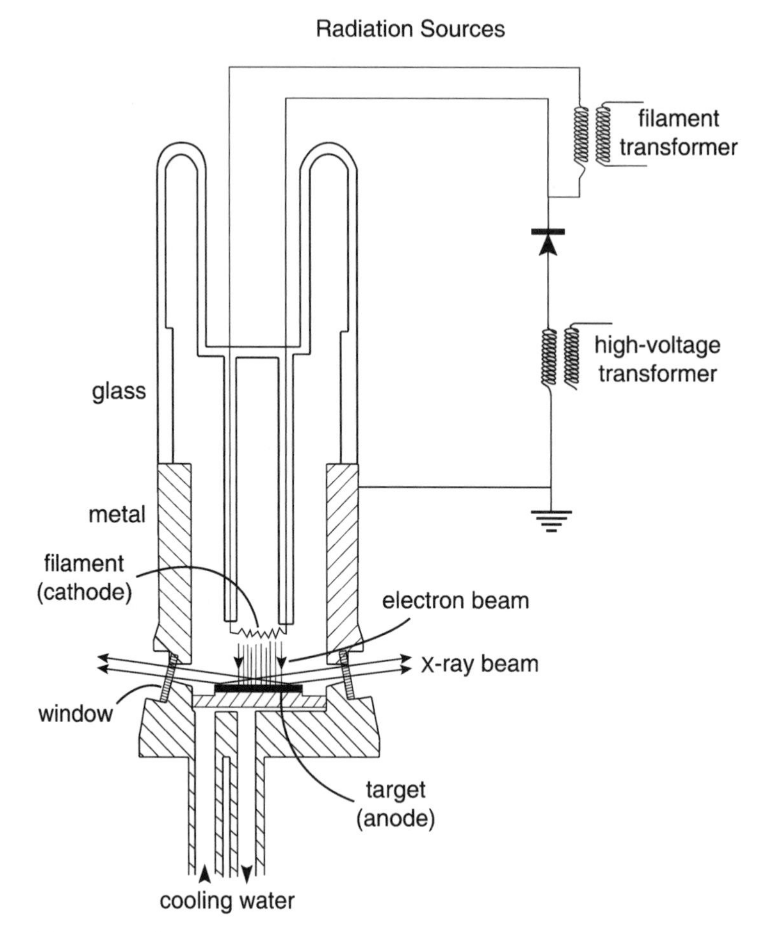

FIGURE 7.1 Schematic diagram of a conventional, broad focus, sealed X-ray tube. The anode material, usually of a pure element, determines the characteristic X-ray spectrum that is produced.

small area of the target material where X rays are produced. Collision of electrons with the atoms of the target yield a continuous spectrum of radiation (i.e., X rays of all wavelengths) as well as discrete lines (i.e., wavelengths at which very high levels of X rays appear). The characteristic wavelengths at which the lines occur are determined by the composition of the target. Such a spectrum is seen in Figure 7.2 where the target is pure copper, the material most frequently used in laboratory experiments. Continuous radiation is produced by the multiple small transfers of energy, or "bremstrahlen" as it is known. The discrete peaks, or quanta, arise from the displacement and return of individual electrons from the inner atomic shells of the target atoms.

Because radiation should be as monochromatic as possible in order to maximize the diffraction effect from atoms in a crystal, the continuous spectrum is either suppressed by selective filters, also illustrated in Figure 7.2, or a discrete wavelength of X ray is isolated with a monochromater of some sort. For most biological structure analyses in the laboratory where a sealed X-ray tube or a rotating anode source (see below) is employed, the anode is of pure copper or occasionally molybdenum. These two elements have strong, characteristic $\mathbf{K}_\alpha$ peaks in their spectra at **0.154 nm (1.54 Å)** and **0.071 nm (0.71 Å)**, respectively. These

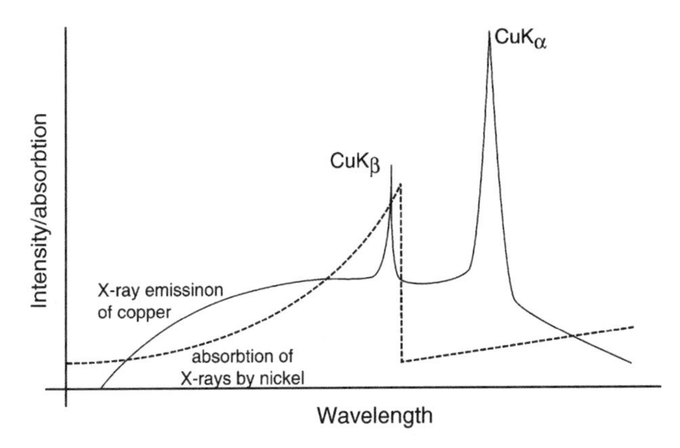

FIGURE 7.2 Bombardment of a pure copper target (the anode) with high-energy electrons emitted by a hot filament (the cathode) produces a broad, continuous distribution of X rays due to energy transfers between the high-velocity electrons and the electrons of the copper atoms. Electrons in the inner shells are also raised to discrete, higher energy levels, and when they fall back to their ground state, they give rise to intense peaks of X rays at discrete, characteristic wavelengths. Shown here are the commonly used $\mathbf{CuK_{\alpha}}$ and $\mathbf{CuK_{\beta}}$ of the copper spectrum. Superimposed on the copper spectrum is the absorption spectrum for X rays of nickel. Insertion of a nickel foil in a beam of X rays produced with a copper target removes the $\mathbf{CuK_{\beta}}$, as well as most of the "white" radiation.

are near ideal because they are wavelengths that are close to the distances between atoms in biological molecules.

Currently there are three commonly used X-ray sources: the first two, sealed X-ray tubes and rotating anode sources, are found in most protein crystallography laboratories. The third source of X rays is synchrotrons, which are available only at specialized facilities, generally national laboratories. X rays produced by synchrotrons, which have a number of unique and highly desirable features, are generated by a completely different principle than that described above for conventional sealed tubes and rotating anode sources.

Sealed tubes (Figure 7.1) are the least complex, both electronically and mechanically, usually least expensive, and most reliable. One simply turns them on. They have been used for the structure determination of most conventional small molecules, and were used as well for most of the early macromolecular structures. They are similar to standard cathode ray tubes. The filament and target are sealed within a glass housing that is evacuated to about $\mathbf{10^{-7}}$ **mm** of mercury. Windows of beryllium, which have insignificant absorption for X rays, allow the radiation to pass to the exterior when the tube is energized. Those X rays are then directed by a collimator system onto the crystal. In conventional tubes, the amount of X rays that can be produced is limited by the rate at which the enormous heat that is generated at the anode surface can be dissipated. The anode, or target, obviously cannot be allowed to melt. Heat removal is severely limiting even when water is pumped through the static target at high pressure.

Rotating anode X-ray generators, one of which is shown in Figure 7.3, have an arrangement of cathode and anode essentially the same as for the sealed tubes described above. The tubes, however, are not evacuated and sealed, but continuously pumped by a diffusion pump backed by a conventional roughing pump. The environment is therefore the same as the inside of the sealed tube. The second important feature of the rotating anode device is exactly that implied by its name. The anode, which is a hollow disc of copper (usually) several inches in diameter, is rotated at 3000 to 6000 rpm by an external motor so that the beam of electrons emitted from the cathode falls on a continuously changing surface. Water,

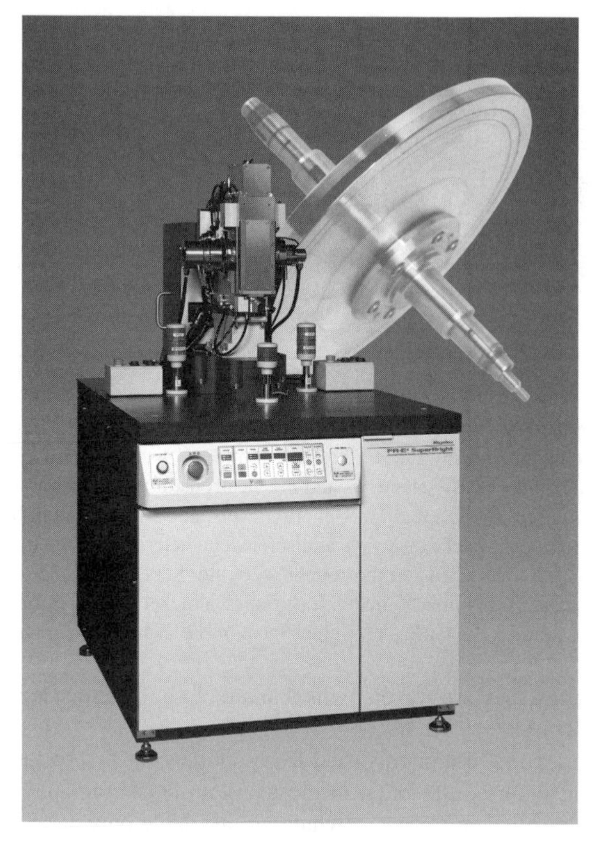

FIGURE 7.3 A modern rotating anode X-ray generator used in most laboratories to produce **CuK_α** radiation. An image of the actual anode is shown in background. (Courtesy of Rigaku America Corp.)

pumped through the interior of the anode through the hollow shaft on which it is mounted, removes the heat generated on the, effectively, much larger copper surface. With rotating anode sources, flux densities that are as much as 100-fold higher than conventional sealed tubes can be achieved.

The most important characteristic of an X-ray source is the radiation flux density that it can generate. This is a function of the power at which it is operated and the size of the focal spot produced on the anode. The ratio of these factors is the specific loading of the source. A standard sealed, broad focus tube is normally operated with a cathode-to-anode voltage difference of **40 kV**, an emission current of **20 mÅ**, and a focal spot size of **1 x 10 mm**, or a specific load of **$80\ \mathrm{w/mm^2}$**. Rotating anode machines can be run at **50 kV** and **50 mÅ**, with a focal spot size of **0.2 x 1.0 mm**. The specific loading under these conditions is **$1.25 \times 10^4\ \mathrm{w/mm^2}$**.

Although new technologies and the further development of micro focus tubes are driving a comeback in the use of sealed tube generators, their inherent limitations, in the long term, make it unlikely that they will ever be competitive with rotating anode sources. Thus most laboratory generators will likely continue to be of the rotating anode variety. Both, however, are fading in popularity and importance as synchrotron sources have come to dominate the field.

A synchrotron generates X rays by a different principle. At a synchrotron, electrons are accelerated to high velocity, approaching relativistic speeds, by a linear accelerator and injected into what is called a storage ring. The storage ring is a circular array of powerful "bending" magnets that act on the electrons to continuously change their direction so that they ultimately circulate around the ring. Although the kinetic energy of the electrons does not change as they travel about the ring, their direction does; hence their momentum is under continuous change. A change in momentum implies acceleration, and we know from fundamental physics that a charged particle undergoing acceleration emits radiation, a form of energy. In the case of the electrons circulating about the ring, the radiation, which is emitted tangentially to the ring, has a broad spectrum of energies (or wavelengths) because of the range of velocities inherent to the electrons. This is completely unlike the X rays from a target bombarded with electrons, where characteristic spectral lines are produced, such as the $\mathbf{CuK}_\alpha$ line from a copper target.

Using monochrometers, which are usually large, bent, graphite crystals, a specific wavelength λ can be selected at will from the X rays produced by the circulating electrons, and these X rays are then directed upon the protein crystal target by a collimation system. Synchrotron radiation offers two major advantages over X rays from conventional sources. First, the wavelength can be selected from a broad spectrum, a feature that is absolutely invaluable to any experiment or procedure that relies on anomalous dispersion effects. The wavelength can be "tuned" to maximize the anomalous scattering from particular kinds of atoms, such as selenium, bromine, or heavy atoms like mercury and platinum. This enhances the anomalous "signal" and provides a means to solve the phase problem. Because of this feature, macromolecular structures containing atoms that scatter anomalously are routinely solved with minimal intervention by the investigator, to precisions seldom attained by other approaches such as isomorphous replacement. The second advantage of synchrotron radiation is that it is produced with an intensity that is two to three orders of magnitude greater than that from conventional, laboratory sources. This is extremely significant because it means that useful diffraction data can be obtained from much smaller crystals, and crystals having very large unit cell dimensions. It generally means that data can be recorded to higher resolution, even from the smaller crystals, thereby yielding more precise structural information.

One consequence of the high intensity of synchrotron radiation, however, is that crystals are quickly destroyed in the X-ray beam under normal conditions. This is not usually true of crystals exposed to the lower intensity X rays from conventional sources, though radiation decay certainly does occur but at a much lower rate. The problem can generally be overcome, however, by maintaining the protein crystals frozen at cryogenic temperatures during the course of data collection (see below). Although radiation damage still occurs, it proceeds at a limited pace. This has a further useful consequence. Generally, one or more complete sets of diffraction data can be collected from a single frozen crystal. The benefit is that it is then unnecessary to scale data sets together from different crystals, as scaling is frequently a major source of introduction of data imprecision. At this time virtually all highly refined, ultra precise macromolecular structures are solved and refined using synchrotron data, as well as all very large macromolecular assemblies such as viruses, ribosomes, and multienzyme complexes.

DETECTORS AND THE RECORDING OF DIFFRACTION INTENSITIES

Generating X rays and directing them on to the macromolecular crystal is only half of the data collection process. Equally important is recording the diffraction intensities produced by the interaction of the X rays with the atoms distributed about the Bragg planes in the crystals. There are a number of ways of accomplishing this, from classical means such as film, to modern detectors that employ solid-state charge-coupled devices. Although film is seldom used today for X-ray diffraction, it is instructive to look at how it works, as this is the basis for some more advanced instruments.

Film is a transparent material, usually plastic coated with an emulsion containing grains of silver. When a photon strikes a grain of silver, the silver turns black. The minimum number of photons that strike a film before some measurable degree of blackness occurs is a measure of the film's sensitivity. The more photons that strike the film, the more silver grains turn black, and the stronger the image. Thus film accumulates an image of the diffraction pattern, like those in Chapter 6, for example, in a linear manner with time. The resolution of the image on the film, like its sensitivity, is a function of the size and number of the silver grains. Note, however, that once a silver grain is struck by a photon and turns black, it will not be changed further or made blacker if it absorbs another X ray. After a time then, or when a large number of diffracted X rays strike the same spot on a film, the film becomes saturated. Its blackness no longer accurately reflects the number of photons it received, and it no longer reflects the true intensity of the diffracted beam. The intensity range over which the response is linear, and the blackness of the film reflects accurately the true intensity, is called its dynamic range. All detectors, including those that do not use film, have some characteristic dynamic range.

Film techniques, complimented by detection methods based on a linear ionization chamber incorporated into what were called diffractometers, provided the basis for nearly all X-ray diffraction analyses carried out before about 1980. Diffractometers are primarily of historical interest now and are no longer used for macromolecular work. By comparison with today's detectors, they were painfully slow, and woefully inaccurate. A single native or isomorphous derivative data set for a protein could require 10 to 50 separate crystals, take 6 months to a year to record, and even then yield intensities with mean errors of the order of 6% to 10%. Recording of anomalous dispersion data was a sometimes thing, and was nowhere near precise enough to be used in the MAD and SAD approaches to phase determination currently in use. Nonetheless, many protein, nucleic acid, and even virus structures were solved. If nothing more, this serves as a testament to the patience, care, and ingenuity of the early generations of protein crystallographers.

A major revolution in detection devices occurred with the development of what were known as multiwire counters in the early 1980s. These were essentially large electronic films, about a foot square in the most popular instrument, comprised of a grid of orthogonal wires maintained inside a chamber filled at low pressure with xenon. When an X ray entered the chamber through a large beryllium window in the front, it ionized a xenon atom, which subsequently struck a wire. The grid was position sensitive so that not only the number of ionized xenon atoms generated by photon collisions were detected but also where on the grid of wires those events occurred. In a few minutes an image of a single two-dimensional plane of a diffraction pattern could be recorded. By rotating the crystal in increments over time, a series of electronic images could be recorded that ultimately encompassed all of reciprocal space, at least to the resolution limit of the crystal. If the crystal was initially oriented carefully with respect to the laboratory coordinate system before image accumulation was begun, then the intensities recorded on the electronic film could be indexed, that is, assigned their proper $\boldsymbol{hk\ell}$ value. Multiwire detectors not only increased the precision of measurement substantially but, more important, reduced the amount of time required to collect a single data set to one or two days. It became possible to collect an entire data set from a single crystal, eliminating inaccuracies due to scaling.

The multiwire devices, though workhorses for about 10 years, were an intermediate in the development of the detectors we use today. They provided excellent data, but even faster, more reliable, more accurate devices became available in the late 1980s. The instruments that we use today fall into two categories: those based on what are called image plates, and those based on charge coupled devices (CCD) detectors. While CCD detectors are now exclusively used at synchrotrons because of their almost instantaneous readout of data, necessary with high intensity sources with short exposure times, most university and private laboratories continue to use detectors based on image plates.

Image plate systems, like that shown in Figure 7.4, are, in a sense, advanced technology film, but instead of a single pellicle coated with silver grains, a fixed plate is coated with a lanthanide salt. The grains of the salt are chemically altered when struck by a photon from, say, an X-ray reflection. As with film, they chemically accumulate an image linearly with time, though with two or three orders of magnitude the efficiency of standard film. More important, when a focused laser beam is scanned, in a raster manner, over the surface of the image plate, the exposed, chemically altered lanthanide salt emits light proportional to the number of photons it absorbed. This light can then be recorded by a photomultiplier system and converted directly into an image of the diffraction pattern. Intensity measurements are accurate and rapid. A plate generally requires only 1 to 5 minutes of exposure to the diffraction pattern, and 4 minutes to scan with the laser–photo multiplier system to convert the image to diffraction intensities. An exposed image plate can then be returned to its original state by a second scanning process. Thus an entire set of diffraction data can

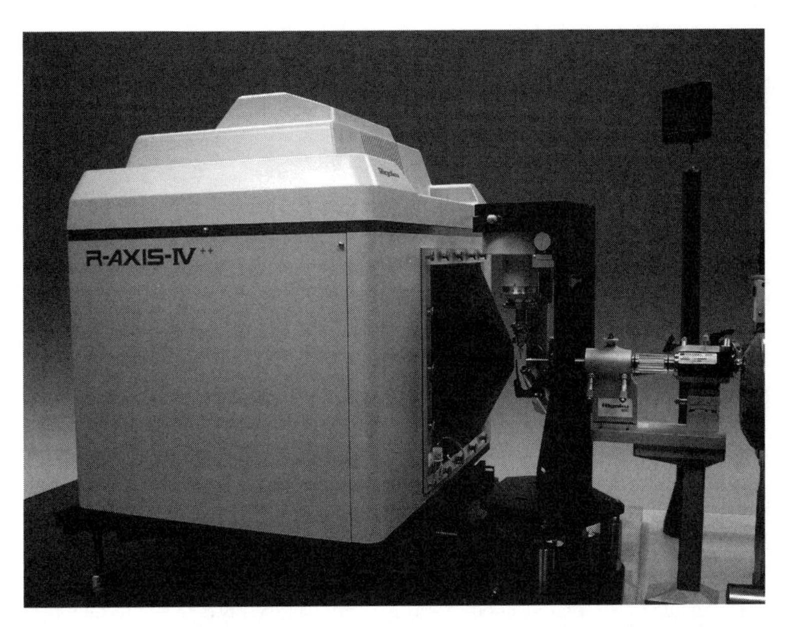

FIGURE 7.4 A multi-image plate X-ray detector. The system contains two image plates on a rotating carrousel so that one can be recording data while the second is read and translated into intensities. (Courtesy of Rigaku America Corp.)

be recorded from a single crystal in roughly a day's time. The sensitivity is excellent, and the image plates have a large dynamic range. Most instruments have not one, but two or even three image plates arranged in a carrousel manner so that while one plate is recording a diffraction image, another is being scanned and read out. This does much to reduce the disadvantage of a somewhat lengthy readout time. A complete laboratory system that includes a rotating anode generator, modern optics, a CCD detector, and and image plate detector is shown in Figure 7.5.

CCD instruments are technically more complicated than can be described here, but the general principles can be sketched out. CCD instruments use much the same technology as the modern electronic cameras that now dominate the consumer market, though built on a grander scale. With this instrument a photon passes through a film from the backside and strikes some luminescent molecule that emits light. The light is led by a bundle of individual optical fibers from a fine grid of points on the film surface to another fine grid of solid-state devices that accumulate charge in proportion to the intensity of the light they receive. The location of the CCD is physically related to the position at which the X ray struck the film by its particular optical fiber. At the end of an exposure, during which time the crystal is rotated through a small angle of **0.5°** to **3°**, the intensities of the diffraction pattern are recorded. At the end of the exposure the entire array of CCDs is read out into a computer. The CCD devices are no more sensitive, in general, than image plates, and their dynamic range is substantially less. In addition they tend to be afflicted with more technical considerations and complexities than are image plates. Nonetheless, their great read-out speed clearly makes them the preferred instrument at high-intensity sources. Currently, using synchrotron radiation and a CCD detector, a single diffraction data set from a crystal of modest cell dimensions, perhaps **60 Å** on an edge, and having, say, orthorhombic symmetry, can be recorded in 30 minutes to an hour.

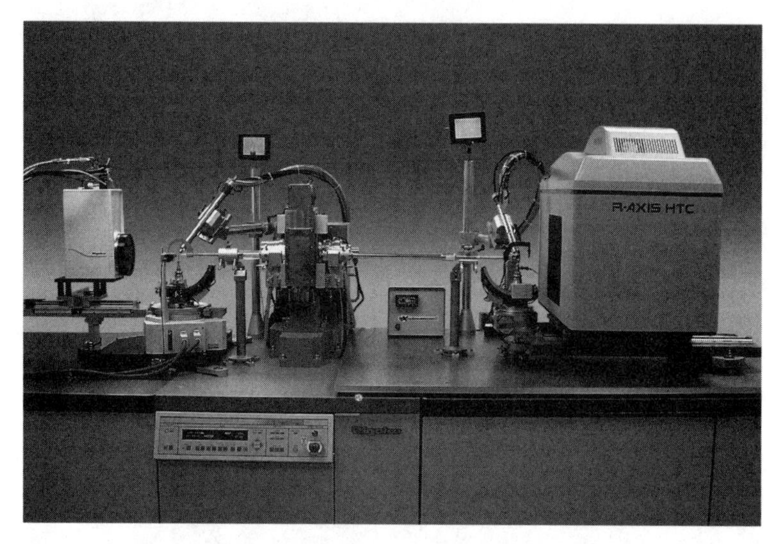

FIGURE 7.5 An integrated rotating anode X-ray generator and X-ray detector system. In this system the detector on the right is a multi-image plate, while that on the left is a CCD detector. The system includes a liquid–gaseous nitrogen cryogenic system that maintains the macromolecular crystal at −173°C. (Courtesy of Rigaku America Corp.)

DATA MANAGEMENT PROCEDURES

Besides the revolutionary developments that have occurred in X-ray sources and detectors, there are two other impressive advances in data collection. These have been in the areas of algorithms and computing, and in the area of how crystals are treated during data collection, that is, cryocrystallography. Both have been essential to realizing the full potential of the technology developments in sources and detectors. To mention only one in the area of computing, the example of indexing reflections comes immediately to mind. As noted above, in the age of the multiwire area detector, and in all times prior to that, it was necessary to begin data collection with the crystal accurately aligned so that reciprocal space axes were precisely related to some laboratory coordinate system, generally that of the camera or goniostat. This was necessary so that the reflections recorded on each image, electronic or otherwise, could be recognized and indexed according to their proper $\boldsymbol{hk\ell}$. This was often a painstaking and time-consuming process, and assumed in advance that one accurately knew the cell dimensions and symmetry of the crystal under investigation.

Currently this is unnecessary. A completely arbitrary plane through reciprocal space can be recorded, or a small number in the worst cases, and computer programs based on powerful algorithms can figure out the cell dimensions, give at least a good idea of the probable Bravais lattice (see Chapter 3), and index all of the reflexions, not only on the planes of reciprocal space examined, but on all subsequent images that will be recorded during data collection. Optimal strategies for data collection based on crystal orientation and symmetry are generated, and as the intensities are recorded, they may be corrected for experimental factors, scaled, and merged. There is no longer much for the investigator to do between freezing the crystal and solving the structure. In fact crystals are now commonly frozen in the laboratory, placed in a metal cassette, shipped in a cryogenic container, a Dewar, to the synchrotron where robots place the frozen crystals on the data collection instrument and the data are recorded. The investigator can monitor the process, make choices and decisions, or otherwise intervene remotely from his laboratory through the Internet

CRYSTAL MOUNTING AND HANDLING

X-ray diffraction data may be collected from crystals at either room temperature, or, as is more common today, at cryogenic temperatures around **−173°C**. At synchrotrons, it is always at cryogenic temperature because of the potential for radiation damage, but in the home laboratory, it may be at either. Cryocrystallography only emerged in the early 1990s with the advent of synchrotron radiation. Prior to that time most data were collected at room temperature, or at **4°C**. The art of mounting macromolecular crystals in glass or quartz capillaries for data collection at room temperature is rapidly being lost. This is unfortunate, as there are some crystals, and likely always will be some, that cannot, for whatever reason, be frozen. Thus there will always be a need to collect data from certain crystals in the conventional way, and it would be wise if budding young crystallographers gave some attention to developing the necessary skills.

Crystals of proteins, nucleic acids, viruses, and macromolecular complexes must be handled with considerable care because they are extremely fragile and contain a high proportion of solvent, principally water. Bernal and Crowfoot demonstrated in 1934 that diffraction patterns from protein crystals quickly degenerate upon dehydration in air. Thus it is essential that crystals be maintained in a fully hydrated and stable environment during the entire course of analysis.

For data collection at room temperature, a crystal in a small amount of mother liquor is mounted by drawing it, using a **1 cc** hypodermic syringe, into a thin walled quartz or glass capillary of **0.25** to **1.5 mm** diameter that is coupled to it by a small section of soft rubber tubing. The crystal is then separated from most of the liquid with a fine glass fiber inserted into the open end of the capillary. The crystal should have a thin film of mother liquor around it, but not a lot. Too much and the crystal will be unstable in the capillary and slip around during data collection, ruining the process. To little liquid and it may dehydrate. The capillary is first sealed at one end with dental wax. A small amount of mother liquor, segregated from the crystal, should be included in the capillary at one end to maintain the proper hydration state of the crystal. The capillary is then broken at the other end and similarly sealed with wax to completely isolate the crystal. All of this is done of course under a microscope.

This procedure ensures that the vapor environment of the crystal will remain invariant during the period of analysis and will be the same as that from which it was grown. The capillary itself absorbs only a negligible amount of radiation, and since the glass or quartz is amorphous, it contributes nothing to the discrete diffraction pattern, only background scatter. The capillary is next joined to what is called a "goniometer head," which is a multistage platform with perpendicular translations (and in the old days, two perpendicular arcs as well). When the crystal is affixed to the goniometer head, it may be made to rotate perfectly about a central axis by adjusting the translations. This must always be done to prevent the crystal from moving out of the X-ray beam as it is rotated during data collection.

A more common procedure for preparing crystals for data collection today is by flash freezing them in liquid nitrogen, or in the cryostream that maintains them frozen on the actual data collection instrument (Garmen and Schneider,1997; Pflugrath, 2004). Data collection devices now almost always include a cryosystem to ensure that temperatures do not rise above **−173°C** during data collection. A frozen crystal is immobile, but it is still wise to eliminate as much mother liquor as possible from around the crystal so as to minimize background scatter, which affects data quality. While frozen crystals suffer little radiation damage compared to those at room temperature, the amount varies depending on the particular protein crystal, it does nonetheless exhibit some sensitivity. At synchrotron sources, even frozen crystals have a limited lifetime.

The most arbitrary aspect of cryogenic data collection arises from the requirement that most macromolecular crystals must be removed from the mother liquor in which they were grown and first exposed to some cryoprotective agent, by briefly dipping them in the liquid, before inserting them into liquid nitrogen. The problem is that there is not yet any universal cryoprotective agent that works with every crystal. Therefore it is often necessary, with a new crystal, to try a series of cryoprotective agents in order to find one that will allow freezing of the crystal without loss or reduction in resolution of the diffraction pattern, or disordering of the crystal. Typical cryoprotective agents are based on the natural mother liquor from which the crystal was grown, say a salt or PEG containing solution, plus 20% to 25% glycerol, MPD, ethylene glycol, or 2M sucrose. Most crystallographers have personal preferences, but there is no one favored agent. As noted above, there are some crystals, it appears, that simply cannot be frozen.

To mount crystals for freezing, metal pins mounted on bases and having a small (**0.5** to **5 mm**) soft fiber loop on the opposite end are used. Examples are shown in Figure 7.6. The standard procedure is to enter the mother liquor with the pin, capture the crystal on the

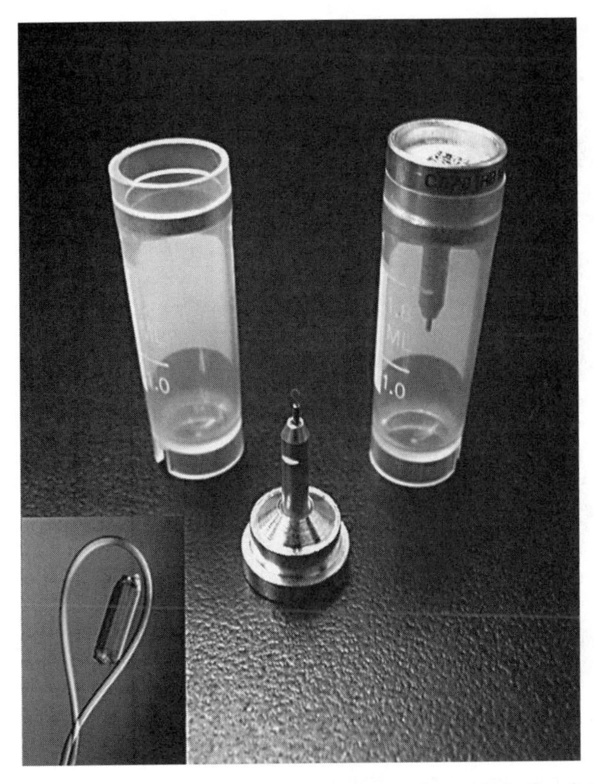

FIGURE 7.6 The standard copper pin used to mount protein crystals for data collection under cryoconditions. At the top of the pin, seen magnified in the inset, is a small fiber loop of **0.1** to **0.5 mm** diameter that is used to "catch" or "scoop up" the crystal from its mother liquor. The magnetic base attaches to the goniostat. Also shown are cryovials where pins carrying crystals can be inserted for protection during storage in liquid nitrogen. (Courtesy of Hampton Research.)

loop, remove the crystal from the mother liquor, pass is briefly through the cryoprotective liquid, and then quickly submerge it, pin and mount as well, into liquid nitrogen. From that time forward, the crystal must not be allowed to warm above **−173°C**. To thaw and refreeze a crystal is generally devastating (though there are procedures, called annealing, that use precisely that process to improve the diffraction qualities of some crystals, Pflugrath, 2004). At many synchrotron beam lines, robotic instruments and procedures have been developed to such an extent that human hands need never to touch a mounting pin once the crystal has been frozen. Such a robotic system at the Stanford University facility is shown in Figure 7.7.

X-RAY DATA PROCESSING

In considering the quality of diffraction data (or phases, electron density, or of any other experimentally derived measurement), it is important to understand the difference between precision and accuracy. This idea is illustrated for a single measurement in Figure 7.8. In general, the precision of any measurement increases (roughly with the square root) of the number of times it is made, that is, the redundancy. That, however, provides no assurance

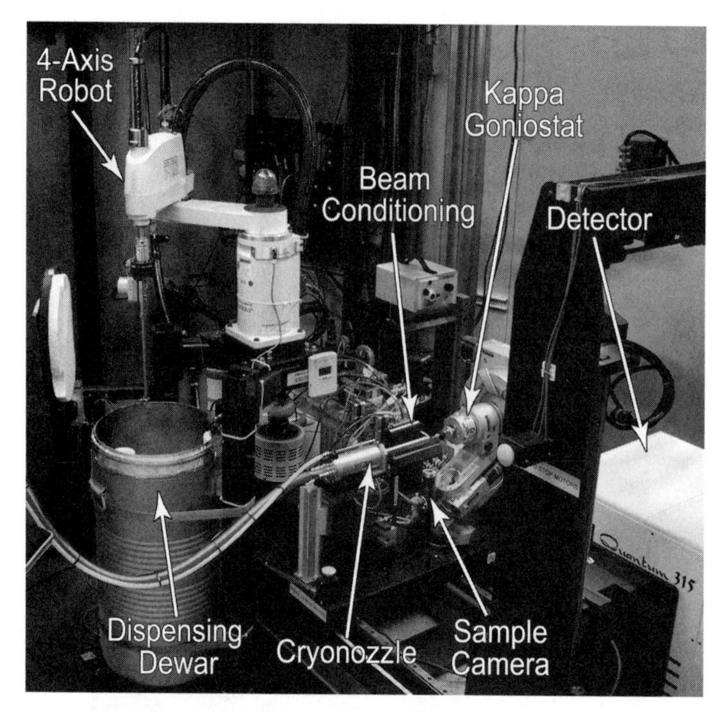

FIGURE 7.7 Shown here is the investigator end of a modern synchrotron X-ray beam line and data collection facility at the Stanford Linear Accelerator (SLAC) at Stanford University. The ensemble of instruments includes the goniostat for precisely positioning and orienting the crystal, a shutter system, a cryo- cooling apparatus to maintain the crystal at **−173°C**, and a robotic system to exchange a series of crystals held in reserve in a dispensing Dewar. All operations are fully automated and under internet control by the investigator who may be at a remote location anywhere on earth. The detector is a fast readout CCD detector. Cameras provide the investigator with a continuous display of the crystal, and allow him to monitor all operations concerning data collection. With a system such as this, a dozen sets of X-ray diffraction data may be recorded in the span of a day.

that the final tally or average is accurate. Indeed, if every observation of a quantity is consistently afflicted with some systematic error (your measuring stick has a flaw, or you use it incorrectly), precision may be very high but accuracy lacking. Accuracy is always the more difficult issue, and the most difficult to evaluate. This is particularly true in X-ray crystallography where the number of experimental variables tends to be large.

There are a number of good programs available that integrate the observed peaks from the detectors and subtract background to create intensities, index the intensities, and ultimately convert them into structure amplitudes, $F_{hk\ell}$, for use in phasing and structure determination. The programs correct the intensities for geometrical factors such as polarization and the Lorentz factor (***Lp*** correction), and for anisotropies in the data if they exist. There are several statistical properties of a set of data that are important in evaluating its quality. The first that one might look at is the resolution of the diffraction data, that is: What are the minimum Bragg spacings in the crystal for which observable reflections were obtained? While this may seem straightforward, it is not always so. This is primarily because crystallographers have not agreed on how to establish a resolution limit. Some would say the limit is where

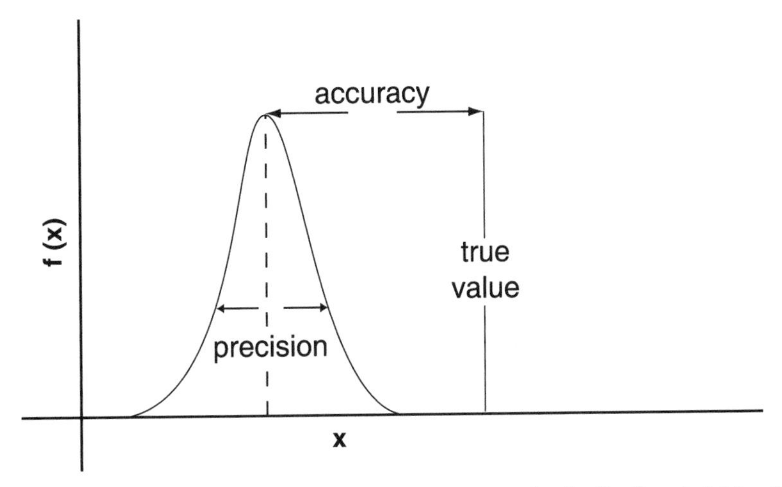

FIGURE 7.8 A set of observations for which the errors are randomly distributed yields a Gaussian curve. The width of that curve is a measure of the "dispersion," or precision of the measurements. The true value, however, may be some distance away from the peak of the Gaussian, and this distance defines the accuracy of the experiments.

the average observed reflection falls to less than the estimated error, others twice the error, and so on. In addition, this is also complicated by the fact that less data might have been collected in the higher resolution ranges than in the lower ranges. Thus, if one has only **30%** of the data in the highest resolution range, how is this to be considered? This question is mostly academic, however, as the final criterion is the quality of the electron density map, and the accuracy of the structure.

The redundancy of measurements is important. Clearly, an $F_{hk\ell}$ will be more accurate if it, or its symmetry equivalents, have been measured many times rather than a single time. Average redundancies of 4 to 8 are desirable, and even more if they can be obtained. It is wise to remember, though, that crystals suffer radiation damage with time and their diffraction patterns gradually change. Thus reflections collected at a later time may be of poorer quality than those measured early on, and it is never a good idea to corrupt good data with poor. A very significant criteria of data quality is how well multiple measures of the same reflection, or symmetry equivalent reflections scale and merge with one another. This is usually indicated by what is known as an R_{sym}, which is the ratio of the sum of the absolute values of the differences between equivalent reflections to the average value of the reflections.

$$R_{sym} = \frac{\sum_{hk\ell} \sum_j |F_{hk\ell-j} - \hat{F}_{hk\ell}|}{\sum_{hk\ell} \hat{F}_{hk\ell}},$$

where $F_{hk\ell-j}$ is the jth symmetry equivalent reflection or multiple measurement of $F_{hk\ell}$ and $\hat{F}_{hk\ell}$ is the mean value for all $hk\ell$ symmetry equivalent reflections.

Normally this ratio is calculated in shells of resolution. The average intensity declines with greater resolution, and the R_{sym} increases because weak reflections tend not to be measured as accurately as strong reflections. Another valuable criterion is the estimated error of the reflections, σ, again computed as a function of resolution, and this generally increases as one proceeds away from the origin of reciprocal space. At higher resolution spot shapes become distorted and backgrounds higher, and reflections simply become more difficult to measure accurately.

The estimated error σ is best determined by averaging redundant and symmetry related reflections, and is given by the formula

$$\sigma = \frac{\sqrt{\sum_j^N (x_j - \hat{x})^2}}{(N-1)},$$

where x_j is the jth measurement of the reflection, or a symmetry equivalent, $\hat{x}$ is the mean value for all measurements j, and N is the total number of measurements.

Because diffraction spots have extent, they are not simply points, the intensity over the entire spot must be summed or integrated. Hence we commonly refer to the raw intensities as integrated intensities. The integrated intensities of each reflection are not, however, the exact values for $I_{hk\ell}$, as there are certain instrumental corrections that must be applied. Principal among these are the Lorentz and polarization corrections, or Lp factor as it is called. These are dependent on the particular geometry of the data collection system and can be calculated and applied in a straightforward manner. This process is done automatically by data collection software, so it should be of little concern to the crystallographer today. Should you have an interest in these corrections and their underlying physical principles, they are discussed in great detail in the classic texts of Buerger (1960, 1963).

SCALING OF X-RAY DIFFRACTION DATA

One of the most important aspects of data collection and data reduction is the necessity of bringing all reflections of the same $hk\ell$, or symmetry equivalent $hk\ell$, onto a common numerical scale. This is generally necessary for all of the intensities $I_{hk\ell}$, or their corresponding structure amplitudes $F_{hk\ell} = \sqrt{I}_{hk\ell}$, from a single crystal, or the $F_{hk\ell}$, from different crystals. In the former case, this generally means scaling together reflections that are related by symmetry in reciprocal space, such as $hk\ell$ and $-hk-\ell$ for a monoclinic crystal, or $hk\ell$ and $-h-k-\ell$ for any crystal due to Friedel's law (if we ignore anomalous dispersion effects). For multiple crystals, this may mean the reflections collected from several ostensibly identical crystals needed to produce a full data set, or similar but not entirely identical crystals as is necessary for isomorphous replacement phase determination. The latter approach will also be required when comparisons are made of protein complexes involving ligands. Scaling is also necessary when the observed diffraction pattern, composed of F_{obs}, is compared with the calculated pattern, F_{calc}, for example, to compute an R factor, an $F_{obs} - F_{calc}$ difference Fourier synthesis, or for use in crystallographic refinement (see Chapters 8 and 10).

Now, one might think that this should not be a serious problem, the scaling of reflections, but it is, and furthermore it often serves as one of the most serious sources of error in the creation of a data set when it is done improperly or carelessly. If we consider, for example, the $hk\ell$ and $-hk-\ell$ symmetry equivalent reflections from a monoclinic crystal, or the $hk\ell$ and $-h-k-\ell$ reflections from any crystal (again ignoring anomalous dispersion) we would expect any measured intensity $I_{hk\ell}$ to be exactly equal to its symmetry or Friedel mate. They seldom are, however, and we may consider the reasons why.

First of all, there is simple random experimental error due to the fact that our measuring devices, our detectors, are not perfect. There is always some inexactness in accumulating the photons that comprise a given $I_{hk\ell}$, an intensity. This is nonetheless the least of our worries. Random experimental error can be minimized simply by measuring the equivalent reflections numerous times and averaging the independent measurements, namely by collecting the data with a high redundancy. By doing this, we can generally reduce the experimental error to about 3% or less of F_{obs} for strong and moderately intense reflections. The error will be higher, of course, for weaker reflections close to the background level, which are more difficult to accurately measure. An estimated error σ can be calculated for each final $F_{hk\ell}$ or $I_{hk\ell}$ based on counting and scaling statistics, and this provides a measure of its reliability. Usually we rate (or weight) reflections in calculations according to the ratio of $F_{hk\ell}/\sigma$; very precise reflections having larger values, and less reliable reflections having low values. Generally, we would view reflections with $F_{hk\ell}/\sigma < 3$ as somewhat suspect but nonetheless useful.

The more significant problems in scaling (or merging) X-ray diffraction data arise not from random error but from systematic error, and here the sources, unfortunately, are legion. Consider first the equivalent reflections from a single crystal that must be merged together to give a collective value for $F_{hk\ell}$. There is the problem of crystal deterioration with exposure time. While $I_{hk\ell}$ might have been measured early on, $I_{-h-k-\ell}$ could have been recorded much later, when crystal decay had become significant. They should ideally have been the same, but they will not be due to the physical change in the crystal.

A crystal is never a perfect sphere, it has some geometric shape. As it rotates in the X-ray beam, different volumes of the whole crystal may be exposed. As a consequence $I_{hk\ell}$ will not be equal to $I_{-h-k-\ell}$ as we supposed. The same effect is produced by absorption because the diffracted X-rays must pass through different volumes of the crystal, or different volumes of the glass capillary or solvent, liquid, or ice, that surrounds the crystal.

The shape of a diffraction spot, or intensity, on the face of a detector is to some extent the projection of the illuminated part of the crystal in a specific direction; that is, the spot shape reflects the crystal shape. As the crystal moves in the beam, the spot shapes on different sectors of the detector face change as well. In addition the spots become increasingly elongated at higher $\sin\theta$. The background also varies as a function of crystal orientation and position on the detector. Finally (but not really), a crystal may not scatter with the same strength, or average intensity, in all directions (along h, k, and ℓ), and this gives rise to a non-isotropic distribution of intensity.

Given all of these effects, it is a wonder that the data sets we obtain are as precise as they are. It is a testament to the patience and persistence of generations of crystallographers that these problems have been dealt with successfully. Current software normally addresses all, or most of these factors, and their effects are now more or less transparent to most practicing crystallographers. Nonetheless, it is wise to know that all these problems are present in the background, and they constantly present a danger to a structure determination if they are mishandled.

Scaling diffraction data from separate crystals, or scaling observed and calculated intensities or structure amplitudes is even more fraught with problems. Consider two separate crystals whose data were collected in two separate experiments, and remember that there are frequently several crystals. First of all, the independent data set from each crystal is subject to all the factors recounted above, and it is unlikely that any two will be identically affected. The two crystals may have had different geometrical properties, different amounts of liquid around them, different rates or allowances of radiation damage, or they

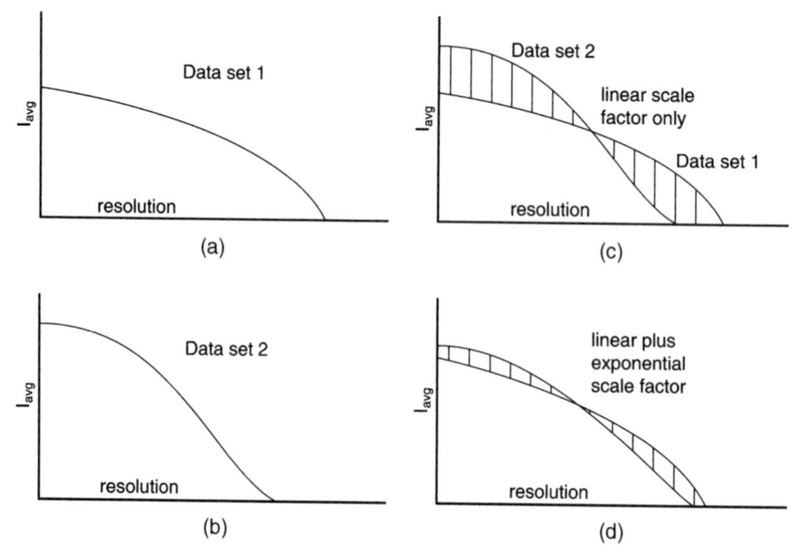

FIGURE 7.9 In (*a*) and (*b*) are the resolution dependencies of two X-ray data sets recorded from two different but isomorphous crystals that show a clear difference in the rates at which average intensity declines with $\sin\theta$. The difference may be due to differences in crystal quality, decay rates, or many other factors. If the two data sets are scaled to one another using only a linear scale factor, the result in (*c*) is obtained, in which the crosshatched area represents the scaling mismatch, or residual. If a linear plus exponential scale factor is used, the falloff of the second data set is made to more closely approximate that of the first set, and the result in (*d*) is obtained in which the residual is substantially reduced.

may have diffracted with different degrees of non-isotropy. The data sets from the two crystals will not, $F_{hk\ell}$ by $F_{hk\ell}$, be identical by any means even if all other factors were equal, and they are not. If the two crystals decline differently in average intensity as a function of $\sin\theta$, reflecting different degrees of inherent order, then they cannot be merged using a simple linear scale factor, meaning $F_{hk\ell-1} = K(F_{hk\ell-2})$. An example is illustrated in Figure 7.9. This can be improved upon by adding an exponential term so that $F_{hk\ell-1} = (K \exp[-K' \sin^2\theta/\lambda^2])F_{hk\ell-2}$. These expressions are almost always approximations, however, and so never exact.

To judge the degree of agreement between multiple, independent measurements of a reflection $F_{hk\ell}$, we can again calculate a residual, or R factor, for the scaling process. This, of course, is done after all data correction has been completed and scale factors applied. The R_{merge}, as it is called, is a summation over all independent reflections $F_{hk\ell}$, and it incorporates measurements from all crystals from which data were collected. It is given by

$$R_{merge} = \frac{\sum_{hk\ell}\sum_j |F_{hk\ell-j} - \hat{F}_{hk\ell}|}{\sum_{hk\ell}\hat{F}_{hk\ell}},$$

where $F_{hk\ell-j}$ is the jth measurements of $F_{hk\ell}$ and $\hat{F}_{hk\ell}$ is the average value for all measurements of $F_{hk\ell}$.

The greatest problem in scaling multiple crystals, however, is non-isomorphism, and this may arise from several sources. If the crystals were not grown from the same batch of protein, grown under the same crystallization conditions, or if they were exposed to different reagents, then non-isomorphism may result. Assumptions of isomorphism upon diffusion of ligands or heavy atoms into crystals are seldom met. Conformation changes occur, side chains shift, and water structure is disturbed. Even if the crystals are from the same sample and are otherwise identical, their defect structures may be different, producing different mosaicities for the diffraction intensities. Even mounting of the crystals for data collection may introduce discrepancies.

The most common procedure for data collection currently in use involves freezing the crystals by plunging them into liquid nitrogen, or into the gas stream of a cryogenic cooling device. No two crystals freeze the same, virtually all suffer some damage, and it is almost too much to expect that multiple, frozen crystals will all diffract with good isomorphism. The lesson to be learned from this is that as much data as is reasonable (judged by statistics) should be collected from a single crystal, and that the number of crystals scaled together and merged be minimized. This often affects, sometimes determines, for example, data collection strategies.

REAL SPACE AND DIFFRACTION SPACE

Although we have repeatedly discussed the interplay of real space and diffraction space, described many of their properties, and seen many examples in Chapter 6, it may be useful to review their relationship once again before proceeding. Table 7.1 outlines the principle relationships, some in terms of precise quantitative or mathematical terms and others in terms of what one observes in diffraction space as a consequence of properties or events in real space. In simplest terms we may think of real space as the physical crystal, the unit cells which make it up, and the distribution of atoms that fill the unit cells. We should in practice think of reciprocal space as diffraction space, or the pattern of diffraction intensities produced by a crystal exposed to X-rays, and the points in space where they may be observed.

Virtually everything that exists or happens in real space has a corresponding property or effect in diffraction space, and vice versa. The correspondences are established through the Fourier transform, which, as we have seen, operates symmetrically in both directions, getting us from real space into reciprocal space and back again. It may occasionally appear that this rule is violated, but in fact it is not. For example, the chirality of molecules and the handedness of their arrangement in real space would seem to be lost in reciprocal space as a consequence of Friedel's law and the addition of a center of symmetry to reciprocal space. If, however, we could record phases of reflections in reciprocal space, we would see that in fact chirality is preserved in phase differences between otherwise equivalent reflections. The phases of $\boldsymbol{F}_{hk\ell}$, for example, are ϕ, but the phase of $\boldsymbol{F}_{-h-k-\ell}$ are $-\phi$. Fortunately the apparent loss of chiral information is usually not a serious problem in the X-ray analysis of proteins, as it can usually be recovered at some point by consideration of real space stereochemistry.

In Table 7.1, where real and reciprocal cell dimensions, or other distances are related, an orthogonal system is assumed for the sake of simplicity. For nonorthogonal systems, the relationships are somewhat more complicated and contain trigonometric terms (as we saw in Chapter 3), since the unit cell angles must be taken into account. Rotational symmetry is preserved in going from real to reciprocal space, and translation operations create systematic absences of certain reflections in the diffraction pattern that makes them easily recognized. As already noted, because of Friedel's law a center of symmetry is always present in diffraction space even if it is absent in the crystal. This along with the absence of

TABLE 7.1 Real Space ——⇒ Fourier Transform ⇐—— Reciprocal Space

Real Space	Reciprocal Space
1. Structure factor equation	Electron density equation
2. Families of planes with Miller with indicies ***hkl***	Reciprocal lattice points indicies ***hkl***
3. Crystallographic unit cell	Reciprocal unit cell
4. Distribution of atoms about planes in the family with indicies ***hkl***	Intensity and phase of the diffraction spot at the reciprocal lattice point with indices ***hkl***
5. Distance ***d***	Distance $\mathbf{1/}\boldsymbol{d}$
6. Angle $\boldsymbol{\alpha}$ in real space	Angle $\mathbf{180} - \alpha$
7. Unit cell dimensions $\|\mathbf{a}\|$, $\|\mathbf{b}\|$, $\|\mathbf{c}\|$	Reciprocal cell dimensions $\mathbf{a^* = 1/\|a\|,\ b^* = 1/\|b\|,\ c^* = 1/\|c\|}$
8. Unit cell volume $\mathbf{V = a \times b \times c}$	Reciprocal unit cell volume $\mathbf{V^* = 1/\|a\| \times 1/\|b\| \times 1/\|c\| = a^* \times b^* \times c^*}$
9. Rotational symmetry	Same rotational symmetry
10. Translational operators	Systematic absences
11. No inversion symmetry at origin	Center of symmetry at origin
12. Chiral symmetry in space group or symmetry molecules	Phase difference of equivalent reflections
13. Orientation of unit cell, directions of unit cell axes **a, b, c**	Orientation of reciprocal cell axes $\mathbf{a^* \parallel a,\ b^* \parallel b,\ c^* \parallel c}$
14. Origin of real space chosen in the crystallographic unit cell	Origin may be anywhere
15. Interplaner spacing **d** for a family of planes ***hkl***	Distance of point ***hkl*** from the origin of reciprocal space
16. Frequency of sampling of the unit cell contents by a family of planes ***hkl***	Resolution of intensity ***hkl***
17. Magnitude of temperature factors **B** of atoms in the unit cell	Rate of decline of average intensity **I** with $\mathbf{sin}\,\boldsymbol{\theta}$
18. Pseudo symmetry of atom distribution in the unit cell	Pseudo symmetry of intensities of reflections
19. Anisotropy of ordering of atoms in crystallographic unit cell	Anisotropy of average intensity
20. Variation in ***d*** spacing for a family of planes due to imperfections of ordering, or crystal defects	Volume of diffraction spot, mosaicity

translational operations in reciprocal space leads to an apparent degeneracy there of space group symmetry. The consequence is that the symmetry of diffraction space, the symmetry that relates the intensities, is the point group of the actual space group, after a center of symmetry, or inversion center, has been added. This degenerate symmetry group is called the Laue group of the crystal.

It is wise to remember that every operation, mathematical or physical, in real or reciprocal space has an equivalent operation in the other. Often these are not obvious, but they are always there. Molecular replacement, for example, can in theory be carried out in either space, though our computational tools for doing so are much more powerful in reciprocal space. Similarly, structure refinement may be carried out using least squares procedures in reciprocal space or, equivalently, difference Fourier methods in real space.

Finally, we see from the Fourier transform equations, for the structure factor $\boldsymbol{F}_{hk\ell}$ and the electron density $\rho(x, y, z)$, that any change in real space (e.g., the repositioning of an atom) affects the amplitude and phase of every reflection in diffraction space. Conversely, any change in the intensities or phases in reciprocal space (e.g., the inclusion of new reflections) affects all of the atomic positions and properties in real space. There is no point-to-point correspondence between real and reciprocal space. With the Fourier transform and diffraction phenomena, it is "One for all, and all for one" (Dumas, *The Three Musketeers*, 1844).

CHAPTER 8

SOLVING THE PHASE PROBLEM

The phase problem might initially seem intractable, since, in the absence of a center of symmetry, the phase of any reflection $\boldsymbol{hk\ell}$ may lie anywhere between **0°** and **360°**, and there is nothing apparent in the diffraction pattern to indicate the values of any of them. If we can only record intensity information, then how can phases ever be obtained? There are, however, some ameliorating circumstances, chief among them being that we do not, in calculating a Fourier summation, require exact values for phases. If we are close, that is, if we have reasonable estimates, usually that is good enough. We generally have so many $\bar{\boldsymbol{F}}_{\boldsymbol{hk\ell}}$ to include in the Fourier summation that we can tolerate rather large phase errors for individual reflections. Many may even be entirely wrong. The Fourier transform has the important property that correct phase (and intensity) information tends to reinforce and sum constructively to present a coherent image. On the other hand, incorrect phase (and intensity) information tends only to sum to a more or less uniform, though fluctuating, background.

How clearly the Fourier image emerges from the background (i.e., the contrast) is principally a function of the quality of the phases, and to a lesser extent the intensities. If phases for only some $\bar{\boldsymbol{F}}_{\boldsymbol{hk\ell}}$ are correct, or if most $\bar{\boldsymbol{F}}_{\boldsymbol{hk\ell}}$ have a large mean phase error, the electron density image produced by a Fourier will be lost in background noise. If, on the other hand, the average phase error is modest, or if a large number of reflections have accurate phases, then the image will stand out above the background. As will be seen below, in almost all cases where we can solve the phase problem, we do so by somehow obtaining rough estimates of the phases for most of the $\bar{\boldsymbol{F}}_{\boldsymbol{hk\ell}}$, and then gradually improving the quality of the phases by reducing the mean phase error.

Ultimately we can solve the phase problem for a number of reasons: (1) We have chemical and physical information about the structure of the molecules making up the crystal that we can use to interpret and improve even a poor electron density image. (2) Information actually does reside in the intensity distribution alone, cryptic information about the relative phases

Introduction to Macromolecular Crystallography, Second Edition By Alexander McPherson

of sets of reflections that may be used in some instances. (3) We can devise methods that allow us, from slightly altered diffraction patterns obtained from experimentally altered crystals, to obtain phase information. (4) We can take advantage of the fact that similar molecular structures yield similar diffraction patterns. (5) We can obtain phase information experimentally by using internal wave interference phenomena that are characteristic of certain kinds of atoms, namely anomalous dispersion. (6) Finally, we can obtain phase information when the asymmetric unit contains noncrystallographic symmetry.

Based on the ideas above, and with the exception of some unique cases that present unusual opportunities, the kinds of approaches available to us for solving the phase problem are as follows:

1. Deconvolution of the Patterson map
2. The heavy atom method
3. Isomorphous replacement
4. Anomalous scattering
5. Molecular replacement
6. Direct methods
7. Noncrystallographic symmetry and density modification

PATTERSON METHODS

The Patterson synthesis (Patterson, 1935), or Patterson map as it is more commonly known, will be discussed in detail in the next chapter. It is important in conjunction with all of the methods above, except perhaps direct methods, but in theory it also offers a means of deducing a molecular structure directly from the intensity data alone. In practice, however, Patterson techniques can be used to solve an entire structure only if the structure contains very few atoms, three or four at most, though sometimes more, up to a dozen or so if the atoms are arranged in a unique motif such as a planar ring structure. Direct deconvolution of the Patterson map to solve even a very small macromolecule is impossible, and it provides no useful approach. Substructures within macromolecular crystals, such as heavy atom constellations (in isomorphous replacement) or constellations of anomalous scatterers, however, are amenable to direct Patterson interpretation. These substructures may then be used to solve the phase problem by one of the other techniques described below.

THE HEAVY ATOM METHOD

The heavy atom method is not applicable to macromolecules, again because of the very large number of atoms involved. It is, however, a very useful approach for the solution of conventional crystal structures having as many as 50 or so atoms in the asymmetric unit. For about the first 40 years of X-ray crystallography it was really the only practical method for deducing crystal structures.

The heavy atom method for structure determination, though not useful for macromolecular crystals, is an easily understandable approach and it is illustrative of the ideas and devices that are incorporated into the other, more applicable techniques. The heavy atom method only works, it should be said at the outset, when the molecule under study contains at least one atom that has significantly more electrons than all the other atoms in the molecule, meaning at the very least two or more times the number of electrons. This is common in organic molecules, which may contain a sulfur or halogen atom, or, for example, a metal coordination complex. Heavy atom coordinates (x, y, z) for all kinds of crystals, both conventional and macromolecular, can be obtained directly by Patterson techniques as will be described in the next chapter. Knowing the coordinates of the heavy atom in the unit cell is crucial to application of the method.

Even for a conventional crystal, one is not generally interested in only knowing the positions of the heavy atoms in the unit cell. The distribution of light atoms is usually the objective of the analysis. Once the heavy atom coordinates are known, however, it is usually possible to derive positions for the lighter atoms in the structure using an iterative, interactive approach, without further reliance on Patterson interpretation.

The structure factor $\bar{F}_{hk\ell}$ for any $hk\ell$ is a wave which can be represented as a vector composed of scattering contributions $\bar{f}_j$ for every atom j in the unit cell, where the individual $\bar{f}_j$ are also vectors. In Figure 8.1, the vector sum of the $\bar{f}_j$ for all of the light atoms, $\sum_j \bar{f}_j$ is usually a resultant vector of modest length. This is because the sum of any collection of small, essentially arbitrary vectors tends to be small (waves of random phase tend to destructively interfere rather than constructively interfere; for vectors, a random walk never

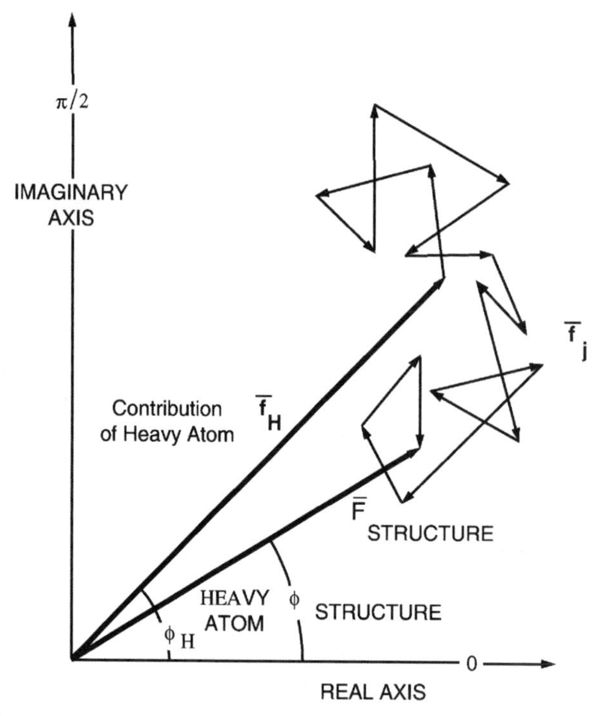

FIGURE 8.1 $\bar{F}$ is the structure factor for an asymmetric unit containing a single heavy atom, along with many light atoms. Thus $\bar{F}$ is the sum of the scattering vectors (diffracted waves) for all the atoms in the structure. The sum of the many short uncorrelated vectors $\bar{f}_j$ is usually small; hence the large contribution $\bar{f}_H$ of the single heavy atom dominates the summation. The resultant vector $\bar{F}$ has a magnitude and phase close to $\bar{f}_H$, the heavy atom contribution alone. Hence the phase of $\bar{f}_H$ serves as a good approximation to that for the ensemble in an initial Fourier synthesis.

leads far from the starting point), and the total structure factor $\bar{F}_{hk\ell}$ will be dominated by the $\bar{f}_H$ of the single heavy atom. The phase ϕ of $\bar{F}$ will be reasonably close to the phase ϕ_H of the heavy atom contribution alone.

Once the position of the heavy atom is known, ϕ_H can be computed from the structure factor equation using only the heavy atom contribution, that is, its coordinates. The ϕ_H can then be used as an approximation to the phase angle for the entire structure ϕ in a Fourier synthesis $\rho(x, y, z)$. This synthesis of the electron density will have coefficients F directly measured, from the diffraction pattern, and phases ϕ_H, calculated from the heavy atom. In this hybrid electron density map the heavy atoms will appear as intense peaks, since their contribution to both the total structure amplitude F and phase ϕ will have been properly accounted for. The light atoms, however, will have contributed only to the observed structure amplitudes but not to the calculated phases. They will therefore appear as only weak electron density peaks, a fraction of their true value, in the Fourier map, and many may not be found at all. Light atoms, whose positions can tentatively be identified in the map (and are not in disagreement with what is known about the structure from chemical considerations), are then combined with the heavy atom in the structure factor equation to derive a better approximation to ϕ, and a new cycle of Fourier synthesis is computed with the improved phases. In succeeding Fourier maps as the calculated estimate of the phase more nearly equals the correct phase ϕ, additional light atoms emerge. They are then included to produce increasingly better phase approximations until the coordinates of all of the atoms in the unit cell can be identified in electron density maps, and a model of the structure created.

This heavy atom method is a "bootstrap" approach where approximate phases, obtained from the heavy atom alone, are gradually improved as light atoms are found and added. The calculations alternate between real and reciprocal space, between $\rho(x, y, z)$ and $\bar{F}_{hk\ell}$, respectively, and improvements made in one space are used to affect improvements in the other. At the end of this process the investigator is in a position to take advantage of one of the most powerful advantages offered by X-ray crystallography, a means of checking the final solution, or model, against the true solution, the actual distribution of atoms in the unit cell.

THE *R* FACTOR AND CRYSTALLOGRAPHIC REFINEMENT

It was demonstrated in Chapter 5, where the structure factor equation was derived, that if the positions of all of the atoms x_j, y_j, z_j and their scattering factors f_j are known for all atoms j in the asymmetric unit, then we can accurately calculate all $\bar{F}_{hk\ell}$ (setting aside for the moment the thermal factor correction). Furthermore we know that $|\bar{F}_{hk\ell}| = \sqrt{I}_{hk\ell}$, and therefore we can directly calculate from a model what the intensity $I_{hk\ell}$ should be for any and every reflection $hk\ell$ in the diffraction pattern. This being the case, what is needed is a statistical measure of how well the actual, recorded diffraction pattern, or the $\sqrt{I} = F_{obs}$ derived from it, compares with the calculated F_{calc} computed from the atomic model. This measure of agreement (or disagreement really) is called the crystallographic residual R, or "R factor" in the vernacular of the street.

The R factor is defined as

$$R = \frac{\sum_{hk\ell} ||\bar{F}_{hk\ell}|_{obs} - |\bar{F}_{hk\ell}|_{calc}|}{\sum_{hk\ell} |\bar{F}_{hk\ell}|_{obs}}.$$

If the model derived from the diffraction data were correct in every way, and the measured data were perfect, then the agreement would be exact: $F_{hk\ell-obs}$ would always equal $F_{hk\ell-calc}$ and $R = 0$. But this is never the case. The data, of course, contain measurement errors, the atomic positions may be accurate and precise, but still not perfect, temperature factors and the ellipsoids of vibration may be only approximate, and so on. In general, even for a very well and correctly determined structure, R will commonly be in the range of 0.05 to 0.10 for a conventional crystal having an asymmetric unit of 50 or so atoms. For macromolecular structure determinations, R is normally in the range of 0.15 to 0.25. The R factor is in most cases the ultimate criterion of model quality at the resolution it was determined.

It so happens that the R factor is useful not only in telling us when we have the correct solution, and how good it is. The R factor is also invaluable in directing us from an imperfect model, or a very approximate structure solution, to the best model that we can obtain. This occurs in the process known as crystallographic refinement (see Chapter 10).

Using any phasing method, an electron density map is ultimately computed that reveals the positions x_j, y_j, z_j of all of the atoms j in the asymmetric unit. The essential physical and chemical features of the structure then become apparent. Detailed information, however, such as accurate bond lengths and angles, positions of hydrogen atoms, and deviations from planarity, may require a higher level of precision. This is readily obtained for a conventional small molecule by refining the coordinates and thermal parameters of each atom in the structure by one of several techniques. Refinement, which is discussed in more detail in Chapter 10, is accomplished in reciprocal space by a comparison of structure amplitudes calculated from the imprecise trial structure F_{calc}, with those actually observed to be produced by the crystal F_{obs}. That is, by computing the R factor. The various structural parameters x_j, y_j, z_j, B_j are then adjusted as a mathematical (least squares) procedure dictates to bring about closer agreement, and the minimization of R.

Refinement also may be accomplished in real space by calculating difference Fourier syntheses using as coefficients $\Delta F = F_{obs} - F_{calc}$ and phases ϕ_{calc} derived from the trial structure. In general, if an atom has been incorrectly placed near its true location, a negative peak will appear at its assigned position and a positive peak will appear at its proper location. That is, in the difference Fourier we have subtracted electron density from where the atom isn't, and not subtracted density from where it is. The atom is then shifted by altering x, y, z, improved atomic parameters are included in a new round of structure factor calculations, and another difference Fourier computed. The process is repeated until the map is devoid of significant features.

In recent times, means have been devised to simultaneously refine not only the crystallographic R factor, which is a comparison in reciprocal space, but real space functions as well. The real space functions may measure the degree by which the atomic model fails to display ideal geometry (Hendrickson and Konnert, 1981; Sussman et al., 1977) or the difference between its potential energy and the minimum energy (Brunger et al., 1987). It is important to note that although these very powerful mathematical procedures have led us to increasingly more accurate structures at a faster rate, they can, at times, obscure errors present in crystallographically determined structures. This will be discussed more fully in Chapter 10.

ISOMORPHOUS REPLACEMENT

Although useful for conventional low-molecular-weight compounds that contain a heavy atom as part of their structure, successive Fourier approximation methods like that described above cannot be used to deduce phase information for structure amplitudes measured from protein crystals. This is due to the fact that the scattering amplitude from a collection of several thousand light atoms (carbon, oxygen, nitrogen) is no longer dominated by the contribution of one or a few heavy atoms. That is, the phase of the structure factor from a heavy atom derivative of a protein crystal is no longer approximately the same as that of the heavy atom component.

The major advance in the determination of protein structures by X-ray diffraction was achieved by M. F. Perutz (Boyes-Watson et al., 1947; Bragg and Perutz, 1954) who recognized that a conventional but little used technique, sometimes employed to solve small molecule crystals, might be applicable to crystals of macromolecules. He realized that protein crystals, composed of about 50% solvent, might, through diffusion, permit reactive heavy atom compounds to reach specific amino acid side chains on the molecules. By using different heavy atom compounds and soaking conditions, such a procedure could allow creation of an isomorphous series of derivatized crystals. The crucial requirement in this method is that each derivative in the series should be structurally identical, isomorphous, with the native crystal structure, with the single exception of the presence of the heavy atom.

The underlying idea behind isomorphous replacement as a means for obtaining estimates of the unknown phases of waves whose amplitudes alone could be measured is quite old. The same principle has been used in acoustics, in many fields of interferometry with visible light, and a host of other wave-dependent applications. At its base is the idea of combining, or interfering a wave of measurable amplitude but unknown phase with a standard wave whose amplitude and phase are both known, and then measuring the amplitude of the result. To see how this might occur in the laboratory, examine the example illustrated in Figure 8.2.

We have an unknown wave passing from left to right through our laboratory and parallel with our laboratory benchtop, or maybe we have moved our bench so that it is. We can measure its amplitude easily. If it is a sound wave, we can just listen to it or measure its loudness with some device (e.g., the one they use on talent shows to measure audience approval). That defines the amplitude of the unknown wave but says nothing about its phase. If the wave is light instead of sound, we can record it on film as we do X rays. The darkening of the silver grains, the intensity, gives us a measure of its amplitude. A scintillation tube would do as well, again yielding the wave amplitude but providing no information about its phase.

At the left end of the bench we set up an oscillator, a device that generates waves of specified frequency and amplitude. We assign the position of the oscillator to be the origin of our system. We must define an origin, since no wave has a meaningful phase except with respect to some origin, or relative to some other wave whose phase is zero. Because our oscillator sits exactly at the origin, the wave it emits not only has a specified amplitude (and the same frequency, which we set, as the unknown wave) but its phase angle is zero. It is a reference wave.

If we have set up our apparatus properly, the reference wave that we generate, and the unknown wave whose phase we wish to measure, will combine and interfere as we now know all waves to do. Now let us assume we have placed the device that records amplitude at the other end of the lab bench and we measure the wave amplitude that results from the two waves interfering. We call this process "beating waves." If the two waves (the reference wave and the unknown wave) were perfectly in phase, the amplitude of the recorded wave would be precisely the sum of the amplitudes of the reference and unknown waves, both of which we know. If the two waves were exactly out of phase, then the observed wave

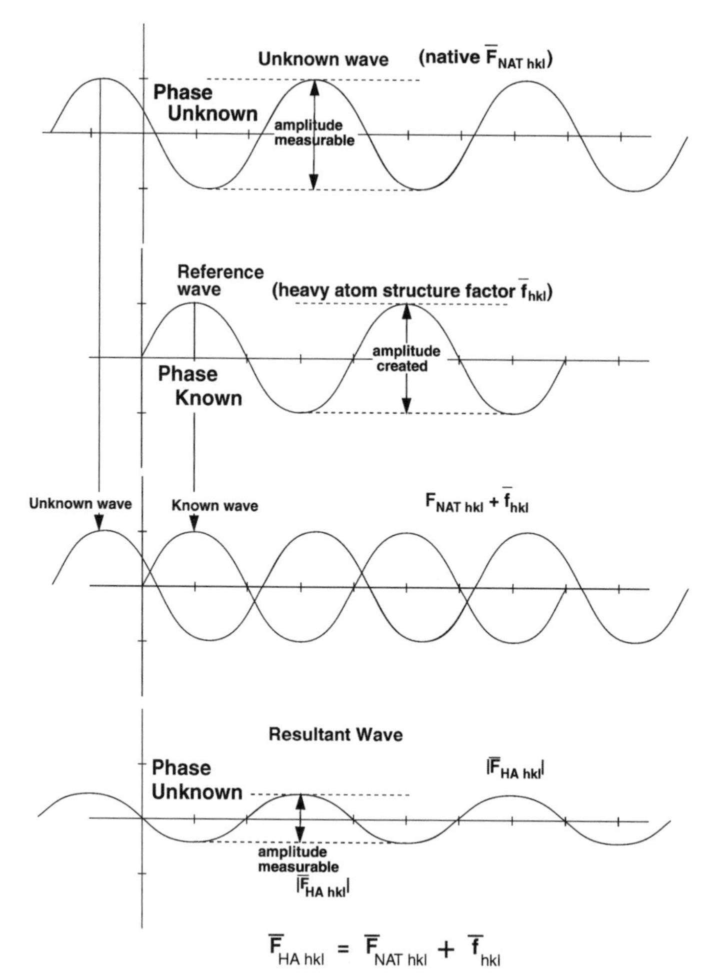

FIGURE 8.2 The unknown phase of a wave of measurable amplitude can be determined by "beating" it against a reference wave (created experimentally) whose phase and amplitude are both known. The unknown wave at top and the reference wave below are allowed to interfere with one another to produce the resultant wave at bottom. Although the phase of the resultant wave cannot be measured, its amplitude can. That resultant amplitude is a function of the amplitudes of the unknown and reference waves (both known or measurable), the phase of the reference wave (experimentally set to zero), and the phase of the unknown wave, which is in question. In isomorphous replacement the unknown wave is the native structure factor (measured), the reference wave is the structure factor of the heavy atom alone (calculated from its position), and the resultant wave is the structure factor of the heavy atom derivatized crystal (measured).

amplitude would be the difference of the amplitudes of the unknown and reference waves. More likely it would be somewhere in between.

In this experiment we have three known amplitudes and one known phase, that of the reference wave. Although we do not know the phase of the observed wave, we can nonetheless figure out what the phase angle of the unknown wave must have been in order to have given us the resultant amplitude. We can deduce what its phase angle must have been so that when it interfered with a wave of zero phase and known amplitude, their sum produced a wave having the measured, resultant amplitude.

Indeed, if we were to make the determination as described here, we would find that there were two possible phases that the unknown wave could have had with respect to the reference wave and yet give the same resultant amplitude. These correspond to phases of ϕ and $-\phi$. Such an ambiguity is always true of this technique for phase determination. It always yields two acceptable solutions at plus and minus ϕ with respect to the phase of the reference wave. It is true for the isomorphous replacement method as well. In our laboratory we could break the phase ambiguity by simply moving our oscillator to another point along our bench and repeating the experiment, but keeping the origin the same. That is, we change the phase of the reference wave from what it was in the first experiment. When we carried out the repetition, we would find that one, and only one, of the phase solutions would be the same as one of those found in the original experiment. This would be the correct phase.

How does the laboratory experiment described above resemble isomorphous replacement with protein crystals and X rays. Actually it is the same. It is illustrated in Figure 8.3. The crystal is our laboratory bench and the X-ray film or detector serves the same purpose as our audience applause meter; it measures the amplitude of waves. What are the other analogous

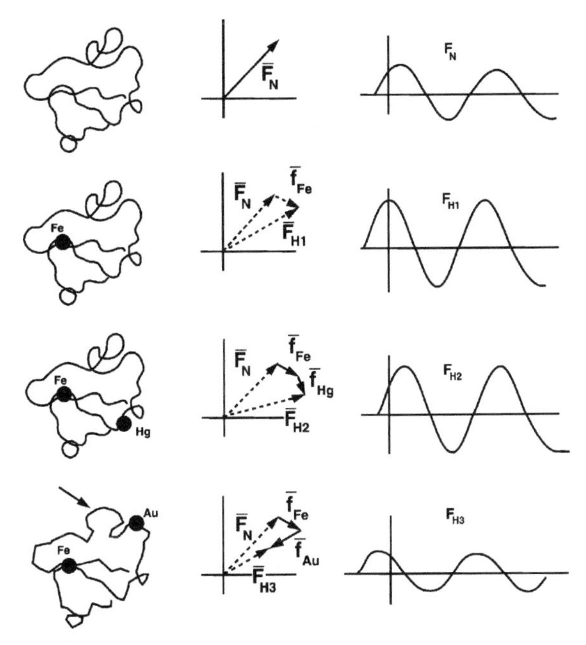

FIGURE 8.3 A hypothetical series of isomorphous heavy atom derivatives for a crystalline macromolecule, represented here by the polypeptide backbone of rubredoxin. (*a*) The apo-protein, stripped of its metal ion, provides native structure factors $\bar{F}_N$, shown in vector and waveform on the right; (*b*) the protein with its naturally bound iron atom and $\bar{F}_{H1}$, the first derivative structure factor; (*c*) the protein with its iron plus an attached mercury atom, and the resultant structure factor $\bar{F}_{H2}$ from the double derivative; (*d*) a second multiply substituted derivative formed by attachment of a gold atom to the protein–iron complex. This last derivative is only marginally useful, however, since the reaction with gold also produces a modification in the tertiary structure of the protein (denoted by an arrow). Since this non-isomorphism is equivalent to introducing a nonnative structure factor contribution, the observed $\bar{F}$'s cannot be properly accounted for, and an erroneous heavy atom contribution $\bar{f}$ results. This final derivative will yield an inaccurate phase estimate ϕ_N for the native protein.

components? The unknown wave whose phase we wish to measure is the native structure factor, or diffracted wave from a particular family of planes $\boldsymbol{hk\ell}$ in the crystal; it is $\bar{F}_{hk\ell}$. Now how can we introduce a source for a reference wave whose amplitude and phase are both known to us, and what would the resultant wave be?

We could, like Perutz, introduce a heavy atom into each unit cell of the crystal, say by attaching a mercury atom to a specific sulfhydril group on every asymmetric unit. If the crystal were then exposed to X rays, the heavy atom would strongly scatter the radiation because of its high atomic number. The wave scattered by this heavy atom $\bar{f}_{HA}$ would interfere with the wave produced by all of the other protein atoms in the crystal, the native $\bar{F}$, and it would produce a resultant wave that was the interference sum of the two. The heavy atom in the unit cell provides the reference wave, and scattering in concert with the native atoms in the crystal, it would produce a wave that has the structure amplitude of the heavy atom derivatized crystal. This resultant amplitude, like that for the native crystal, can be measured by recording the diffraction pattern of the derivitized crystal. In vector notation we have

$$\bar{F}_{hk\ell-deriv} = \bar{F}_{hk\ell-nat} + \bar{f}_{hk\ell-HA},$$

where $\bar{f}_{-HA}$ is the structure factor (the wave contribution) of the heavy atom alone in the unit cell.

A problem remaining is that we must know the phase and amplitude of the reference wave provided by the heavy atom. We can, however, calculate both of these directly from the structure factor equation if we know the atomic number of the heavy atom, which we can read off a periodic table of the elements, and if we know its x, y, z, position in the unit cell. Given this information, then we have at our disposal the amplitude of the unknown wave F_{nat}, the amplitude of the resultant wave F_{deriv}, the amplitude of the reference wave f_{HA}, and the phase of the reference wave ϕ_{HA}. We have the information required to solve for the phase angle of the native structure factor, the unknown wave, just as we did for the benchtop experiment above.

It may not be obvious how we would locate the x, y, z coordinates of the heavy atom in the unit cell. Indeed it is sometimes not a simple matter to find those coordinates, but as for the heavy atom method described above, it can be achieved using Patterson methods (described in Chapter 9). As we will see later, Patterson maps were used for many years to deduce the positions of heavy atoms in small molecule crystals, and with only some modest modification they can be used to locate heavy atoms substituted into macromolecular crystals as well. Another point. It is not necessary to have only a single heavy atom in the unit cell. In fact, because of symmetry, there will almost always be several. This, however, is not a major concern. Because of the structure factor equation, even if there are many heavy atoms, we can still calculate $\bar{f}_{hk\ell}$, the amplitude and phase of the ensemble. This provides just as good a reference wave as a single atom. The only complication may lie in finding the positions of multiple heavy atoms, as this becomes increasingly difficult as their number increases.

In our laboratory investigation a phase ambiguity arose when the experiment was performed only once. It was resolved simply by moving the oscillator to a new location and changing the phase of the reference wave. The analogous operation in X-ray crystallography is not quite so simple but still within reach. We must prepare a second (or even more, to be sure) heavy atom derivative bound at some other location in the unit cell. This second "oscillator position" generates then a second reference wave that yields, in the phase calculation, one phase solution in common with the first experiment.

FORMULATION OF ISOMORPHOUS REPLACEMENT IN PROTEIN CRYSTALLOGRAPHY

The structure factors $\bar{F}_{hk\ell}$ of a crystal, and therefore the observed structure amplitudes $F_{hk\ell-obs}$, depend only on the distribution of scattering material. Each $\bar{F}_{hk\ell}$ is the sum of the scattered waves from the individual atoms in the unit cell. If additional "heavy atoms" are introduced into the unit cell, and all else remains constant, the new resultant $\bar{F}_{hk\ell-deriv}$ will be the sum of the old, native $\bar{F}_{hk\ell-nat}$ plus the contribution of the wave scattered by the heavy atom. This is illustrated by a hypothetical case in Figure 8.3 and graphically in Figure 8.4. We are dealing with a phase-dependent interference phenomenon. Hence, when waves are added, the new $\bar{F}_{hk\ell-deriv}$ of the derivative structure may have an amplitude either greater or less than the $\bar{F}_{hk\ell-nat}$ for the native structure.

To simplify notation, $\bar{F}_{nat}$ will designate the set of native structure factors, $\bar{F}_{deriv}$ is the set of structure factors for the derivative structure, and $\bar{f}$ is the contribution of the heavy atom. The indexes $hk\ell$ are understood and considered the same for $\bar{F}_{nat}$, $\bar{F}_{deriv}$, and $\bar{f}$.

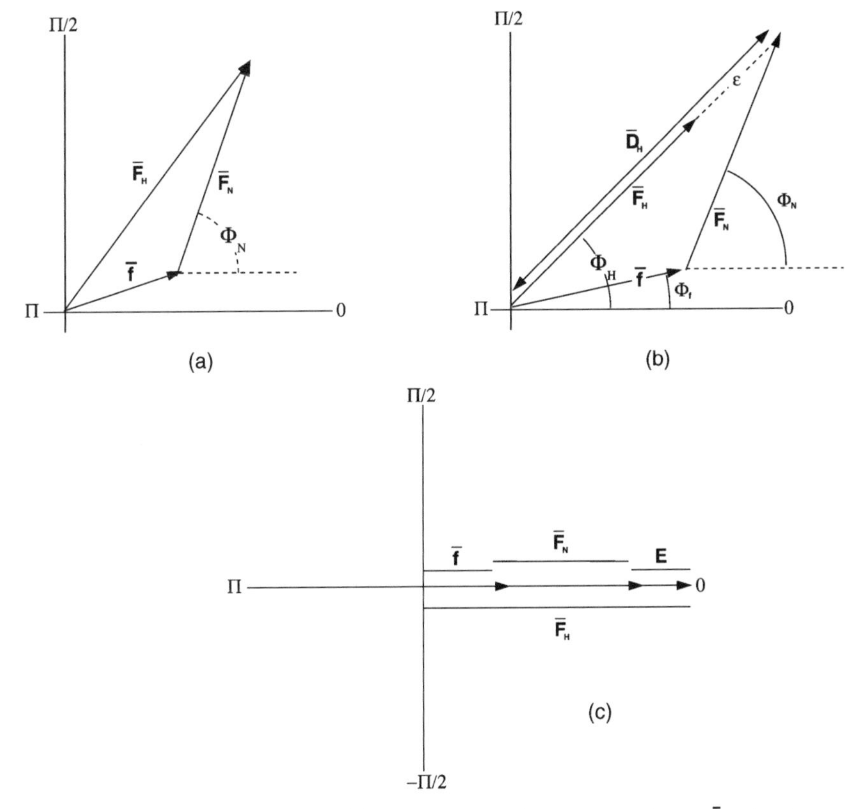

FIGURE 8.4 (*a*) The vector relationship among the native structure factor $\bar{F}_N$, the heavy atom contribution $\bar{f}$, and the isomorphous derivative structure factor $\bar{F}_H$. The phase angle ϕ_N of $\bar{F}_N$ is unknown. (*b*) Because of errors in the data, the phase triangle cannot be closed. The lack of closure error is denoted by ε, and $\bar{D}_H$ is the vector required to close the triangle. The most probable phase for ϕ_N occurs when ε is minimized. The terms ϕ_N, ϕ_f, and ϕ_H are the phase angles of $\bar{F}_N$, $\bar{f}$, and $\bar{F}_H$, respectively. (*c*) The centric case for which the circles are degenerate, since the phases must all be **0** or π. The lack of closure in centric zones of reflections can only be due to inaccuracies in the data, since ϕ_N is always exact. This provides a means for estimating the errors, E.

The diffraction patterns of an isomorphous pair are similar, but not identical, differing from one another, on average, by about 10% to 15%. The small differences reflect the scattering contributions, $\bar{f}$, of the heavy atom. The structure factor $\bar{F}_{nat}$ for a given native reflection $hk\ell$ may be represented by a vector of magnitude $|\bar{F}_{nat}|$ and phase angle ϕ_{nat}; the derivative structure factor $\bar{F}_{deriv}$ is similarly represented. Now $\bar{F}_{deriv}$ must be the vector sum of the native crystal structure factor and the structure factor contribution of the attached heavy atom $\bar{f}$. Figure 8.4 shows this vector relationship between $\bar{f}$, $\bar{F}_{nat}$, and $\bar{F}_{deriv}$. From the coordinates of the heavy atom in the unit cell (x, y, z) and its scattering factor, $\bar{f}$ can be calculated from the structure factor equation

$$\bar{f}_{hk\ell} = f[\cos 2\pi(hx + ky + lz) + i \sin 2\pi(hx + ky + lz)].$$

To solve this vector relationship, David Harker proposed the simple graphical method shown in Figure 8.5 for determining the native protein phase ϕ_{nat}, given $\bar{f}$, F_{nat}, and F_{deriv}

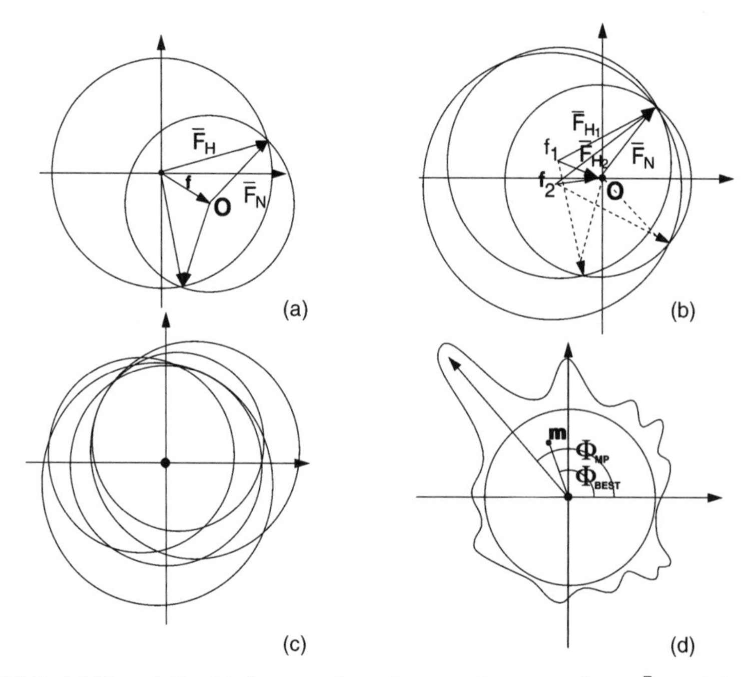

FIGURE 8.5 (*a*) The relationship between the native crystal structure factor $\bar{F}_N$ and that of an isomorphous derivative crystal $\bar{F}_H$, which contains a heavy atom that contributes a scattering component $\bar{f}$. There are two points of intersection of the circles in this Harker diagram, thereby yielding two different solutions for the phase of $\bar{F}_N$. In (*b*) two isomorphous derivatives are available with structure factors $\bar{F}_{H1}$ and $\bar{F}_{H2}$. There is only one point at which all three circles intercept; hence there is no ambiguity in the determination of the phase for $\bar{F}_N$. Both (*a*) and (*b*) are ideal cases. In (*c*) is a more realistic situation where several isomorphous derivatives each contain errors from measurement and other sources. These random errors in the data require implementation of a probabilistic determination of the correct phase of $\bar{F}_N$. (*d*) The statistical likelihood of a phase angle ϕ being correct is plotted around a **360°** unit circle. The most probable phase is ϕ_{MP}, but the electron density map with the least overall error is computed using coefficients weighted with m, the figure of merit, and phases ϕ_{BEST}, where m, ϕ_{BEST} are the polar coordinates of the centroid of the probability distribution shown in the figure.

(Harker, 1956). Note that although $\bar{F}_{nat}$ and $\bar{F}_{deriv}$ are unknown, F_{nat} and F_{deriv} can be measured directly. On a vector diagram in Figure 8.5*a* $-\bar{f}$ is plotted and circles (which can be thought of as vectors having all possible phases, or unknown phases) are drawn of radius F_{nat} and F_{deriv}, using the end point and origin of $-\bar{f}$, respectively, as the center. ϕ_{nat} is the phase of a vector drawn from the origin to one of the two intersection points of the circles. Both represent phases that satisfy the vector relationship.

The ambiguity we saw in our earlier experiment reappears, since there are again two solutions corresponding to ϕ_1 and ϕ_2, symmetrical with respect to $-\bar{f}$, where the vector relationship $\bar{F}_{deriv} = \bar{F}_{nat} + \bar{f}$ is valid. If the process is repeated, however, with a second heavy atom derivative having a different $-\bar{f}'$ and F'_{deriv}, two more intersection points will be found. Ideally one of these will be coincident with either ϕ_1 or ϕ_2, and the other will not. This double intercept will establish the correct phase for the native crystal. This is seen in Figure 8.5*b*.

ISOMORPHOUS REPLACEMENT IN PRACTICE

The expected change in intensity $I_{hk\ell}$ produced by the attachment of a heavy atom to a crystalline macromolecule was estimated by Crick and Magdoff (1956) to be given by the formula

$$\frac{\sqrt{\Delta I_{avg}^2}}{I_{avg}} \cong \sqrt{2}\left(\frac{N_H}{N_N}\right)^{1/2}\left(\frac{f_H}{f_N}\right),$$

where N_H is the number of heavy atoms introduced into the unit cell, f_H their scattering power, N_N the number of native atoms, and f_N their average scattering power. For a single gold atom attached to an asymmetric unit of M_r = 25,000 Daltons, an average difference might be expected of approximately 20% of the mean structure amplitude. While this appears substantial (experience has clearly shown that it is much overestimated), complications arise from measurement and scaling errors between the two sets of data of, on average, 2% to 5%. Slight structural changes produced by the introduction of the heavy atom may generate intensity change due to non-isomorphism, and there may be incomplete substitution at the available sites on all of the molecules in the crystal. All these complications contribute to a reduction in the phasing effectiveness of the isomorphous derivative.

In the discussion of single and multiple isomorphous replacement (called SIR and MIR respectively), we have to this point tacitly assumed all things to be ideal. In reality this is never the case, certainly not in X-ray crystallography. In terms of Harker diagrams this means that for virtually all $\bar{F}_{hk\ell}$, the circles for the first heavy atom derivative and the second heavy atom derivative never really intersect the native circle at a common point. In the best of circumstances they come close, but for many $hk\ell$ they don't. Figure 8.5*c* represents a more typical case, and Figure 8.5*d* how we have traditionally dealt with it (for details, see Dickerson, 1968; Blundell and Johnson, 1976; Blow and Crick, 1959; Drenth, 1999).

The reason for the nonideality and the necessity of a more sophisticated, probabilistic treatment is that the measured data $I_{hk\ell}$ contain random errors of several percent or more. There is invariably some non-isomorphism so that the shifted native atoms contribute to the derivative structure amplitude. $\bar{f}_{hk\ell}$ is imprecise because its occupancy and temperature factor may not be properly known, and even more important, its x, y, z coordinates may be uncertain. Scaling is never perfect; thus differences between native and derivative structure amplitudes may be systematically in error. Heavy atoms of low occupancy, called minor sites, may not have been identified and included in the calculation of $\bar{f}_{hk\ell}$. All in all, it would seem that the X-ray crystallographers have quite a mess on their hands.

To cut through this accumulation of innocent and/or unavoidable errors and tenuous assumptions, a number of clever approaches have been devised that take into account, and partially circumvent, the problems. If reasonable estimates of the errors associated with all quantities, such as intensity data, heavy atom coordinates, and non-isomorphism, can be obtained, then these approaches become particularly powerful (Read, 1997; de la Fortelle and Bricogne, 1997).

Methods for treating isomorphous replacement data, in practice, are mathematical in nature, employ probability and statistics to deduce the best possible phases $\phi_{hk\ell}$ for each $\bar{F}_{hk\ell}$, and assign to that phase some measure of its precision. In the Fourier syntheses used to produce electron density maps, the individual terms are then weighted with their likelihood of being accurate or according to their precision.

The first useful approach to formulating isomorphous replacement in probabilistic terms was that of Blow and Crick (1958), as modified by Dickerson and his colleagues (Dickerson et al., 1968). Since those days new ideas have entered the mix, and today there are a host of competing approaches, and most are improvements upon the early methods. For extensive treatments of the newer techniques, how they differ, and when they are most applicable, the reader is referred to Furey and Swaminathan (1997).

In practice, Harker diagrams, like those described in Figure 8.5 are not actually constructed. Instead, the probability of any phase on the circle being correct, given the structure amplitudes of two or more derivatives and the calculated structure factors of the heavy atoms, is computed at increments around the circle, say at every 5 or 10 degrees. Blow and Crick (1958) showed that the probablility formulation, given certain assumptions regarding the distribution of errors, for any reflection is

$$\rho(\phi) = \frac{\exp - \varepsilon(\phi)^2}{2E^2},$$

where $\varepsilon(\phi)$ is the lack of closure of the vector relationship $\bar{F}_{deriv} = \bar{F}_{nat} + \bar{f}$ for the particular phase angle ϕ, and E is an estimate of the error in the measured quantities. The lack of closure is illustrated in Figure 8.6 for a reflection $hk\ell$.

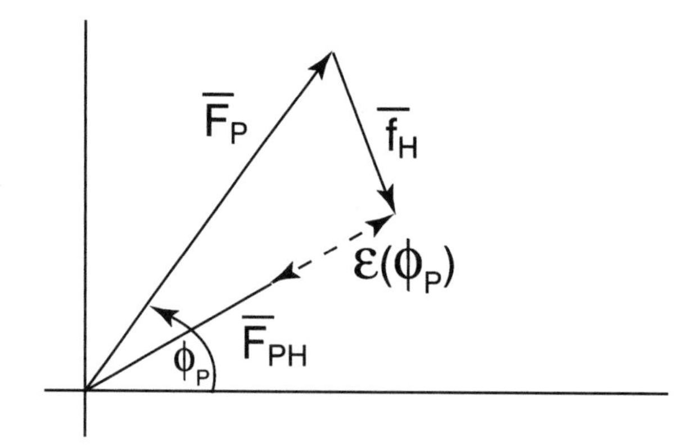

FIGURE 8.6 For any native phase angle ϕ_p the vector triangle $\bar{F}_{PH} = \bar{F}_P + \bar{f}_H$ fails to close by an amount $\varepsilon(\phi_p)$, which is termed the "lack of closure" error. The phase angle of maximum probability, that for which $\varepsilon(\phi_p)$, is minimized when considered over all isomorphous heavy atom derivatives.

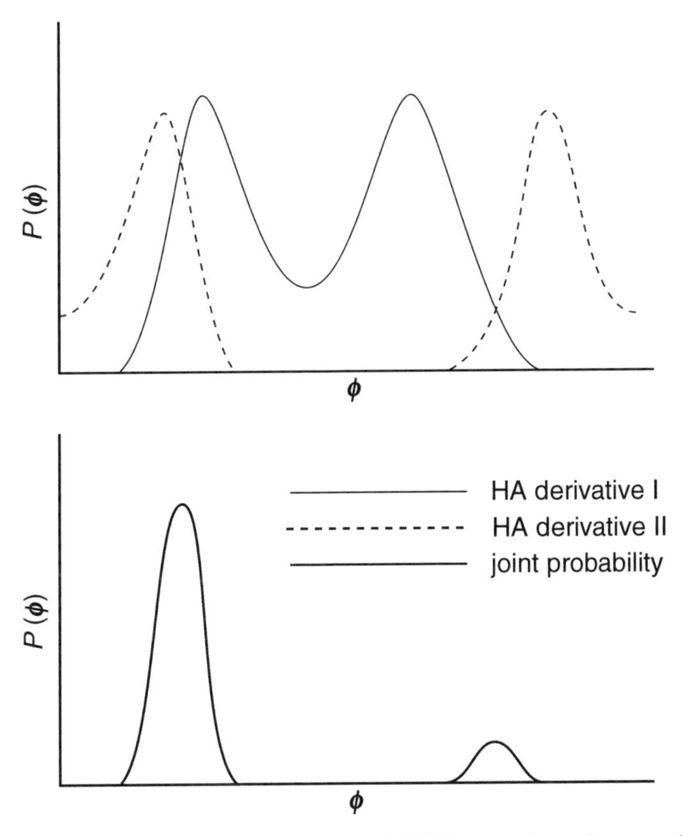

FIGURE 8.7 The top graph shows the separate probabilities over the entire range 0 to 1 for two isomorphous heavy atom derivatives **I** and **II**. Both are bimodal and each separately predicts the two most likely phase angles for the native structure factor. At the bottom the joint probability distribution strongly predicts a single most probable phase.

Figure 8.7 shows the phase probability distributions for two independent heavy atom derivatives, and how they combine to yield a joint probability distribution, and a relatively unambiguous choice of the correct phase angle. Because this determination is entirely a matter of computing and combining phase probabilities, it can readily be carried out for each reflection ***hkℓ*** in a digital computer.

MOLECULAR REPLACEMENT

We have seen that the diffraction pattern of a crystal is the convolution of the contents of a unit cell with that of the crystal lattice (or product of the diffraction patterns, or Fourier transforms). As we have seen, and as illustrated in Figure 1.8 of Chapter 1, and in Figure 8.8, the lattice determines the points in reciprocal space where the transform of the molecules can be observed, and the arrangement of atoms within the unit cell specify the intensity, or value of the combined transform at each point. The asymmetric units in the unit cell are related by space group symmetry, and this symmetry is carried over, except for translations, into reciprocal space. Thus the diffraction pattern reflects the rotational symmetry elements of

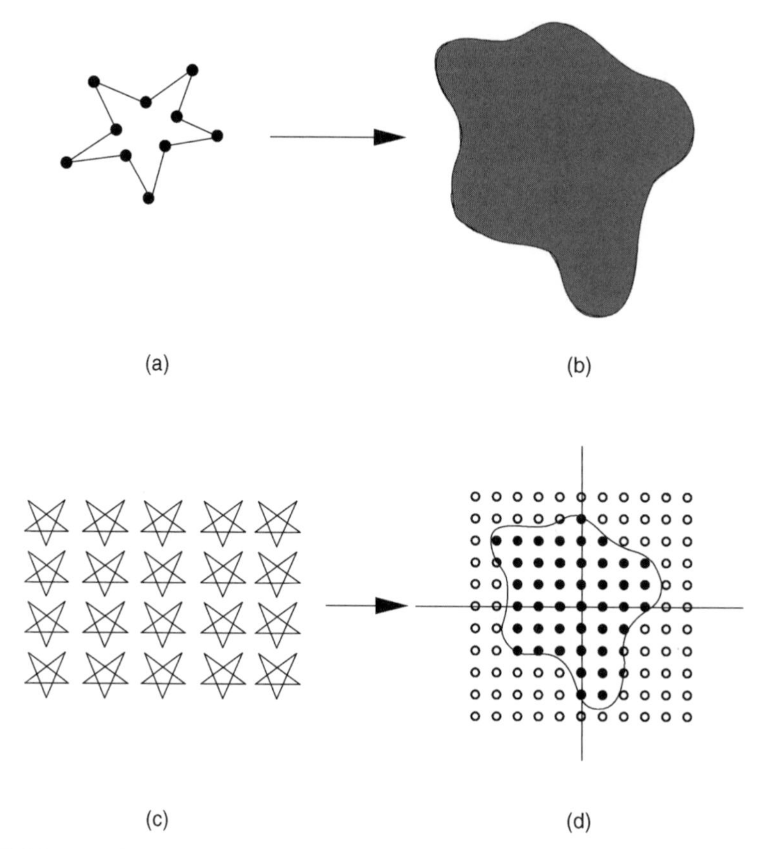

FIGURE 8.8 The distribution of atoms making up the molecule in (*a*) has a continuous Fourier transform as in (*b*). When the molecule in (*a*) is periodically arranged according to a crystal lattice as in (*c*), the diffraction pattern of the array in (*d*) provides sampling of the continuous transform at discrete points consistent with Bragg's law. A discontinuous image of the molecular transform therefore is embedded in the crystal's diffraction pattern. The value of the diffraction intensity of any point in reciprocal space is the intensity of the molecular transform (its diffraction pattern) at that reciprocal lattice point. If the continuous transform of a molecule like that in (*b*) can be calculated, then its image can be sought in (*d*) using the rotation function.

the space group, with their translational components expressed as systematic absences. This was discussed in detail in Chapter 6. Space group symmetry, also called crystallographic symmetry, is perfect and universal throughout the crystal, and throughout the diffraction pattern.

Assume, however, that the asymmetric unit is composed of two or more identical copies of a molecule. These may be related by exact rotational symmetry, such as a dyad or triad, that is not coincident with any crystallographic operator and, hence, is not a part of the space group symmetry. The molecules may also be related by some completely general rotation plus translation. The asymmetric unit will give rise, in either case, to a continuous transform that is essentially a superposition of the transforms of the two independent molecules. This is illustrated schematically in Figure 8.9. The two molecular transforms are identical because the two molecular structures are the same, but the two transforms will be rotated relative to

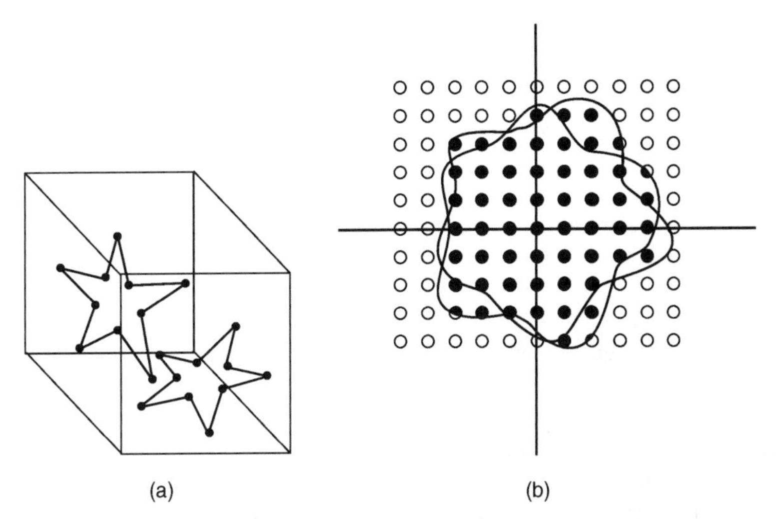

FIGURE 8.9 Two identical molecules comprise the asymmetric unit but are rotated and translated with respect to one another in (*a*). The transforms of the two molecules must also be identical, and both will appear embedded, though overlapping, in the diffraction pattern of the crystal. The two transforms will be rotated with respect to one another according to their disposition in real space. In molecular replacement procedures the crystal diffraction pattern, or its corresponding Patterson map, is rotated through all possible orientations, superimposed upon itself, and a correlation function calculated for each orientation. When the discontinuous transform of one copy of the molecule superimposes upon that of the other, the correlation function (generally called the rotation function) will assume its maximum. The angular value at which the maximum occurs reveals the relative orientations of the two molecules in the asymmetric unit.

one another in diffraction space the same way that the molecules are rotated with respect to each other in real space. The same is true within the Patterson map of the unit cell that can be computed directly from the diffraction intensities, as we will see in Chapter 9. It will contain the Patterson image of one molecule superimposed upon the Patterson image of the other, but rotated according to the relative dispositions of the two molecules in real space.

It is possible, as shown by Rossmann and Blow (1962), to search for redundancies in Patterson space that correspond to the multiple copies of molecular transforms. Rossmann and Blow show, however, that the Patterson map does not need to be computed and used in any graphical sense, but that an equivalent search process can be carried out directly in diffraction or reciprocal space. Using such a search procedure, called a rotation function, they showed that noncrystallographic relationships, both proper and improper rotations, could be deduced in many cases directly from the X-ray intensity data alone, and in the complete absence of phase information. Translational relationships (only after rotations have been established) can also be deduced by a similar approach. Rotation functions and translation functions constitute what we call molecular replacement procedures. Ultimately the spatial relationships among multiple molecules in an asymmetric unit can be defined by their application.

While no attempt will be made here to do so, it can be shown that if the spatial relationships of multiple identical components within the asymmetric unit of a crystal are known, these can subsequently be used to obtain information about the phases of reflections (Rossmann and Blow, 1963). The physical basis is that because the molecular transforms are rotated with respect to one another in diffraction space, the lattice transform (reciprocal lattice) samples these identical molecular transforms at different points. In other words, we are allowed to observe the molecular transforms at more points than Bragg's law would ordinarily permit. Twice as many times for a twofold molecular redundancy, three times for a threefold, and so forth.

A theorem exists in Fourier mathematics that when transforms can be sampled at least one additional time between normally allowed reciprocal lattice points, phases for the allowed reflections can be determined. Use of this "non-integral sampling" approach is steeped in complex mathematical formulations and processes, but it has nonetheless been put to good use in a number of cases, particularly with viruses and other highly redundant complexes whose symmetry arrangement exceeds the space group symmetry.

Molecular replacement (see Rossmann, 1972) has another very useful application, one that has for the most part superseded that described above for directly deducing phase information from noncrystallographic relationships. It derives from the tendency of macromolecules to fall into classes with shared structural motifs, in that they are homologous and often evolutionarily related. Macromolecules of similar function from different species often share structural features and are frequently almost identical. Viruses of the same or similar families do as well. Protein domains of common occurrence may serve as modules to be assembled in different combinations to make a variety of proteins having redundant structural features.

If we know the structure, the relative $\boldsymbol{x}$, $\boldsymbol{y}$, $\boldsymbol{z}$ coordinates in space of one member of a homologous pair, or series, then we can compute its continuous molecular transform, or diffraction pattern, as well as its Patterson function. If we additionally have an observed diffraction pattern recorded from crystals of another, homologous molecule, then we can compare the computed transform of the known molecule with the observed transform from the unknown crystalline homologue. This can be accomplished even though the former transform is continuous and the latter discreet. The computed transform will initially, of course, have an arbitrary and undefined orientation with respect to that observed, but the spatial relationship can ultimately be resolved using the same rotation function search procedure described above. There are some other complexities that must also be addressed in this approach, but they are usually not insurmountable as is witnessed by the remarkable number of successful applications currently swelling the literature.

Once a rotation and translation search has been successfully achieved, then it is possible to place the known molecule in the crystallographic unit cell of the unknown molecule so that its atoms assume the approximate coordinates of the corresponding unknown atoms. By the structure factor equation, the spatially transformed coordinates of the known molecule can be used to calculate phases for the $\bar{\boldsymbol{F}}_{hk\ell}$ of the unknown crystal. These phases, of course, will only be approximate because the molecules are not truly identical, yet, because they are structurally similar, the calculated phases may provide adequate estimates. These approximations can then be used as a starting point for improvement and refinement of the unknown molecules in both real and reciprocal space. As described in Chapter 10, this knowledge can ultimately guide us to the correct structure for the unknown.

As more macromolecular structures become known through X-ray crystallography, then this form of molecular replacement will see ever greater application. With sequence information to guide us, we may eventually be able to accurately and confidently predict what known model structure should be chosen to determine the approximate phases for any new but still unknown macromolecular crystal.

PHASE EXTENSION USING NONCRYSTALLOGRAPHIC SYMMETRY

If an asymmetric unit possesses high symmetry, as is often the case for large macromolecular complexes such as the pyruvate dehydrogenase enzyme complex, or icosahedral viruses, then another approach to solving the phase problem becomes available. Viruses, in particular, are amenable because their symmetry operators are very precise and their orientations are well defined. Asymmetric units also occassionally contain redundant copies of a protein that are related by noncrystallographic symmetry through proper or improper rotations and translations.

The essential point to bear in mind is that the symmetry elements of the virus or multimolecular complex do not generally coincide with crystallographic symmetry operators. Some, if not all, are strictly local, not global symmetry elements. That is the basis for the redundancy within the asymmetric unit. In general, in order to utilize the noncrystallographic symmetry for phase determination it is necessary to have some initial phase information at low resolution, say between **8 Å** and **40 Å** resolution. The noncrystallographic symmetry (found using rotation and, if necessary, translation functions) is then used to extend the phasing to higher resolution structure factors.

The idea of phase extension (Bricogne, 1976; Johnson, 1978) is based on the following: if an electron density map is transformed to produce structure factors, then the phase of no structure factor calculated at resolution higher than the original map has meaning. That is, you can calculate **2 Å** resolution structure factors from a **3 Å** resolution electron density map, but the structure factors between **3 Å** and **2 Å** resolution will be meaningless. The phase angles of those $\bar{F}_{hk\ell}$ calculated at "too high" a resolution will be random. Meaningful phases can only be generated within the same resolution as the $\bar{F}_{hk\ell}$ originally used to produce the map. There is no free lunch.

If, on the other hand, the electron density map calculated at some resolution r is somehow improved in quality in real space, then if it is transformed to produce structure factors, phases at somewhat higher resolution, $r + \Delta r$, can be computed that do have some measure of validity. Improvement of an electron density map thus allows gradual extension of phases in reciprocal space to higher resolution, and ultimately to an electron density map of sufficient quality and detail that a model can be constructed. This is another example of those "bootstrap," incremental procedures so common to X-ray crystallography.

The means for improving the real space, electron density map may involve a combination of ideas, but by far the most powerful is application of noncrystallographic symmetry. If the asymmetric unit contains noncrystallographic symmetry, then its electron density map does as well. If the dispositions of the noncrystallographic symmetry operators are known, then the electron density map can be self-averaged using these operators.

For example, the asymmetric unit of a $T = 1$ icosahedral virus must contain pentagonally related protein subunits. There can be no crystallographic fivefold symmetry operators; thus pentagonally related subunits must be related in a crystal by noncrystallographic symmetry within the asymmetric unit. Often the virus crystallographic asymmetric unit is a multiple of 5, and 15 is common. The orientation of the fivefold axes can usually be identified from a self-rotation function, as can the orientations of twofold and threefold elements. An electron density map of the viral asymmetric unit, calculated from estimated phases at some low resolution r, is averaged about the fivefold symmetry operators. Invariably this substantially improves the quality of the map. Phases are then calculated by Fourier transforming the averaged map to a resolution of $r + \Delta r$.

Extension of the phases from $\boldsymbol{r}$ to $\boldsymbol{r} + \Delta\boldsymbol{r}$ then allows calculation of a new, higher resolution map that can then again be averaged. This way, by continually cycling between real space, where map averaging is applied, and Fourier transforming into reciprocal space, the phases can be improved and extended to higher and higher resolution. Ideally the map would ultimately be extended to sufficient resolution to permit a structural model to be built, and from that point forward, phases calculated from the model would be used.

The low-resolution phases initially required may be obtained in a variety of ways, but frequently these depend on other imaging techniques outside of X-ray crystallography. These may include transmission electron microscopy, cryo-electron microscopy, or atomic force microscopy. Low-resolution phases are even more often obtained by placing the known structure of a closely related virus, or complex, in the correct disposition in the unit cell (determined by rotation and translation functions) and using its low resolution calculated phases.

Real space averaging about noncrystallographic symmetry operators may be combined with other techniques that also serve to improve the electron density map (and that can be used alone to the same end). Chief among these is solvent flattening (Wang, 1985). When noncrystallographic symmetry averaging is combined with this method, not only is the electron density map averaged but a molecular boundary is defined. Outside of that boundary, for example, the central void of a virus or the space beyond its outside limit, the map is set to a uniform low value consistent with disordered solvent. On each cycle the electron density map is improved both by symmetry averaging and solvent flattening, and the higher resolution phases improved as a consequence.

ANOMALOUS SCATTERING APPROACHES

Excellent and detailed treatments of the use of anomalous dispersion data in the deduction of phase information can be found elsewhere (Smith et al., 2001), and no attempt will be made to duplicate them here. The methodology and underlying principles are not unlike those for conventional isomorphous replacement based on heavy atom substitution. Here, however, the anomalous scatterers may be an integral part of the macromolecule; sulfurs (or selenium atoms incorporated in place of sulfurs), the iron in heme groups, Ca^{++}, Zn^{++}, and so on. Anomalous scatterers can also be incorporated by diffusion into the crystals or by chemical means. With anomalous dispersion techniques, however, all data necessary for phase determination are collected from a single crystal (but at different wavelengths); hence non-isomorphism is less of a problem.

As with the isomorphous replacement method, the locations $\boldsymbol{x}, \boldsymbol{y}, \boldsymbol{z}$ in the unit cell of the anomalous scatterers must first be determined by Patterson techniques or by direct methods. Patterson maps are computed in this case using the anomalous differences $\boldsymbol{F_{hk\ell}} - \boldsymbol{F_{-h-k-\ell}}$. Constructions similar to the Harker diagram can again be utilized, though probability-based mathematical equivalents are generally used in their stead.

In addition to the elimination of non-isomorphism, there are other factors that make anomalous scatterers mighty in terms of phasing power, though the $\boldsymbol{\Delta F}$ anomalous may be small. Scaling between crystals, a common portal for the introduction of error, is usually eliminated. Anomalous scattering does not decline with $\sin\theta$ as does the real scattering component; hence it maintains its power at higher resolution where phase information is usually more difficult to obtain. Finally, phase information from anomalous scattering experiments can readily be integrated with phase information provided by other sources such as MIR, SIR, or molecular replacement to produce improved electron density maps.

It has been argued that anomalous scattering should better be called resonance scattering because it arises, not from the anomalous behavior of electrons producing Thompson scattering of X rays, but by their absorption and reemission of X rays having an energy near to the electron's own nuclear binding energy. When the frequency of the oscillating electric field of the X ray is close to that of the frequency of the orbiting electron, a resonance is established. The electron absorbs energy from the X ray, undergoes a quantum transition, and reemits the energy as a photon of the same wavelength, but altered in phase with respect to the normal scattered waves. This resonance effect is negligible for light atoms with weakly bound electrons, including those in most organic macromolecules. For atoms of higher atomic number the effect can be significant, and it becomes substantial for some heavy atoms like mercury or uranium. The effect of this resonance scattering, combined with the conventional Thompson scattering, has a curious consequence. It causes $\bar{F}_{hk\ell}$ and $\bar{F}_{-h-k-\ell}$ to no longer be equal in amplitude for noncentric crystals. That is, it produces a violation of Friedel's law.

For more than 50 years it has been known that the barely measurable differences between $F_{hk\ell}$ and $F_{-h-k-\ell}$ contained useful phase information. For macromolecular crystals lacking anomalous scattering atoms, this phase information was impossible to extract and use because it was below the measurement error of reflections. Anomalous dispersion was, however, sometimes useful in conjunction with isomorphous replacement where the heavy atom substitutent provided a significant anomalous signal. The difference between $F_{hk\ell}$ and $F_{-h-k-\ell}$ was, for example, employed to resolve the phase ambiguity when only a single isomorphous derivative could be obtained (known as single isomorphous replacement, or SIR) or used to improve phases in MIR analyses.

In the past 15 years a number of technical advances have made it possible to maximize and precisely measure anomalous dispersion differences, and more powerful mathematical approaches have been devised to optimize its use for phase determination. The experimental problem was always to amplify the difference between $F_{hk\ell}$ and $F_{-h-k-\ell}$ and obtain its most accurate measure. This has become possible with the advent and increased use of extremely intense synchrotron X-ray beams, and more important, with the ability to tune the wavelength of the sources to virtually any desired value. The tuning of λ (which is also a measure of X-ray energy) to the absorption edge of whatever anomalous scatters are introduced into (or are naturally present in) the macromolecular crystals greatly enhances the method's power. Improvements in X-ray detectors and collection methods have also contributed greatly to the measured accuracy of otherwise small anomalous effects. Finally, an effective approach for introducing anomalously scattering atoms into macromolecules has resulted from the introduction of selenium methionine into recombinant proteins in place of normal sulfur containing methionine. The selenium atoms are relatively efficient anomalous scatterers.

Currently, use of selenium methionine, recombinant protein crystals, or crystals of wild type protein into which have been introduced heavy atom anomalous scatterers, has made single and multiple wavelength anomalous scattering the method of choice for phase determination. In addition, the ways it can be used, and the increasing opportunities for its application are dramatically expanding its popularity.

The underlying principle of anomalous scattering methods is related to the isomorphous replacement concept of creating reference waves within crystal unit cells that interfere in some way with the resultant wave from all of the light atoms belonging to the macromolecules. The scattering power at a few defined locations in the unit cell is changed to provide reference waves. In isomorphous replacement, this is accomplished by the introduction of heavy atoms at specific locations. In anomalous scattering, the contribution to the structure factor of preexisting atoms, the selenium atoms, for example, is varied by altering the wavelength of the X rays. Variation of wavelength produces measurable changes in anomalous differences between structure amplitudes, and these changes can be used to estimate phases by a procedure that is mathematically similar to that used in isomorphous replacement.

There are two principal methods for phase determination using anomalous dispersion. The first is utilized when a single isomorphous heavy atom derivative is available. In this case the native and derivative X-ray data are collected and used to produce SIR phases. The wavelength for the data collection, however, is chosen to correspond with or approach the adsorption edge of the heavy atom substitutent. This maximizes the anomalous scattering differences, and these differences are then applied to resolve the phase ambiguity. The second approach, called multiple wavelength anomalous diffraction (MAD) does not rely on any isomorphous derivative scattering contribution, but only on anomalous dispersion (Hendrickson et al., 1985). Here two sets of data, both containing $\boldsymbol{F}_{hk\ell}$ and $\boldsymbol{F}_{-h-k-\ell}$ pairs of reflections, are collected at two different wavelengths λ_1 and λ_2. One wavelength lies at an absorption edge of the anomalous scattering atoms in the structure, and the other wavelength some distance away. The two sets of anomalous differences, like two different heavy atom derivatives, provide an algebraic solution to the phase problem and, in principle, leave no ambiguity. The anomalous scattering atoms may be selenium in selenomethonines, or they may be atoms that occur naturally in the macromolecule such as the iron in heme groups. They may include other common prosthetic metal atoms such as Zn and Cu as well. While two wavelengths are an absolute minimum in MAD phasing, as are two heavy atom derivatives with MIR, collection of data at even more wavelengths is better. Hence experiments are now often carried out on not just two, but several different wavelengths.

As with the isomorphous replacement technique it is necessary to identify the positions, the x, y, z coordinates of the anomalous scatterers. This can be done by anomalous difference Patterson maps, which are Patterson syntheses that use the anomalous differences $\boldsymbol{F}_{hk\ell} - \boldsymbol{F}_{-h-k-\ell}$ as coefficients (Blow and Rossmann, 1961). These maps are interpreted identically to isomorphous difference Patterson maps (see Chapter 9). Rapidly surpassing Patterson approaches, particularly for selenomethionine problems and others where the number of anomalous scatterers tends to be large, are direct methods (see below). These are strictly mathematical methods that have proved to be surprisingly effective in revealing the constellation of anomalous scatterers in a unit cell.

Phase determination using anomalous dispersion measurements involves more complex vector relationships. In principle, however, and in the way MAD is formulated, it closely resembles the treatment we use for MIR phase determination. For a single heavy atom derivative having significant anomalous dispersion components, the method is not too complicated in principle, and is illustrated in Figure 8.10. There the effect of anomalous scattering on Friedel related reflections is evident. The anomalous effects (represented by their vectors in the complex plane) have been exaggerated in order to emphasize the outcome. As seen in the figure, when the anomalous components of the scattering are added to the real scattering component as

$$f_H = f_O + \Delta f' + i\Delta f'',$$

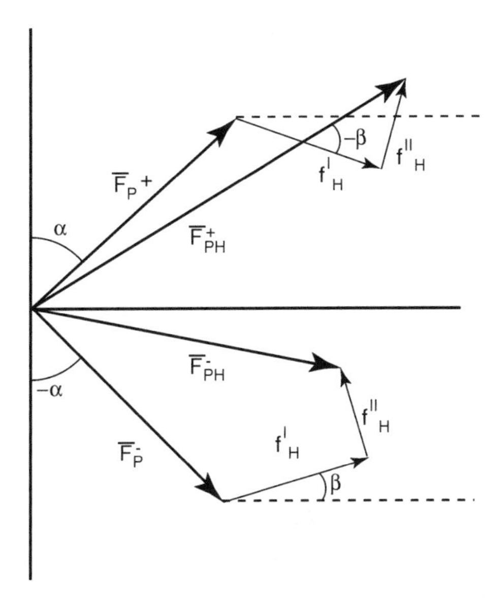

FIGURE 8.10 The effect of anomalous dispersion on the measured structure amplitude is shown here. The normal scattering contribution has the same magnitude, though phase β of opposite sign, for both $\bar{F}_P^+$ and $\bar{F}_P^-$. The real component f'_H of the anomalous scattering atom has the same features. The imaginary component of f''_H, however, always leads f'_H by **90°** for both F_P^+ and F_P^-. The consequence is that the resultant vectors $\bar{F}_{PH}^+$ and $\bar{F}_{PH}^-$ have different lengths, which depend on the phase of $\bar{F}_P^+$.

the magnitude of $\bar{F}_{hk\ell}$ is greater than $\bar{F}_{-h-k-\ell}$, though it might well be the other way around. Furthermore, the difference in length of the vectors for the Friedel related reflections is dependent on the phase $\phi_{hk\ell}$ of $\bar{F}_{hk\ell}$. Because we can calculate $\bar{f}_H$, if we know the coordinates of the heavy atoms, we can again use a method analogous to the Harker procedure for MIR to solve for the unknown phase.

As with MIR, we do not actually draw Harker diagrams but compute and combine probability distributions. Phase determination by anomalous dispersion has a probability form very similar to that for MIR:

$$\rho(\phi) = \frac{\exp - \delta(\phi)^2}{2D^2},$$

where D is an estimate of errors in the measurements and $\delta(\phi)$ is a measure of the lack of closure for phase ϕ. The isomorphous replacement and anomalous phase probabilities for a single heavy atom derivative often are combined so that phases can be sufficiently well estimated from only a single isomorphous heavy atom derivative to solve the structure.

DIRECT METHODS

What are traditionally termed direct methods (Harker and Kasper, 1947; Hauptman and Karle, 1952; Sayer, 1952), also referred to as ab initio methods, had until recently made rather little contribution to macromolecular crystallography, though Hauptman and Karle were awarded the Nobel Prize for the application of direct methods to conventional molecules in 1985. This is true because direct methods were generally not applicable to structures having more than 100 atoms. Further they require, for application to small-molecule crystals, X-ray diffraction data that extends to atomic resolution, namely about **1.0 Å** resolution or better. The latter necessity derives from the fundamental assumptions of direct methods, atomicity (the principle that molecules are composed of point atoms), and that the electron density must everywhere be positive. In conventional crystallography, however, where these limitations pose few problems, direct methods have come to dominate phase determination. Indeed it is now almost routine to solve complicated crystallographic problems of less than 100 atoms in a few hours time using highly automated programs based on direct methods.

If atomicity is assumed, it can be shown that special relationships among the phases of specific classes of reflections, such as among $\bar{F}_h$, $\bar{F}_{h'}$, and $\bar{F}_{h+h'}$ must exist (Sayer, 1952; Hauptman and Karle, 1952). This leads to an extensive network of interrelating phase constraints. The collection of relationships can then be analyzed, using very sophisticated mathematical approaches, to deduce phases for specific $\bar{F}_{hk\ell}$.

While the number of atoms in protein crystals has restricted the use of direct methods, the need for atomic resolution data has been equally prohibitive, since only rarely do macromolecular crystals diffract to such a resolution. Some proteins do diffract to beyond **1.2 Å**, in great part due to our now intense synchrotron X-ray beams and ultrasensitive detectors, and some smaller proteins of about a thousand atoms have in fact been solved using conventional direct methods.

The more important applications of direct methods, however, are currently a consequence of their potential to reveal substructures in unit cells whose crystals yield data to a more modest resolution of **1.5** to **2.2 Å**. By substructures we mean constellations of heavy atoms in isomorphous replacement, or constellations of selenometionines for anomalous scattering approaches. This obviates the often difficult task of determining the positions of these scattering centers using Patterson techniques. In these applications direct methods are successful because the total number of atoms in the substructure is roughly the same as in a conventional small-molecule asymmetric unit, and because the heavy atoms are not bonded to one another but are spread arbitrarily throughout the volume of the unit cell. Their distances of closest approach are therefore much larger. This latter feature endows the substructures with a kind of pseudoatomicity that direct methods find acceptable.

Direct methods are likely to increase in importance and have more widespread application for several reasons. As experimental tools become more powerful, and we learn to grow better crystals, atomic resolution data for proteins of increasing size will become more common. As the use of anomalous scattering approaches expands, the opportunities for their application to deduce the constellations and substructures will increase as well. Finally, as the algorithms strengthen and computing methods become even more powerful, the direct methods themselves will become more effective.

CHAPTER 9

INTERPRETING PATTERSON MAPS

In order to exploit the heavy atom method with crystals of conventional molecules, or to utilize the isomorphous replacement method or anomalous dispersion technique for macromolecular structure determination, it is necessary to identify the positions, the x, y, z coordinates of the heavy atoms, or anomalously scattering substituents in the crystallographic unit cell. Only in this way can their contribution to the diffraction pattern of the crystal be calculated and employed to generate phase information. Heavy atom coordinates cannot be obtained by biochemical or physical means, but they can be deduced by a rather enigmatic procedure from the observed structure amplitudes, from differences between native and derivative structure amplitudes, or in the case of anomalous scattering, from differences between Friedel mates.

The mathematical function that must be interpreted in order to deduce the heavy atom coordinates, a puzzle really, is called a Patterson function or Patterson synthesis (Patterson, 1935). It has a form similar to the equation for electron density except that all phases are effectively zero. It yields, also in a similar manner, a three-dimensional density distribution. The peaks in this map, however, do not correspond to electron density centers but mark the interatomic vectors relating those centers.

The Patterson function has been employed since its formulation in 1935 for determining the locations of heavy atoms in crystals of conventional compounds. This alone made possible application of the heavy atom technique (see Chapter 8) for structure determination. For conventional molecules the information for the heavy atom positions is contained entirely within the native diffraction data, unlike macromolecules, where the information is embedded in differences between two independent data sets, or differences between Friedel mates. Aside from the coefficients employed, use of the function is virtually identical in all cases. Perhaps the major difference arises from the fact that diffraction data from macromolecular crystals, and therefore corresponding difference Patterson maps, contain more noise than

Introduction to Macromolecular Crystallography, Second Edition By Alexander McPherson

do conventional data and their Patterson maps. In addition, when structure amplitude differences are used as Patterson coefficients, the function is only an approximation rather than a rigorously correct synthesis. As a result, Patterson syntheses involving protein crystals are frequently more difficult to interpret, and the solutions are less certain.

It should be noted that there are alternatives in some cases to Patterson syntheses for locating heavy atom positions, and these are direct methods (see Chapter 8). Indeed direct methods have achieved sufficient power, in a mathematical sense, that they are now frequently competitive with Patterson methods. In cases where vast constellations of selenium atoms must be found for MAD analyses, often equivalent itself to solving a challenging conventional molecule structure, direct methods have become the methods of choice; occasionally they are the only choice.

Unless direct methods are used to locate heavy atom positions, an understanding of the Patterson function is usually essential to a full three-dimensional structure analysis. Interpretation of a Patterson map has been one of two points in a structure determination where the investigator must intervene with skill and experience, judge, and interpret the results. The other has been the interpretation of the electron density map in terms of the molecule. Interpretation of a Patterson function, which is a kind of three-dimensional puzzle, has in most instances been the crucial make or break step in a structure determination. Although it need not be performed for every isomorphous or anomalous derivative used (a difference Fourier synthesis using approximate phases will later substitute; see Chapter 10), a successful application is demanded for at least the first one or two heavy atom derivatives.

WHAT IS A PATTERSON MAP

The Patterson synthesis was not derived by a mathematician and then put into use by an experimenter. On the contrary, it was calculated by a perceptive experimenter, who was initially uncertain of its meaning, who examined its features, and deduced its properties. Only some time later, and after considerable discussion among interested investigators was its actual physical basis deduced and its meaning mathematically established (see Glusker et al., 1987). It seems reasonable therefore to follow that same path and simply state how it is calculated, what it means in terms of atom distributions, and then later show how it may be used in structure determination.

The Patterson equation is written as follows,

$$P(u, v, w) = \frac{1}{V^2} \sum_{hk\ell} F_{hk\ell}^2 \cos 2\pi(hu + kv + lw).$$

Remembering that $F_{hk\ell}^2$ is simply the measured intensity $I_{hk\ell}$, a scaler quantity, then the expression can be recognized as simply the electron density equation (see Chapter 5) with squared coefficients and all phases $\phi_{hk\ell}$ set equal to zero. The normalization constant is here $1/V^2$ because of the squared coefficient, where V is the volume of the unit cell. The units it implies for the function, something per volume squared, immediately indicates that $P(u, v, w)$ is not electron density but some other spatial function. Because the equation yields something other than electron density, existing in some unique space, we cannot denote it by $\rho(x, y, z)$ in x, y, z (real space); we must designate it by $P(u, v, w)$ in some alternative coordinate space whose variables are u, v, w. Otherwise, $P(u, v, w)$ is the equation for a periodic function in u, v, w space. The Patterson function, or Patterson wave between $\boldsymbol{u} = \mathbf{0}$ and $\mathbf{1}$, $\boldsymbol{v} = \mathbf{0}$ and $\mathbf{1}$, and $\boldsymbol{w} = \mathbf{0}$ and $\mathbf{1}$ repeats systematically throughout all space. $\boldsymbol{P(u, v, w)}$ is the value of the function at any point $\boldsymbol{u, v, w}$ within what we can consider to be the Patterson unit cell. $\boldsymbol{P(u, v, w)}$ is called the Patterson function, and the coordinate space defined by $\boldsymbol{u, v, w}$ is called Patterson space.

If we examine a Patterson map calculated from the observed diffraction intensities of a crystal it looks superficially like an electron density map. Figures 9.1 and 9.2 are examples. The unit cell vectors of the Patterson unit cell, $\boldsymbol{u, v, w}$, are parallel with the unit cell vectors $\boldsymbol{a, b, c}$, for the real lattice. Inspection will show a Patterson map to have a unique, monumentally large value, or peak, at exactly $\boldsymbol{u} = \mathbf{0}$, $\boldsymbol{v} = \mathbf{0}$, $\boldsymbol{w} = \mathbf{0}$, its origin. More significantly, the Patterson density within the Patterson unit cell will have large values, or peaks of Patterson density, emerging at seemingly arbitrary points, against a generally lower background value of the function. The Patterson map $\boldsymbol{P(u, v, w)}$ may be contoured to emphasize the peaks just as we contour an electron density map (see Chapter 10). It must again be remembered that these maxima are not electron density; they do not appear at atomic positions. They are Patterson peaks.

Now what are these peaks, why do they have the values we observe, and what implications do their map positions $\boldsymbol{(u, v, w)}$ carry for the true distribution of atoms at $\boldsymbol{x, y, z}$ within the real crystallographic unit cell?

What Patterson theorized, and with the help of others eventually proved, was the following: if we choose any point in the Patterson map at $\boldsymbol{u, v, w}$, that point specifies a vector

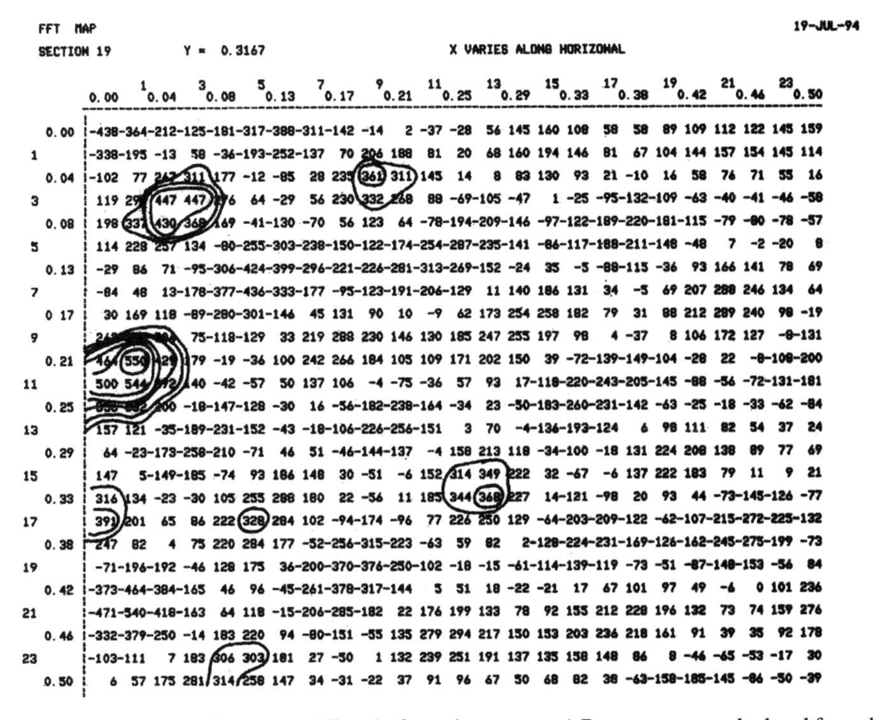

FFT MAP 19-JUL-94

SECTION 19 Y = 0.3167 X VARIES ALONG HORIZONAL

	0.00	1	0.04	3	0.08	5	0.13	7	0.17	9	0.21	11	0.25	13	0.29	15	0.33	17	0.38	19	0.42	21	0.46	23	0.50
0.00	-438	-364	-212	-125	-181	-317	-388	-311	-142	-14	2	-37	-28	56	145	160	108	58	58	89	109	112	122	145	159
1	-338	-195	-13	58	-36	-193	-252	-137	70	206	188	81	20	68	160	194	146	81	67	104	144	157	154	145	114
0.04	-102	77	[illegible]	311	177	-12	-85	28	235	361	311	145	14	8	83	130	93	21	-10	16	58	76	71	55	16
3	119	[illegible]	447	447	[illegible]	64	-29	56	230	332	268	88	-69	-105	-47	1	-25	-95	-132	-109	-63	-40	-41	-46	-58
0.08	198	337	430	368	169	-41	-130	-70	56	123	64	-78	-194	-209	-146	-97	-122	-189	-220	-181	-115	-79	-80	-78	-57
5	114	228	257	134	-80	-255	-303	-238	-150	-122	-174	-254	-287	-235	-141	-86	-117	-188	-211	-148	-48	7	-2	-20	8
0.13	-29	86	71	-95	-306	-424	-399	-296	-221	-226	-281	-313	-269	-152	-24	35	-5	-88	-115	-36	93	166	141	78	69
7	-84	48	13	-178	-377	-436	-333	-177	-95	-123	-191	-206	-129	11	140	186	131	34	-5	69	207	288	246	134	64
0 17	30	169	118	-89	-280	-301	-146	45	131	90	10	-9	62	173	254	258	182	79	31	88	212	289	240	98	-19
9	[illegible]	[illegible]	[illegible]	75	-118	-129	33	219	288	230	146	130	185	247	255	197	98	4	-37	8	106	172	127	-8	-131
0.21	[illegible]	550	[illegible]	179	-19	-36	100	242	266	184	105	109	171	202	150	39	-72	-139	-149	-104	-28	22	-8	-108	-200
11	500	544	[illegible]	140	-42	-57	50	137	106	-4	-75	-36	57	93	17	-118	-220	-243	-205	-145	-88	-56	-72	-131	-181
0.25	[illegible]	[illegible]	200	-18	-147	-128	-30	16	-56	-182	-238	-164	-34	23	-50	-183	-260	-231	-142	-63	-25	-18	-33	-62	-84
13	157	121	-35	-189	-231	-152	-43	-18	-106	-226	-256	-151	3	70	-4	-136	-193	-124	6	98	111	82	54	37	24
0.29	64	-23	-173	-258	-210	-71	46	51	-46	-144	-137	-4	158	213	118	-34	-100	-18	131	224	208	138	89	77	69
15	147	5	-149	-185	-74	93	186	148	30	-51	-6	152	314	349	222	32	-67	-6	137	222	183	79	11	9	21
0.33	316	134	-23	-30	105	255	288	180	22	-56	11	185	344	368	227	14	-121	-98	20	93	44	-73	-145	-126	-77
17	391	201	65	86	222	328	284	102	-94	-174	-96	77	226	250	129	-64	-203	-209	-122	-62	-107	-215	-272	-225	-132
0.38	247	82	4	75	220	284	177	-52	-256	-315	-223	-63	59	82	2	-128	-224	-231	-169	-126	-162	-245	-275	-199	-73
19	-71	-196	-192	-46	128	175	36	-200	-370	-376	-250	-102	-18	-15	-61	-114	-139	-119	-73	-51	-87	-148	-153	-56	84
0.42	-373	-464	-384	-165	46	96	-45	-261	-378	-317	-144	5	51	18	-22	-21	17	67	101	97	49	-6	0	101	236
21	-471	-540	-418	-163	64	118	-15	-206	-285	-182	22	176	199	133	78	92	155	212	228	196	132	73	74	159	276
0.46	-332	-379	-250	-14	183	220	94	-80	-151	-55	135	279	294	217	150	153	203	236	218	161	91	39	35	92	178
23	-103	-111	7	183	306	303	181	27	-50	1	132	239	251	191	137	135	158	148	86	8	-46	-65	-53	-17	30
0.50	6	57	175	281	314	258	147	34	-31	-22	37	91	96	67	50	68	82	38	-63	-158	-185	-145	-86	-50	-39

FIGURE 9.1 A section from a raw (directly from the computer) Patterson map calculated from the structure amplitudes recorded from a protein crystal. Contour lines have been drawn around areas of high Patterson function values to emphasize positions of Patterson peaks. Because of inaccuracies in the data and other sources of error, fluctuations in the background value of the Patterson function may be appreciable.

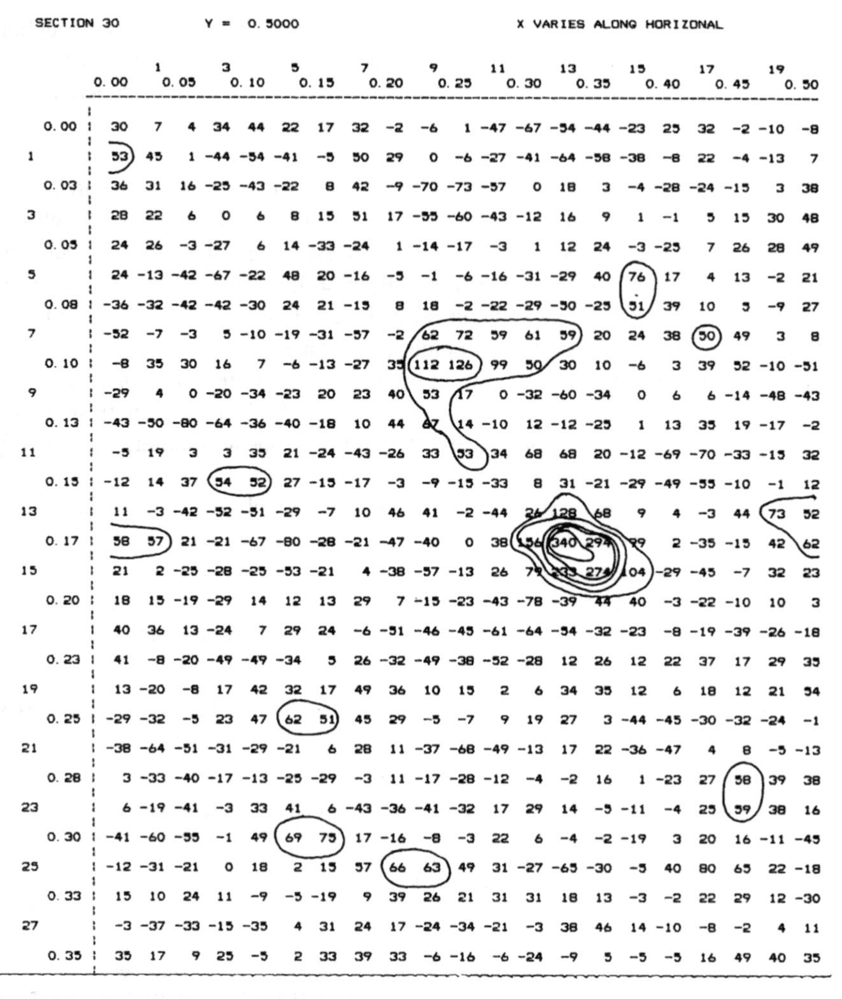

SECTION 30 Y = 0.5000 X VARIES ALONG HORIZONAL

	0.00	1	0.05	3	0.10	5	0.15	7	0.20	9	0.25	11	0.30	13	0.35	15	0.40	17	0.45	19	0.50
0.00	30	7	4	34	44	22	17	32	-2	-6	1	-47	-67	-54	-44	-23	25	32	-2	-10	-8
1	53	45	1	-44	-54	-41	-5	50	29	0	-6	-27	-41	-64	-58	-38	-8	22	-4	-13	7
0.03	36	31	16	-25	-43	-22	8	42	-9	-70	-73	-57	0	18	3	-4	-28	-24	-15	3	38
3	28	22	6	0	6	8	15	51	17	-55	-60	-43	-12	16	9	1	-1	5	15	30	48
0.05	24	26	-3	-27	6	14	-33	-24	1	-14	-17	-3	1	12	24	-3	-25	7	26	28	49
5	24	-13	-42	-67	-22	48	20	-16	-5	-1	-6	-16	-31	-29	40	76	17	4	13	-2	21
0.08	-36	-32	-42	-42	-30	24	21	-15	8	18	-2	-22	-29	-50	-25	51	39	10	5	-9	27
7	-52	-7	-3	5	-10	-19	-31	-57	-2	62	72	59	61	59	20	24	38	50	49	3	8
0.10	-8	35	30	16	7	-6	-13	-27	35	112	126	99	50	30	10	-6	3	39	52	-10	-51
9	-29	4	0	-20	-34	-23	20	23	40	53	17	0	-32	-60	-34	0	6	6	-14	-48	-43
0.13	-43	-50	-80	-64	-36	-40	-18	10	44	67	14	-10	12	-12	-25	1	13	35	19	-17	-2
11	-5	19	3	3	35	21	-24	-43	-26	33	53	34	68	68	20	-12	-69	-70	-33	-15	32
0.15	-12	14	37	54	52	27	-15	-17	-3	-9	-15	-33	8	31	-21	-29	-49	-55	-10	-1	12
13	11	-3	-42	-52	-51	-29	-7	10	46	41	-2	-44	26	128	68	9	4	-3	44	73	52
0.17	58	57	21	-21	-67	-80	-28	-21	-47	-40	0	38	156	340	294	99	2	-35	-15	42	62
15	21	2	-25	-28	-25	-53	-21	4	-38	-57	-13	26	79	233	274	104	-29	-45	-7	32	23
0.20	18	15	-19	-29	14	12	13	29	7	-15	-23	-43	-78	-39	44	40	-3	-22	-10	10	3
17	40	36	13	-24	7	29	24	-6	-51	-46	-45	-61	-64	-54	-32	-23	-8	-19	-39	-26	-18
0.23	41	-8	-20	-49	-49	-34	5	26	-32	-49	-38	-52	-28	12	26	12	22	37	17	29	35
19	13	-20	-8	17	42	32	17	49	36	10	15	2	6	34	35	12	6	18	12	21	54
0.25	-29	-32	-5	23	47	62	51	45	29	-5	-7	9	19	27	3	-44	-45	-30	-32	-24	-1
21	-38	-64	-51	-31	-29	-21	6	28	11	-37	-68	-49	-13	17	22	-36	-47	4	8	-5	-13
0.28	3	-33	-40	-17	-13	-25	-29	-3	11	-17	-28	-12	-4	-2	16	1	-23	27	58	39	38
23	6	-19	-41	-3	33	41	6	-43	-36	-41	-32	17	29	14	-5	-11	-4	25	59	38	16
0.30	-41	-60	-55	-1	49	69	75	17	-16	-8	-3	22	6	-4	-2	-19	3	20	16	-11	-45
25	-12	-31	-21	0	18	2	15	57	66	63	49	31	-27	-65	-30	-5	40	80	65	22	-18
0.33	15	10	24	11	-9	-5	-19	9	39	26	21	31	31	18	13	-3	-2	22	29	12	-30
27	-3	-37	-33	-15	-35	4	31	24	17	-24	-34	-21	-3	38	46	14	-10	-8	-2	4	11
0.35	35	17	9	25	-5	2	33	39	33	-6	-16	-6	-24	-9	5	-5	-5	16	49	40	35

FIGURE 9.2 A section from a difference Patterson map calculated between a heavy atom derivative and native diffraction data (known as a difference Patterson map). This map is for a mercury derivative of a crystal of bacterial xylanase. The plane of Patterson density shown here corresponds to all values of u and w for which $v = \frac{1}{2}$. Because the space group of this crystal is $P2_1$, this section of the Patterson map is a Harker section containing peaks denoting vectors between 2_1 symmetry related heavy atoms.

whose tail is at the origin of Patterson space $(0, 0, 0)$ and whose head is at the point. That vector in turn defines a vector in the crystallographic unit cell. That is, a point in Patterson space, or its corresponding vector, represents a vector in real space. The value of the Patterson synthesis $P(u, v, w)$ at the point u, v, w in Patterson space is the sum of the products of the electron density at one end of the vector in the real unit cell with the electron density at its other end, when the tail of the vector is sequentially placed at all points x, y, z in the real unit cell. The set of all points within the Patterson unit cell therefore represents the collection of all possible interatomic vectors in the real crystallographic unit cell, but weighted with the sum of the products of the electron densities at the points they connect.

In addition to being confusing, this might at first also appear quite useless. The value of $P(u, v, w)$ at any point u, v, w in Patterson space is the sum of products obtained by placing its corresponding vector at all points x, y, z in real space. Remember that the value of $P(u, v, w)$ for any point or vector u, v, w depends on the products of the electron densities it connects. Most Patterson vectors placed at most positions in the real unit cell will have tail or head, or both where there is no atom, hence zero electron density. In all these cases, representing the vast majority, $P(u, v, w) = 0$. The only time this will not be true is when a vector u, v, w falls in the real unit cell so that it connects two atoms. Hence the peaks in a Patterson synthesis $P(u, v, w)$ comprise the map of all interatomic vectors.

The Patterson function $P(u, v, w)$ includes all possible interatomic vectors in the crystallographic unit cell; hence it must also contain the vector between every atom and itself. This vector always has components $u = 0$, $v = 0$, and $w = 0$. It is for this reason that the origin peak in a Patterson map is so large. It is the sum of the squares of the atomic numbers of all the atoms in the real unit cell.

The Patterson map, or Patterson unit cell within the bounds of **0** to **1** along $u, v,$ and w, must contain not only vectors from any atom x, y, z to x', y', z' but also the vectors in the opposite directions from atoms x', y', z' to x, y, z. Hence every atomic relationship in real space is represented by a pair of centrosymmetrically related vectors $u, v, w\ (x - x', y - y', z - z')$ and $-u, -v, -w\ (x' - x, y' - y, z' - z)$. This tells us that Patterson maps, like reciprocal space and the diffraction pattern, always contain a center of symmetry at their origin even when the real crystallographic unit cell does not.

CREATING A PATTERSON MAP FROM A CRYSTAL

Let us now review what has been said and see how it would work in practice. The three-dimensional function $P(\bar{u}) = P(u, v, w)$, as deduced by Patterson, is the product, taken over the entire unit cell, of the electron density $\rho(x, y, z)$ at each point (x, y, z) and at the point related by vectors $\bar{u} = (u, v, w)$ for all vectors $\bar{u}$ that can be drawn in the unit cell. Formally, it may be written as

$$P(u, v, w) = \sum_{unitcell} \rho(x, y, z) \times \rho(x + u, y + v, z + w).$$

The Patterson function is yet another example of a convolution function (see Chapter 3), and it maps vector relationships in real space into a second coordinate system, which is Patterson space. It will be instructive here to examine the Patterson function's relationship to a real atom distribution by asking how it may be physically generated if the distribution of atoms in real space is known. Examples are illustrated in Figures 9.3 and 9.4. The Patterson function of a known structure is formed in the following way.

A Patterson coordinate system is defined based on unit vectors $u, v,$ and w, which are parallel with the axes of the real unit cell of the crystal. Each point ($u, v,$ and w) in Patterson space defines the end point of a vector $\bar{u}$ having a unique direction and length from the origin of Patterson space to that point, that is, $\bar{u} = (u, v, w)$. Every point, or vector, in Patterson space (u, v, w) will have associated with it a value $P(\bar{u})$.

For every point j at (x_j, y_j, z_j) in the unit cell of the crystal, a vector $\bar{u}$ is systematically set down so that it connects the point (x_j, y_j, z_j) with a second point $(x_j + u, y_j + v, z_j + w)$. For all pairs of points related by the vector $\bar{u} = (u, v, w)$, the product of the electron densities at the two points is computed, $\rho(x, y, z) \times \rho(x_j + u, y_j + v, z_j + w)$.

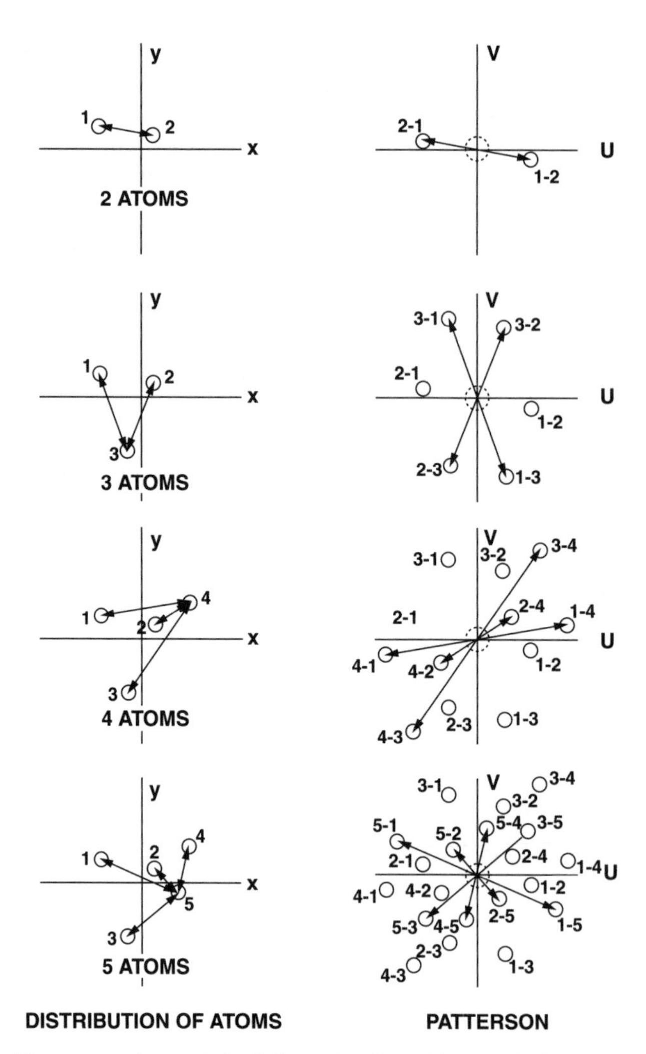

FIGURE 9.3 (From top to bottom) On left are the dispositions of **2, 3, 4,** and **5** atoms, and on right, the corresponding Patterson peak distribution. Only vectors arising from the addition of the most recent atom are indicated at right, although all peaks are shown. In each Patterson map the number of peaks (excluding the origin peak indicated by dashed lines) is $N(N-1)$, where N is the number of atoms. Note how rapidly the complexity of the Patterson map and its interpretability increases with the number of atoms.

For a real crystal this product will be nonzero only when $\bar{u}$ connects two nonzero scattering elements, namely atoms. Any Patterson vector that connects an atom with nothing (a point of no density) will be zero. Any vector connecting nothing to nothing will also be zero. Thus the Patterson function should be zero everywhere except where a Patterson point defines a Patterson vector that connects two atoms, namely an interatomic vector. The sum of all these products for a given $\bar{u}$ is $P(\bar{u})$, and this sum is entered at point (u, v, w) in the Patterson coordinate system. In carrying out this exercise, it becomes clear that the function would be of little value were the scattering matter in the unit cell not composed of discrete atoms and the electron density zero everywhere except near atomic centers.

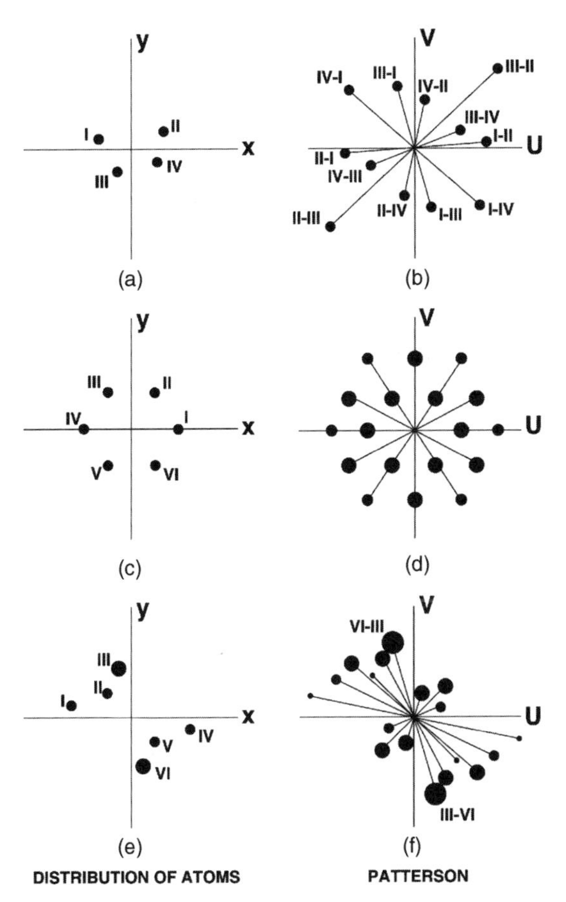

FIGURE 9.4 In (*a*) is a two-dimensional structure composed of four atoms, and in (*b*) its corresponding Patterson map. The four atoms give rise to $N(N-1) = 12$ vector peaks plus the origin peak, which is not shown. The Patterson contains a center of symmetry, even though the real structure does not. If one were presented with the Patterson map on the right, the objective (called deconvolution) would be to deduce a set of atomic positions whose interatomic vectors predict all the Patterson peaks but require no others. A set of six atoms with a center of symmetry is shown in (*c*) along with the corresponding Patterson map in (*d*). Many vectors are identical, although they relate completely different atoms. This gives rise to overlap in Patterson space and produces peaks of multiple weight, as well as a reduction in the total number. Note that the symmetry present in real space is preserved in Patterson space. In (*e*) is a dyad-related pair of three atom structures, containing one atom considerably more electron dense than the others. The corresponding Patterson map is in (*f*). The vectors between the symmetry related heavy atoms are outstanding in the Patterson because of the greater weight of their corresponding peaks. The heavy atom–heavy atom Patterson vectors precisely fix the coordinates of the heavy atoms, with respect to the twofold axis, in real space.

The process is now repeated for every vector $\bar{u}$ that can be drawn within the unit cell of the crystal until the value of $P(\bar{u})$ for every point (u, v, w) in Patterson space has been calculated and entered. The collection of all points (u, v, w) with their associated value $P(u, v, w)$ is the Patterson map of the crystallographic unit cell. The Patterson map yields, at least directly, no information regarding the absolute positions of scattering matter in the unit cell, atoms, but it does provide a map of all interatomic vectors in the crystal. The

Patterson function of the unit cell of a crystal is then the set of all atom-to-atom vectors emanating from a common origin.

Occasionally more than one pair of atoms in a structure may be related by identical vectors, for example, atoms that lie on opposite sides of a benzene ring. Therefore overlap can occur in Patterson space. For some points $\boldsymbol{u}, \boldsymbol{v}, \boldsymbol{w}$ the Patterson value $\boldsymbol{P(u, v, w)}$ will be the sum of more than one nonzero product. This is what makes interpretation of Patterson maps particularly challenging, and obviously this becomes more so as the number of atoms increases. In general, however, every pair of atoms in the unit cell will give rise to two centrosymmetrically related, nonzero vectors in Patterson space. Hence, if there are $\boldsymbol{N}$ atoms in the unit cell, there will be $\boldsymbol{N(N-1)}$ peaks in the Patterson map, as well as the vector from every atom to itself, but this vector is always $\bar{\boldsymbol{u}} = \boldsymbol{(0, 0, 0)}$.

PATTERSON MAPS AS MOLECULAR COVOLUTIONS

The Patterson map of a structure can be viewed in yet another way: as a set of identical images of the distribution of atoms in the molecule, or set of molecules. This is because the end points of all the vectors (the Patterson peaks) from any one atom to all others (including that to itself, which appears at the origin) form an image of the structure in the Patterson map. The number of self-images of the structure in the map will be equal to the number of atoms in the structure. Each image will have the same orientation as any other, but a different atom in the structural image will appear at the Patterson origin.

This alternative way of looking at a Patterson map is illustrated by a four atom and a five atom structure in Figures 9.5 and 9.6. This is sometimes a useful way of considering the Patterson map because it provides the basis for various kinds of Patterson search methods where the objective is to find the image of a known part of a molecule in Patterson space. It is also the basis for the rotation and translation functions used in molecular replacement procedures (see Chapter 8).

DECONVOLUTING PATTERSON MAPS

If the real unit cell contains only a very few atoms, as is often the case with an ionic crystal or salt, the Patterson map, calculated by including its diffraction intensities as coefficients in the Patterson synthesis, may be treated as a puzzle. The object is to contrive a distribution of atoms whose interatomic vectors yield the highest peaks in the Patterson map. This direct approach in fact provided the means by which many of the first small molecules and simple ionic crystals were solved. It is not practical for larger, more complicated structures.

The surprising utility of the Patterson function stems from two implicit features. Since the magnitude of $\boldsymbol{P(\bar{u})}$ at $\bar{\boldsymbol{u}} = \boldsymbol{(u, v, w)}$ in Patterson space is the product of the electron density at the ends of the vector $\bar{\boldsymbol{u}}$, the vectors between heavy atoms (large atomic numbers $\boldsymbol{Z}$) in the unit cell, will yield much larger peaks in Patterson space than those between light atoms. Thus a peak corresponding to a bromine-bromine vector would be $35 \times 35 = 1225$, that from a bromine to a carbon would be $35 \times 6 = 210$, and a nitrogen-carbon peak would have a magnitude of only $6 \times 7 = 42$. The vectors between heavy atoms will therefore contribute the most outstanding peaks and dominate a Patterson map. Such vectors, in the Patterson map of a simple compound, can often be identified simply by inspection.

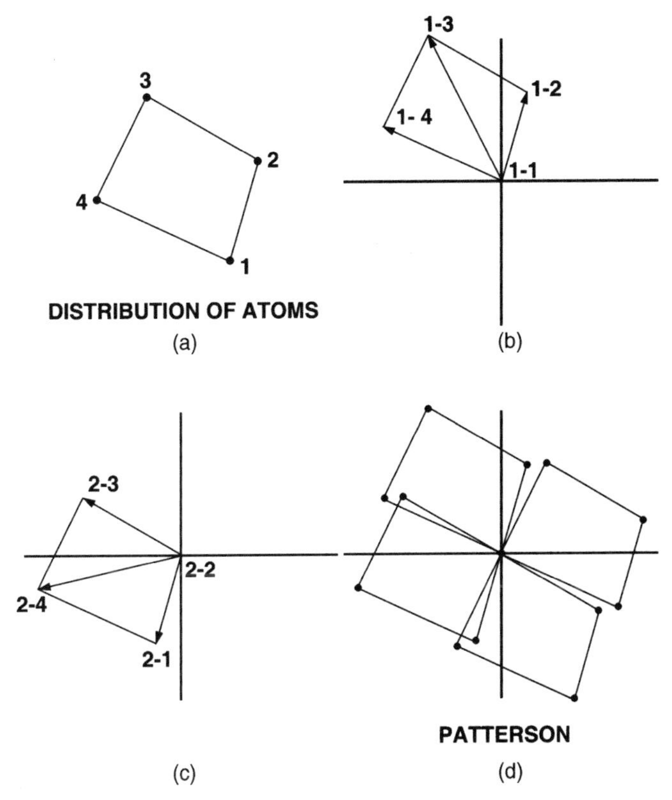

FIGURE 9.5 A molecule in (*a*) is composed of four atoms. In (*b*) we see the molecular image produced in the Patterson map when the vectors from atom 1 to all other atoms are drawn, then, in (*c*) when those from atom 2 are drawn, and finally in (*d*) the set of molecular images that appear in the Patterson map when all intramolecular vectors are compiled.

HARKER PLANES OR SECTIONS

Because crystallographic unit cells contain space group symmetry related asymmetric units (except for space group ***P*1**), certain prescribed patterns are imposed on the distribution of Patterson peaks. This was the invaluable deduction of the great crystallographer David Harker (1936). Unique planes and lines in Patterson maps, now called Harker sections and Harker lines, are heavily populated with peaks (defining the ends of interatomic vectors) as a consequence of the particular symmetry operators within unit cells. This idea is illustrated in Figures 9.7 and 9.8. A key to understanding the use of Patterson maps is to comprehend the idea of Harker peaks, and to be able to interpret the Harker planes produced as a consequence of the symmetry operators of particular space groups.

We know that any symmetry operator, such as a twofold axis or a sixfold screw axis, is equivalent to a set of general fractional coordinates in the unit cell where, by symmetry, corresponding atoms on different asymmetric units are found (see Chapter 3). Thus to specify that there is a twofold axis along $\boldsymbol{y}$ (and it is always essential to specify the direction of the operator) is the same as saying that for an atom at any $\boldsymbol{x}, \boldsymbol{y}, z$, there must be an identical atom at $-\boldsymbol{x}, \boldsymbol{y}, -z$. To say there is a $\mathbf{3_1}$ axis along z is equivalent to stating

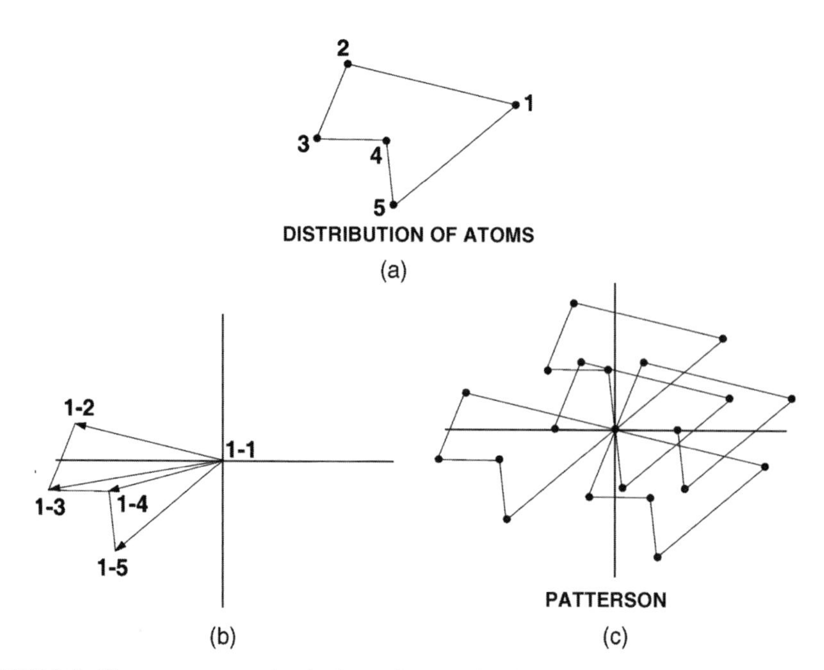

FIGURE 9.6 The same progression is shown for a molecule composed of five atoms as was done for a molecule of four atoms in Figure 9.5. These two figures also illustrate that the Patterson map of a molecule is the convolution of the molecule with itself.

that if there is an atom at any position, x, y, z, then there must be equivalent atoms at $-y, x-y, z+\frac{1}{3}$ and $y-x, -x, z+\frac{2}{3}$ as well.

A Patterson map is the collection of all possible interatomic vectors in the unit cell drawn from a common origin. Therefore it will include, among others, peaks arising from vectors between identical atoms lying at symmetry-equivalent positions. Although we usually cannot predict where vectors between nonsymmetry related atoms will fall, we can predict where vectors between symmetry equivalent atoms will occur. This is especially significant if you bear in mind that peaks in the Patterson map have magnitudes essentially equal to the products of the atomic numbers of the atoms that the corresponding vectors connect. Thus two heavy atoms related by symmetry yield a Patterson peak equal to the square of the atomic number of the heavy atom. That, in most cases, produces a strong, and generally recognizable peak. Equally important, because of David Harker, we know where in the Patterson map to look for it.

Consider the case in Figure 9.7*a* of a twofold axis along z which gives rise to two symmetry equivalent positions at x, y, z and $-x, -y, z$. The vector between these two sites, no matter what the actual values of x, y, z would be $(x, y, z) - (-x, -y, z)$, which equals $\mathbf{2x, 2y, 0}$. Thus for a twofold axis along z, the vector between any twofold symmetry related atoms will lie on the two-dimensional plane through the Patterson map (the Harker section) for which $\boldsymbol{w} = \mathbf{0}$. The $\boldsymbol{w}$ direction in Patterson space corresponds to the z direction in real space. In practice, then, we need calculate only the $\boldsymbol{uv}$ section of the Patterson map for which $\boldsymbol{w} = \mathbf{0}$ and examine it for large peaks. If the asymmetric units contain an outstanding atom, then we might reasonably expect the most prominent peak on that section to correspond to the vector between the twofold related heavy atoms. The value of knowing this vector will be seen shortly.

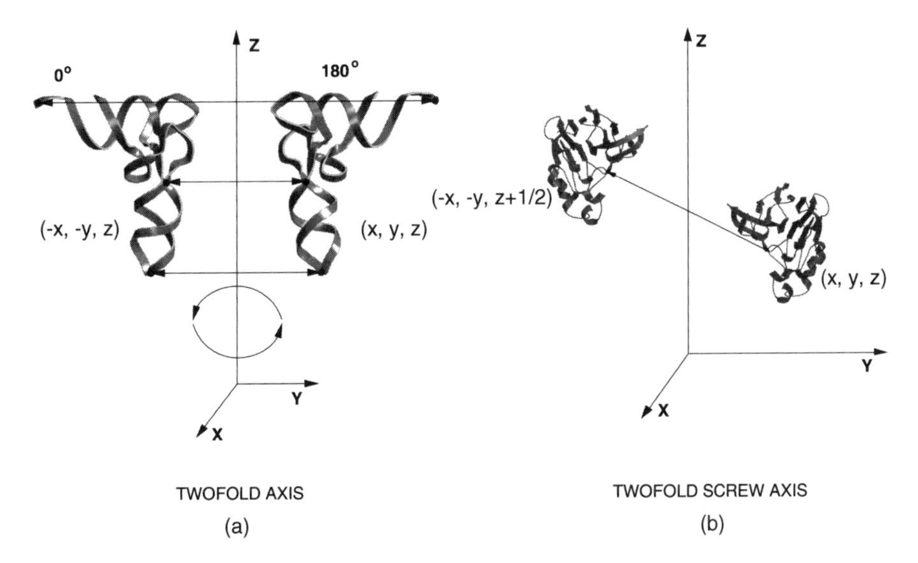

FIGURE 9.7 Two molecules of tRNA in (*a*) are related by a twofold symmetry axis along z in the crystal. A point $\boldsymbol{x},\ \boldsymbol{y},\ z$, which could be the site of a heavy atom in one molecule, has an identical corresponding site in the dyad-related molecule at $-\boldsymbol{x},\ -\boldsymbol{y},\ z$. The vector that connects the two sites will be $(\boldsymbol{x},\ \boldsymbol{y},\ z) - (-\boldsymbol{x},\ -\boldsymbol{y},\ z) = \mathbf{2}\boldsymbol{x}, \mathbf{2}\boldsymbol{y}, \mathbf{0}$. This vector, a Harker vector, must appear on the $\boldsymbol{w} = \mathbf{0}$ section of the corresponding Patterson map computed from the intensities of the diffraction pattern. In (*b*) the heavy atom site on the protein molecule at $\boldsymbol{x},\ \boldsymbol{y},\ z$ appears on the $\mathbf{2_1}$ screw axis (along z) related asymmetric unit at $-\boldsymbol{x},\ -\boldsymbol{y},\ z - \frac{1}{2}$. But $-\frac{1}{2}$, the unit translation, is the same as $+\frac{1}{2}$, so the difference vector is $\mathbf{2}\boldsymbol{x},\ \mathbf{2}\boldsymbol{y},\ \frac{1}{2}$. This Harker vector would appear on the plane of the Patterson map containing points for which $\boldsymbol{w} = \frac{1}{2}$.

Consider another example, that seen in Figure 9.7*b*. A $\mathbf{2_1}$ axis along z would correspond to equivalent positions in the unit cell for all atoms of $\boldsymbol{x},\ \boldsymbol{y},\ z$ and $-\boldsymbol{x}, -\boldsymbol{y}, z + \frac{1}{2}$. The vector between any two symmetry equivalent points in space will have $\boldsymbol{u},\ \boldsymbol{v},\ \boldsymbol{w}$ components equal to the difference of their coordinates. Thus vectors between equivalent positions will be $\boldsymbol{u} = \mathbf{2}\boldsymbol{x}, \boldsymbol{v} = \mathbf{2}\boldsymbol{y}, \boldsymbol{w} = \frac{1}{2}$. Here the Harker section containing the corresponding peaks between symmetry related atoms, screw axis related atoms, will be the two-dimensional plane for which $\boldsymbol{w} = \frac{1}{2}$.

These examples suggest a strategy for the analysis of the Patterson map of any crystal. First, consider the space group and ask; What symmetry elements does it contain, and what are the corresponding equivalent positions? These are conveniently presented in the *International Tables for X-ray Crystallography*, Volume 1 and are not things that need to be derived or kept in the head (though after working with a space group for some time, they tend to become indelible upon one's mind).

Second, algebraic differences between the equivalent positions for the space group are formed. For each pair of equivalent positions, one coordinate difference will turn out to be a constant, namely $\mathbf{0}, \frac{1}{2}, \frac{1}{4}, \frac{1}{3}, \frac{1}{6}$, depending on the symmetry operator. These define the Harker sections for that space group, which are the planes having one coordinate $\boldsymbol{u},\ \boldsymbol{v},$ or $\boldsymbol{w}$ constant, and that will contain peaks corresponding to vectors between symmetry equivalent atoms. In focusing attention only on Harker sections, the Patterson coordinates $\boldsymbol{u}, \boldsymbol{v}, \boldsymbol{w}$

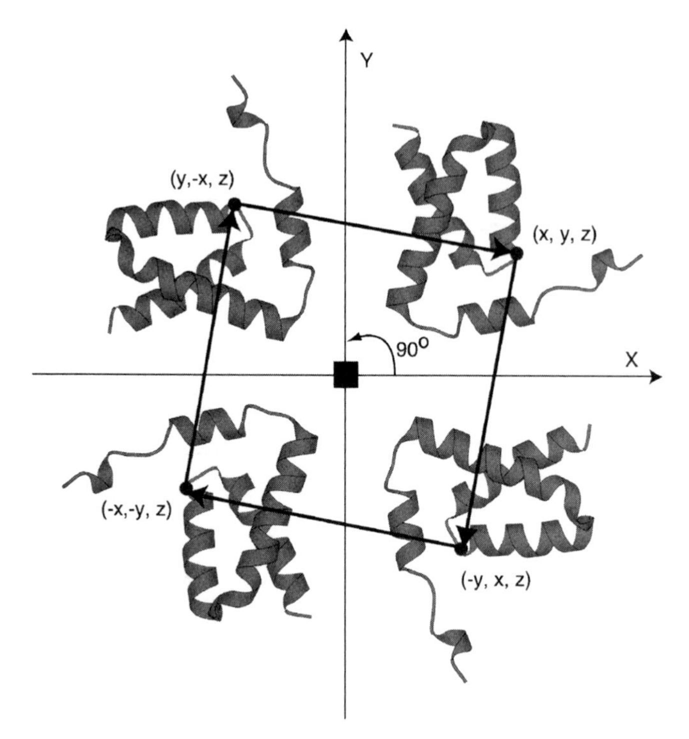

VECTORS BETWEEN EQUIVALENT POINTS
FOURFOLD AXIS

FIGURE 9.8 Four protein molecules are related by a fourfold axis along z that is perpendicular to the common plane of the molecules. The equivalent positions of the four molecules are indicated. Note that vectors connecting equivalent points on the molecules are also in the same plane with their $w = z_2 - z_1$ and $(z_3 - z_1,$ or $z_4 - z_1)$ components equal to zero. The vectors themselves, plus their equivalents in the opposite direction, form a constellation of Patterson peaks that also exhibit fourfold symmetry. This fourfold arrangement also provides additional Harker vectors that are not shown in the diagram, the vectors connecting atoms on diagonal corners of the square.

corresponding to vectors between symmetry equivalent atoms of greatest atomic number are identified.

These Harker peaks, it so happens, can ultimately be used to reveal the actual coordinates (x, y, z) of the heavy atoms in the real unit cell. Let us again look at a case, illustrated in Figure 9.9, where the unit cell contains a twofold axis along y. The corresponding Harker peaks are at $2x, 0, 2z$. Suppose that a peak is found on the $v = 0$ plane at $U, 0, W$ (we will use capital U, V, W to represent specific numerical coordinates in the Patterson map, or Patterson space) which is of outstanding magnitude, and which is likely to represent the vector between two atoms of high atomic number. We know this vector arises from a coordinate difference in real space of $2x, 0, 2z,$ but we do not know what x and z are. Because the coordinates in Patterson space for this vector are U, O, W, which we have read directly from the Patterson map, U must equal $2x$, and W equal $2z$. Therefore $x = U/2$ and $z = W/2$. Hence the observed coordinates for the Patterson peak of the vector connecting two heavy atoms, U and W, yield the actual atomic coordinates in the unit cell of the crystal, x and z.

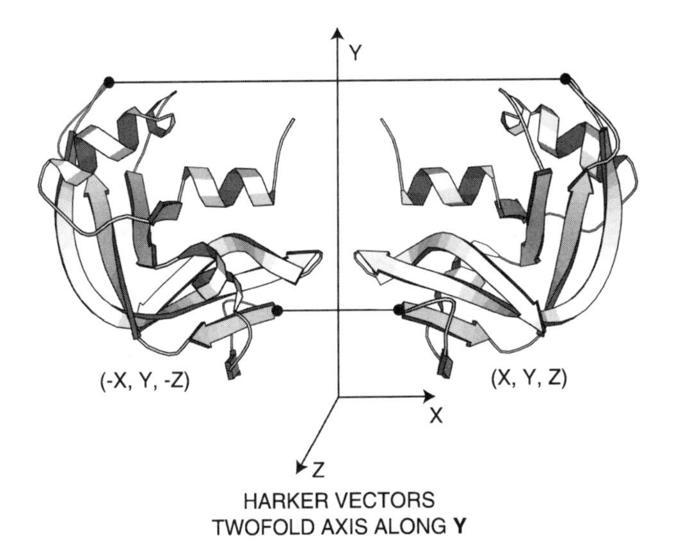

FIGURE 9.9 Two RNase A molecules are related by a twofold axis along y in a crystal. Assume that the points indicated correspond to sites of heavy atoms. The vector between any pair of symmetry-equivalent heavy atom sites is $\boldsymbol{U}$, $\mathbf{0}$, $\boldsymbol{W}$ in Patterson space and is $x_2 - x_1, y_2 - y_1, z_2 - z_1$ in real space. Thus $\boldsymbol{U} = x_2 - x_1$, $\boldsymbol{V} = y_2 - y_1 = \mathbf{0}$, and $\boldsymbol{W} = z_2 - z_1$. $\boldsymbol{U}$, $\boldsymbol{V}$, and $\boldsymbol{W}$ can be obtained by inspection of the Patterson map calculated from the diffraction intensities of the crystal and x and z, two coordinates of the heavy atoms, be determined unambiguously.

Consider the example in Figure 9.10 of a $\mathbf{2_1}$ axis along $\mathbf{z}$ in the unit cell of a crystal. The corresponding Harker peaks will be at $\boldsymbol{2x}, \boldsymbol{2y}, \frac{1}{2}$. If, in examining the $\boldsymbol{w} = \frac{1}{2}$ section of the Patterson map, we find a propitious peak at $\boldsymbol{U}, \boldsymbol{V}, \frac{1}{2}$, then we can properly conclude that the real space coordinates of the heavy atom are at $\boldsymbol{x} = \boldsymbol{U}/\mathbf{2}$, $\boldsymbol{y} = \boldsymbol{V}/\mathbf{2}$. Again, we are able to deduce actual, real space positions of the heavy atoms in the unit cell, $\boldsymbol{x}$ and $\boldsymbol{y}$, from coordinates of peaks on the appropriate Harker section of the Patterson map, $\boldsymbol{U}$ and $\boldsymbol{V}$.

In the examples described here, two of the three atomic coordinates were deduced and one remained indeterminate. In space groups that have only a single symmetry axis, such as ***P*2** and ***P*2₁**, a third coordinate is not established physically by the space group symmetry operators. The origin could be anywhere along the third direction, that is, the third coordinate is arbitrary and can be chosen as zero. In the majority of space groups, where symmetry elements operate along all three directions in space, such as $\boldsymbol{P2_12_12_1}$ or ***I*422** and most others, the physical origin of the unit cell is uniquely specified by the ensemble of symmetry operators, and the third coordinate cannot be arbitrarily chosen as zero. In these cases, however, the additional symmetry operators give rise to additional Harker sections (e.g., space group $\boldsymbol{P2_12_12_1}$ has three). From similar analyses of multiple Harker sections all three coordinates can be deduced. Thus for any space group there are always enough Harker sections to allow determination of the position of an outstanding, heavy atom.

Let's consider, for example, a unit cell having space group $\boldsymbol{P222_1}$ that contains both twofold and $\mathbf{2_1}$ screw axes. The equivalent positions for this space group (from the *International Tables*) are

1. (x, y, z)
2. $(x, -y, -z)$

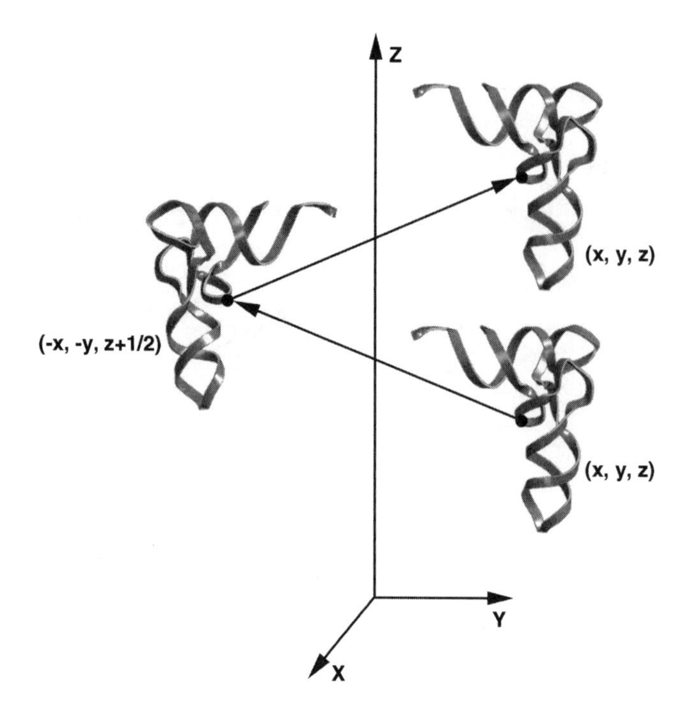

FIGURE 9.10 Two tRNA molecules are related by a $\mathbf{2_1}$ screw axis along the z direction in a crystal, and each bears a heavy atom at the indicated site. The vector between two symmetry ($\mathbf{2_1}$) related heavy atom sites is $U, V, \frac{1}{2}$ in Patterson space and is $x_2 - x_1, y_2 - y_1, z_2 - z_1 = \frac{1}{2}$ in real space. Thus $x_2 - x_1 = U/2$ and $y_2 - y_1 = V/2$.

3. $(-x, -y, \frac{1}{2} + z)$
4. $(-x, y, \frac{1}{2} - z)$

The differences between the general position [1] at x, y, z and each of the other equivalent positions, which define the Harker vectors, are

[1–2] $(0, 2y, 2z)$
[1–3] $(2x, 2y, -\frac{1}{2})$
[1–4] $(2x, 0, 2z - \frac{1}{2})$

Thus we see that for space group $P222_1$ there are three planes of the Patterson map that must contain Harker peaks, and they are $(0, v, w)$, $(u, v, \frac{1}{2})$, and $(u, 0, w)$.

Interpreting Patterson maps for a particular space group may appear daunting at first, but it becomes easier with experience. One must become familiar with the space group and understand what constellation of vectors will result, the relationships between the vectors, and where the vectors should be expected to appear in the Patterson map. The difficulty, in practice, is not in calculating the Patterson map itself, or deducing atomic positions from Harker peaks, but in properly identifying Patterson peaks corresponding to vectors between the correct symmetry equivalent atoms. This is often, in reality, complicated by the inherent noise of the Patterson map arising from errors in the measured structure amplitudes, an overlap of Patterson peaks, and series termination errors in the computations. It is complicated as well by the fact that vectors between nonsymmetry equivalent atoms may fall fortuitously on Harker sections and mislead the investigator.

Consider also the real possibility of an asymmetric unit that contains not one but two atoms of high atomic number. These two atoms are crystallographically independent of one another and are not related by space group symmetry. On the appropriate two-dimensional Harker section (or sections) we should expect to find two unique peaks representing the vectors between each atom and its space group symmetry related atom. From these two peaks we might deduce a pair of coordinates for each of the independent atoms. If the space group contained only a single symmetry element, then the relative third coordinate (relating the two heavy atoms to one another) would remain indeterminate. It can, however, be found by examining the remainder of the three-dimensional Patterson map, the non–Harker sections, for what are known as "cross vectors" between the two nonsymmetry equivalent heavy atoms within the asymmetric unit. Although this requires examination of the entire three-dimensional Patterson map, it is not so difficult as it might first appear, because two of the coordinates for each atom, say x_1, y_1 and x_2, y_2 are already known from the Harker sections. Thus one only needs to examine a single line in the Patterson, in this case along w, for which $U_{1,2} = x_1 - x_2$, and $V_{1,2} = y_1 - y_2$ to find the cross peak that yields the difference in the $\mathbf{z}$ coordinates. If the space group contains multiple symmetry elements, then there will be additional Harker sections. These sections can then be consulted to obtain the missing coordinate.

A Patterson map, different for each space group, is a unique puzzle that must be solved to gain a foothold on the phase problem. It is by finding the absolute atomic coordinates of a heavy atom, for both small molecule and macromolecular crystals, that initial estimates (later to be improved upon) can be obtained for the phases of the structure factors needed to calculate an electron density map.

USING THE PATTERSON MAP FOR ISOMORPHOUS REPLACEMENT

How does a practicing protein crystallographer set about finding the positions of the heavy atoms in a derivative unit cell, having obtained the X-ray diffraction data for the native and (hopefully) isomorphously derivatized crystals, that is, the sets of F_{nat} and F_{deriv}. The first thing is to form differences $\Delta F = F_{deriv} - F_{nat}$ for all $hk\ell$. The squared differences ΔF^2 are then used in the Patterson equation, just as F^2 was for conventional crystals, to compute the three-dimensional Patterson map on an appropriately spaced grid. This is done section by section, just as one computes an electron density map. One can even use the same Fourier program by setting all phases to zero and substituting the Laue group (see Chapter 6) for the space group.

Next, the Harker sections of the Patterson map are contoured. An example is shown in Figure 9.11. As for electron density maps, this is done by drawing contours, lines of equal density, at regular intervals around specified values of the Patterson function, as discussed in Chapter 10, to obtain a topographical map of the "Patterson density." In this manner the major peaks on the section are defined. Remember, however, that a peak in the Patterson map actually represents the end of a vector from the origin of the Patterson map, and this vector, when it appears on a Harker section, is the vector between some atom at x, y, z and the corresponding atom at a symmetry equivalent position.

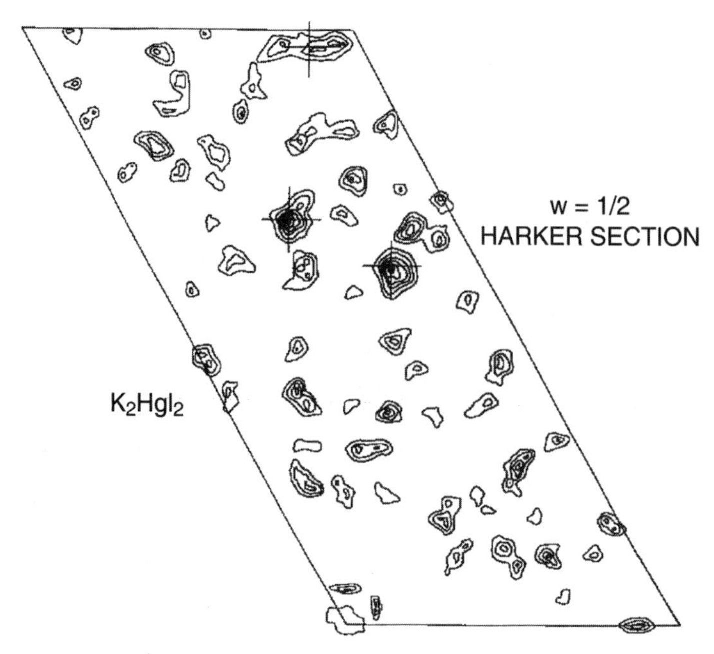

FIGURE 9.11 The $w = \frac{1}{2}$ plane of the difference Patterson map for the K_2HgI_4 heavy atom derivative of the hexagonal crystal form of the protein canavalin. The space group is $P6_3$, so $w = \frac{1}{2}$ is a Harker section. The derivative crystal contained two major K_2HgI_4 substitution sites and one minor substitution site per asymmetric unit. The Patterson peaks corresponding to those sites are marked with crosses. Note that the Patterson peak corresponding to the minor site cannot be discriminated from noise peaks in the Patterson map as is often the case.

The space group of the crystal is then examined and pairs of differences between x, y, z, and all other equivalent positions corresponding to the space group symmetry are formed, as was done for the example above. There is no value in dealing with equivalent positions arising from centering operations (C, I, or F), as they will yield no coordinate information, always producing a set of three fractional constants such as $\frac{1}{2}, \frac{1}{2}, \frac{1}{2}$, or $0, \frac{1}{2}, \frac{1}{2}$. The body-centering operation (denoted by I), for example, creates the equivalent position $(x + \frac{1}{2}, y + \frac{1}{2}, z + \frac{1}{2})$ so that when x, y, z is subtracted, the difference is $\frac{1}{2}, \frac{1}{2}, \frac{1}{2}$. This tells us that all Patterson peaks arising from translationally equivalent sites fall at the same point, such as $\frac{1}{2}, \frac{1}{2}, \frac{1}{2}$, and thereby simply create a second origin peak. Space groups give different numbers of useful Patterson relationships (i.e., Harker sections), depending on the number of symmetry elements they contain, their type, and the way they are combined in the space group.

For each symmetry relationship (and corresponding difference between equivalent positions), it will be found that one of the three Patterson coordinates is a constant, 0 or $\frac{1}{2}$ or $\frac{1}{4}$, and so on. These define the Harker sections of the Patterson map. For example, if one of the Patterson relationships is $u = 2x$, $v = \frac{1}{2}$, and $w = 2z$, then the Harker section $u, \frac{1}{2}, w$ is first calculated and investigated. On this section, like the real unit cell of the crystal whose periodicity it reflects, u and w will vary from 0 to 1 and v will be constant, $\frac{1}{2}$. It is generally useful to separately calculate the individual Harker sections, review them, and only then calculate the full map. The appearance of the Harker sections and their content of peaks, the contrast so to speak, usually provide an immediate impression of the quality of the entire Patterson map.

The Harker sections are inspected for the presence of maxima corresponding to the vectors between symmetry equivalent heavy atoms, and the Patterson coordinates of the peak, or peaks recorded. Maxima positions $\boldsymbol{U}, \boldsymbol{V}, \boldsymbol{W}$ are then correlated with expected real space vectors and real space coordinates $\boldsymbol{x}, \boldsymbol{y}, z$, as in the examples above. Generally, from a Harker plane we obtain only two coordinates of the three necessary to specify the position (or positions) of the heavy atom. If, as in the examples above, the cell is monoclinic, or has only a single symmetry axis, then the third coordinate doesn't matter (at least for the first heavy atom), and we can assign it to be zero. Uniaxial space groups don't have a symmetry-defined origin along the unique axis, so we can simply assume that it is at some arbitrary point lying on the symmetry axis. Once we do assign to a heavy atom a third coordinate of zero, then all subsequent heavy atoms must be defined in terms of this now fixed origin. Correlation of the relative third coordinate can occasionally be problematic but generally can be accomplished by accessory Patterson techniques (see Blundell and Johnson, 1976).

When the unit cell contains several symmetry elements, then each of these will each give rise to a Harker section. The additional Harker sections will in turn yield information about one of the coordinates already known (but not both), as well as information about the missing coordinate. A second Harker section will not only fill out the $\boldsymbol{x}, \boldsymbol{y}, z$ coordinate set for the heavy atom, but will confirm, or serve as a cross check on one of the coordinates deduced from the first Harker section. A third Harker section will, in its turn, provide a cross check on heavy atom coordinates derived from both of the others.

If there is only one heavy atom in the unit cell and the space group is, for example, $\boldsymbol{P2_12_12_1}$, which has three perpendicular screw axes, then three Harker sections exist, each containing coordinate information partially redundant with another Harker section. Thus the absolute crystallographic coordinates can be found from any two of the Harker sections, and whichever is taken as the third serves as a cross check on the interpretation of the others. That is, the $\boldsymbol{x}, \boldsymbol{y}, z$ deduced from any two sections must predict the Patterson coordinates of the peak on the third Harker section. This illustrates the most common challenge, to find convincing peaks that give a consistent interpretation among all Harker sections. If there is more than one heavy atom in the unit cell, then the problem of identifying and correlating peaks becomes, of course, more problematic but by no means insurmountable. It is impossible to describe all of the interesting and demanding puzzles that Patterson maps can present, or the traps they can set, but it is worth mentioning a few common ones here as a warning to the unsuspecting.

If heavy atoms bind at two sites on the asymmetric unit, meaning it is doubly substituted, then there will be two sets of Harker peaks. They will lie on the same Harker sections, however, because the underlying space group symmetry relationships are independently the same for both heavy atoms. Interpretation becomes more perplexing when the number of sites increases beyond two. Internal consistency of solutions with the Patterson map become particularly important. In addition, some space groups, such as $\boldsymbol{P2_12_12_1}$ (the most commonly occurring space group for macromolecular crystals), allow origins to be chosen at several different locations within the unit cell. Thus, even though it may be possible to obtain $\boldsymbol{x}, \boldsymbol{y}, z$, coordinates for both, or several of the heavy atoms independently, it is essential to relate them to a common origin. This, again, requires examination of the full three-dimensional Patterson map and the consideration of cross vectors, that is, peaks that occur at general positions.

Correlating heavy atom positions within the same unit cell, or between an initial and subsequent heavy atom for a series of isomorphous derivatives in a uniaxial space group, can be accomplished by considering cross vectors between nonsymmetry related atoms in the Patterson map. Cross vectors arise between independent atoms, and among their symmetry-related mates. One normally assumes an origin for the first heavy atom, considers each of the other possible origins for the second, and evaluates which choice is most consistent with the non–Harker peaks of the Patterson. This is somewhat more complicated for the correlation of two different heavy atoms in a uniaxial space group, but a similar approach may be taken using a modified difference Patterson synthesis where the coefficients are $\mathbf{\Delta}\boldsymbol{F}$ taken between the two heavy atom derivative data sets (Rossmann, 1960).

Another problem that frequently arises with multiple isomorphous derivatives is that of handedness. In space group $\boldsymbol{P2_12_12_1}$, Patterson maps for two independent derivatives may be interpreted to yield a set of symmetry related sites for one derivative and, independently, a second set for the other. Because handedness is completely absent in a Patterson map (because it contains a center of symmetry), there is an equal chance that the heavy atom constellation for the first will be right handed, and the constellation for the other will be left handed, and vice versa. This won't do. The two heavy atom sets will not cooperate when used to obtain phase information. There are ways of unraveling this problem too, and once again, it involves difference Patterson maps between the two derivative data sets and cross vectors. This case can also be resolved by calculating phases based on only one derivative and then computing a difference Fourier map (see Chapter 10) for the other.

In principle, Patterson maps are not difficult to interpret for a single derivative in a low to moderately high symmetry space group. They are a bit more demanding for high symmetry cells, for example, cubic cells. The greatest challenges arise when there are multiple heavy atom sites within the asymmetric unit for a single derivative and when there are multiple derivatives to be correlated. Frequently Patterson maps are also very noisy because of the great number of interatomic light atom vectors, and these may obscure the true heavy atom Harker peaks. For example, if an asymmetric unit was composed of 1000 light atoms (***C*, *N*, *O***) and one heavy atom, say ***Hg***, then an ***Hg* − *Hg*** peak would be $80 \times 80 = 6400$, but it is superimposed on a map containing 999×1000 light atom Patterson peaks, each of which is about $6 \times 6 = 36$.

The problem of Patterson maps having high background is particularly difficult, and unfortunately common. High background noise arises not only because of light atom vectors but because there is random error in the X-ray data or systematic error from other sources such as scaling. Non-isomorphism is also usually a problem, and statistical disorder or low occupancy of the heavy atoms is another. Misfortunes such as heavy atom Harker peaks overlapping with one another or occurring at special positions seem to appear in much higher proportion than chance should charitably allow.

CHAPTER 10

ELECTRON DENSITY, REFINEMENT, AND DIFFERENCE FOURIER MAPS

To produce an electron density map of the unit cell contents of a crystal from experimentally measured structure amplitudes, and approximate phase angles derived by isomorphous replacement, or any other method, the value of the Fourier synthesis $\rho(x, y, z)$ must be computed at all points (x, y, z) in the unit cell. The electron density is a three-dimensional function and is continuous within the unit cell and periodic throughout the crystal. For visualization, however, a good approximation to a density continuum can be generated by computing the value of $\rho(x, y, z)$ on a grid of points whose separations are sufficiently small. The value of $\rho(x, y, z)$ might be calculated, for example, on a grid of points separated by distances Δx and Δy lying on a particular plane of constant z passing through the unit cell. Initially one might choose the xy plane, which contains all points with coordinates $(x, y, 0)$. The z coordinate is then incremented by $\Delta z, 2\Delta z, 3\Delta z$, and so on, and $\rho(x, y, z)$ is computed on all grid points in each plane $(x, y, n\Delta z)$.

Figure 10.1 is a raw image of one plane of an electron density map, and Figure 10.2 a detail, as they appear directly from the computer, but also after passing through the hands of a crystallographer. By systematically increasing the final coordinate by Δz, the electron density map is built up from a series of two-dimensional electron density planes perpendicular to z. The individual sections are transferred to some transparent film after contour lines have been drawn around areas within selected density limits. In the process the underlying numerical values are eliminated in favor of the contour lines alone. This yields sections of electron density like those shown in Figure 10.3. Each section is a topographical map of the electron density on one plane of the unit cell. When the individual sections are stacked in consecutive order, three-dimensional electron density images, like those shown in Figures 10.4 and 10.5, are formed.

Prior to the advent of computer graphics systems, electron density maps were, in actuality, produced as described above. Each section was printed on paper as a field of numbers, each

Introduction to Macromolecular Crystallography, Second Edition By Alexander McPherson

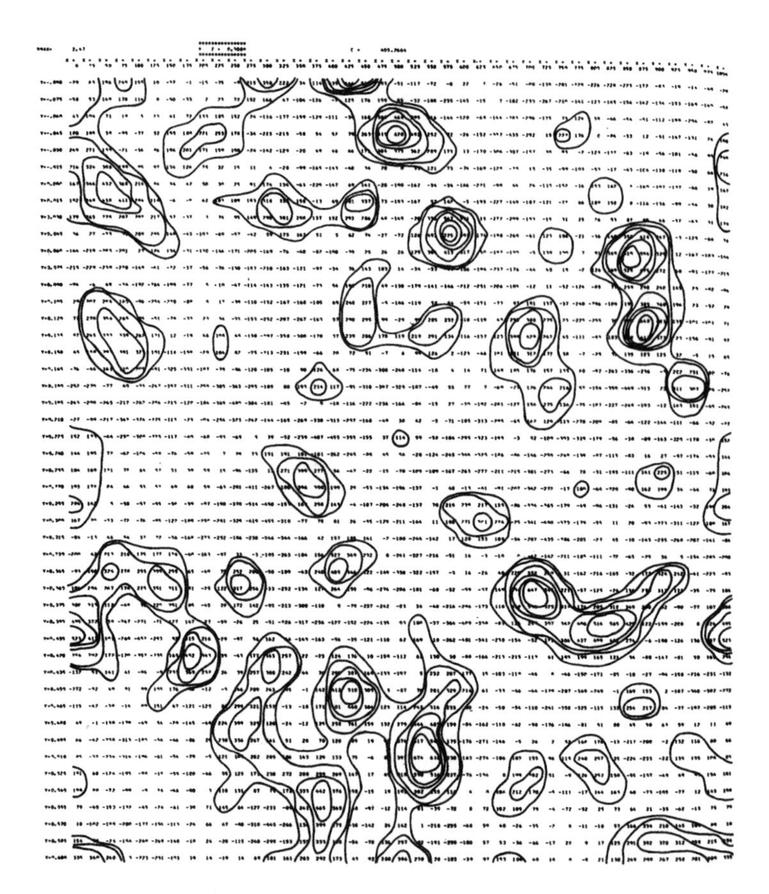

FIGURE 10.1 An electron density map section from a yeast phenylalanine tRNA crystal showing contour lines drawn about areas of high electron density. In the map can be seen what appear to be broken lengths of continuous chains of density. When electron density sections above and below are included, the chains assume greater continuity.

number representing the value of $\rho(x, y, z)$ at a particular point x, y, z in the unit cell. The sheets of numbers were then contoured by eye and hand, reduced in size, transferred to plastic sheets, and displayed in stacks atop laboratory light boxes for inspection (Dickerson, 1992). This entire process is now, of course, entirely automated in the computer (Jones, 1978; McRee, 1999). The investigator may simply collect a sheaf of plastic transparencies containing the contoured sections, ready for display, or he may not even examine the physical map at this point. Currently, most electron density maps are not even presented as contoured sections, but as three-dimensional nets or dot patterns on the computer screen that offer an even better representation of the electron density in three-dimensional space. It is important, however, to understand the origin of those images on the screen of the computer monitor.

In practice, both modes of presentation may be employed. A physical map on a light box, like that in Figure 10.4, often called a minimap, is still very useful for visualizing long-range features of molecules and complexes, that is, to get a global view of the structure. Presentation by computer graphics is nonetheless the only reasonable way to analyze the fine details of a map and construct accurate models consistent with the density. Examples are shown in Figures 10.6 and 10.7.

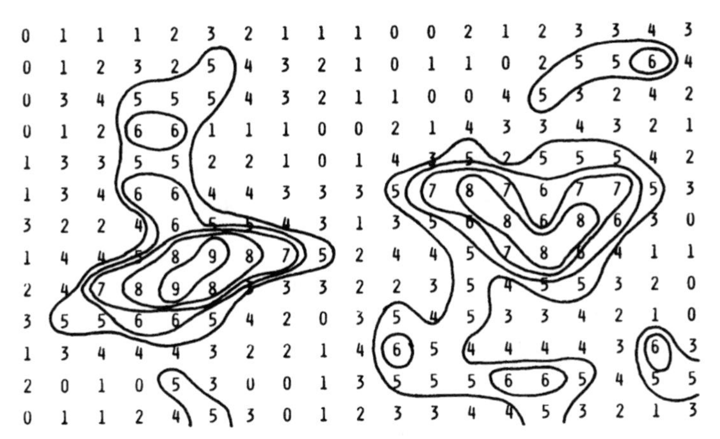

FIGURE 10.2 A small area of a raw electron density map of a protein, directly from the Fourier calculation as it comes off the computer. The location of each number on the plane corresponds to a specific x, y, z fractional coordinate in the unit cell. In general, one of the three coordinates will be constant for the entire plane, and rows and columns will correspond to the other two coordinates. The value of the number at each position is $\rho(x, y, z)$, the electron density at that point. Contours are incrementally drawn around areas having $\rho(x, y, z)$ greater than certain values. This yields a topological map of the electron density on each plane of the unit cell.

RESOLUTION OF ELECTRON DENSITY MAPS

The level of detail that can be discriminated in an electron density map is reflected visually by how fine a computation grid (i.e., Δx, Δy, and Δz) has been chosen. The ultimate resolution of molecular detail contained in a particular map does not depend simply on the grid spacings. It does not matter how finely one samples a blurry image, it is still blurry. What really determines the quality of an electron density image is the number and the accuracy (both amplitude and phase) of the individual $\bar{F}_{hk\ell}$ included in the Fourier summation that produced it. Although the mathematical limits of this summation are over all integers $hk\ell$, the physical restriction that no reflection can occur for which $\sin\theta = n\lambda/2d > 1$ implies that even the highest resolution terms in the series, that is, the $\bar{F}_{hk\ell}$ having the largest $hk\ell$, must arise from reflecting planes with spacings greater than $\lambda/2$. This is the theoretical limit of a diffraction pattern.

In Figure 10.8 appears one of the best known of all demonstrations of the effect of resolution on the image obtained by Fourier transformation. The Taylor and Lipson (1964) illustration shows how the transform of a general object, a duck, becomes increasingly less distinct and featureless as the diffraction information included in the Fourier transform decreases. In Figure 10.9 is another example of Fourier termination effects, but there applied, again using optical diffraction, to a more relevant case, a two-dimensional crystal.

For conventional molecules with relatively small unit cells and low thermal parameters, all the theoretically possible data may be collected and included in the Fourier synthesis. The value $\lambda/2$ is then the practical as well as the theoretical resolution limit. For macromolecular crystals, $\lambda/2$ is never the practical limit, simply because the consistency of structural detail from molecule to molecule, and unit cell to unit cell throughout the crystal is not adequate. Thus, beyond a certain Bragg spacing, usually considerably short of the theoretical limit of **0.77 Å** for CuK_{α} radiation, the intensities decline and ultimately become unobservable. In

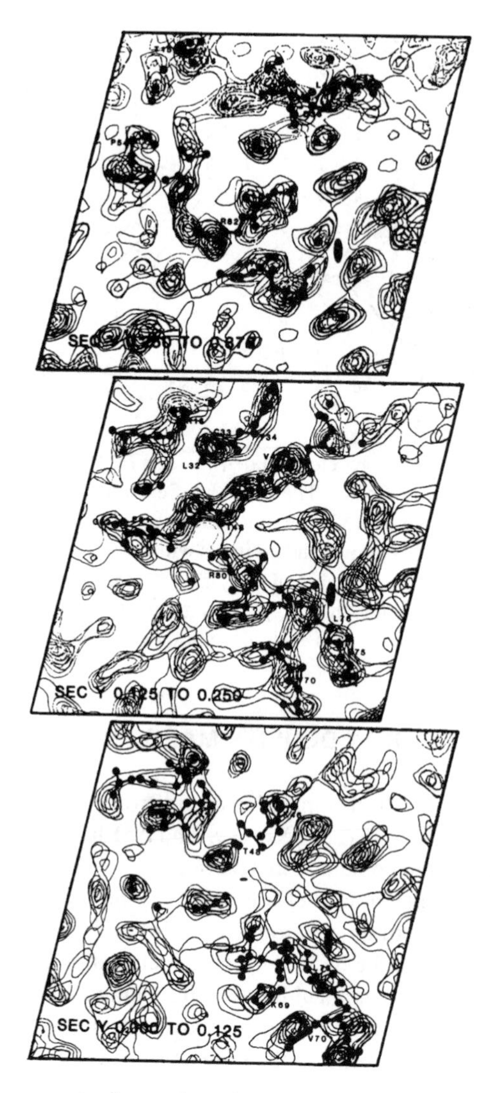

FIGURE 10.3 Three consecutive layers through an electron density map of the Gene 5 DNA Unwinding Protein crystal after contouring. By stacking the sections one atop the other, after transfer to transparencies and placing on a light box, a three-dimensional image of the electron density is obtained.

general, the highest resolution reflections for protein crystals correspond to Bragg spacings on the order of **1.2** to **3.0 Å**, but most commonly to no better than about **1.7 Å** spacings. Hence electron density maps of macromolecules do not often allow one to directly resolve detail of better than **0.6** to **1.5 Å**. As a rule of thumb, electron density maps should be computed with an interval between computation points Δx, Δy, and Δz of about one-third the maximum resolution of the terms included in the Fourier synthesis. It is important to remember that the resolution limit of the electron density map is not necessarily the resolution of detail in the final structure. Refinement methods imposed on putative models generally ensure a precision in atomic coordinates of better than **0.5** to **0.1 Å**.

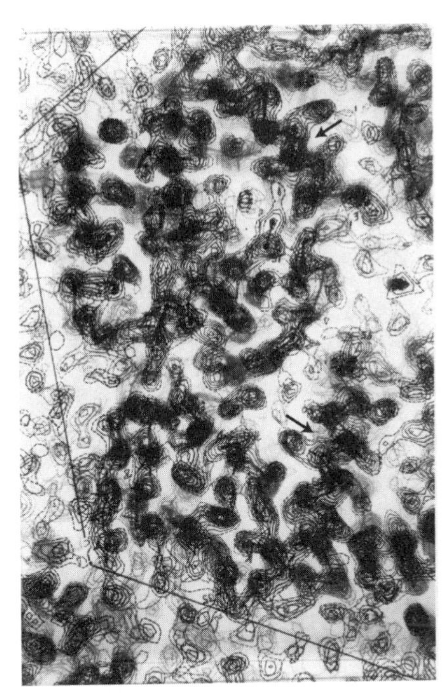

FIGURE 10.4 Several consecutive sections of the electron density of the hexon capsid protein from adenovirus are stacked and displayed on a light box. Here the continuity of the polypeptide begins to emerge. Some recognizable features such as the two α helices marked by arrows are also evident to the practiced eye.

It should be emphasized that the stated resolution of a particular structure determination does not necessarily reflect the clarity of the features present in the electron density map, or the accuracy of the model based on the map. That is, many **3.0 Å** resolution electron density maps are far more interpretable in terms of protein structure, and the amino acid side groups more recognizable, than are some maps that include higher resolution terms. This is because resolution says nothing about the accuracy of the structure factors used to compute the map. In particular, it gives no indication of the quality of the phase determination, the mean phase error, which is the essential factor that influences contrast between the electron density of the protein and the background noise level in the Fourier map.

No Fourier synthesis of a macromolecule phased strictly by isomorphous replacement, multiple anomalous dispersion, or any other experimental technique is likely to permit resolution of the constituent atoms. One must be cautious, therefore, when presented with an electron density map interpretation, that deductions are not made concerning detailed interactions or positions that are unwarranted by the extent or accuracy of the data used in the calculation. One should furthermore be aware that an electron density representation is always only an approximate image of the molecule. The phase angles employed in the synthesis usually contain significant errors that are inherent in the experimental technique, and some phases may be entirely wrong. It is currently estimated, for example, that the deviation of the average isomorphous replacement phase from its true value is generally greater than **45°**. In addition, inaccuracies are introduced by non-isomorphism, scaling errors, and series termination effects. All these problems are, however, addressed by refinement procedures.

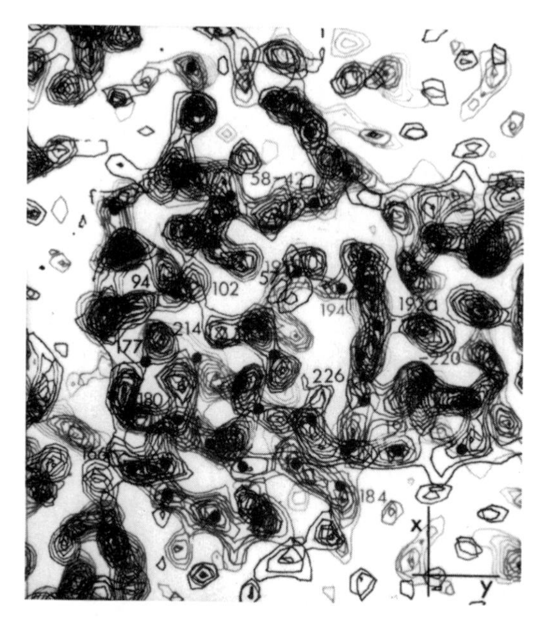

FIGURE 10.5 Several consecutive electron density sections, separated by a constant increment along the z axis, are stacked and displayed on a light box. The continuity of the protein, here the protease from streptomyces, is clear in many places. The dots are used in the early interpretation of the map to mark points, such as putative α carbon positions, along the chain. The first objective in interpreting an electron density map is generally to deduce the overall fold of the polypeptide. Only later are side chains identified and oriented.

INTERPRETATION OF ELECTRON DENSITY MAPS

How is an electron density map interpreted? In the case of a conventional small molecule compound, where the electron density map is computed at or near atomic resolution, atoms appear as distinct, well-separated peaks in the map. Chemical constraints, such as reasonable bond lengths and angles, are then applied to the distribution of atoms, and based on the relative peak heights, geometry, and distances of separation, the structure of the molecule deduced. Since there is no explicit information specifying which atoms are directly bonded, connectivity must be implied. In most instances the interpretation is unambiguous and the structure clear.

Similar principles are applied in the interpretation of the Fourier synthesis of a macromolecule, that is, correlation of the electron density distribution with existing chemical and physical information. In this case, however, the chemical information constraining the structure is of a somewhat different nature. In the case of a protein, the objective is to arrive at a model composed of a known sequence of amino acids linked together in a continuous chain, which is then organized into recognizable secondary structural elements, such as α-helices and β-sheets, that are then joined together and arranged in a topologically sensible pattern. In reality the amount of structural information regarding macromolecules has now become so large (Berman et al., 2000) that the quantity and quality of chemical and physical constraints available in fixing the structure of a newly determined macromolecule is approaching that for conventional small molecules.

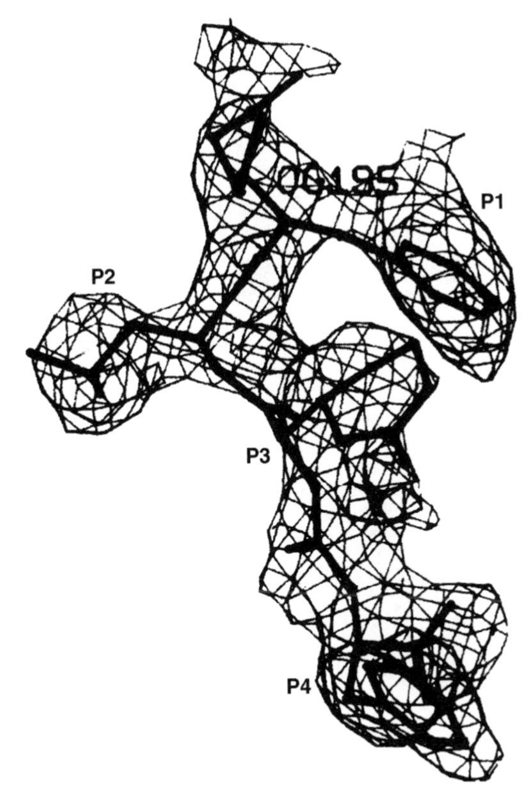

FIGURE 10.6 A tetrapeptide inhibitor of a bacterial serine protease is seen here as a stick model fitted to its corresponding difference electron density generated by computer graphics. The three-dimensional mesh representation for the density is clearly an advance over the multiple, two-dimensional density sections in terms of fine detail.

CONSTRUCTING A MODEL

The penultimate stage of structure determination consists of building a physical or virtual model into the electron density obtained from a Fourier synthesis. The synthesis may be obtained by application of multiple or single isomorphous replacement, MAD or SAD, molecular replacement, direct methods, or other possible approaches to solving the phase problem (see Chapter 8). This is generally the point at which you begin regretting that you did not give greater attention to data collection and phase determination. The electron density map will reflect the exact features of the true structure (e.g., the conformation of the polypeptide backbone or the dihedral angles of the amino acid side chains) as a function of the accuracy and precision of the measured intensities, and the quality of the experimental phases.

Model building requires fitting, as carefully as possible, a polypeptide or nucleotide chain into the strongest density in the map while maintaining chemical reality, geometric and stereo chemical proprieties, and using simple common sense. Model building is currently done using computer graphics systems, like those used to play video games in many laboratories, or those residing on simple laptop computers. There are currently two principal graphics programs for model building, the venerable program ***O*** that has been in use

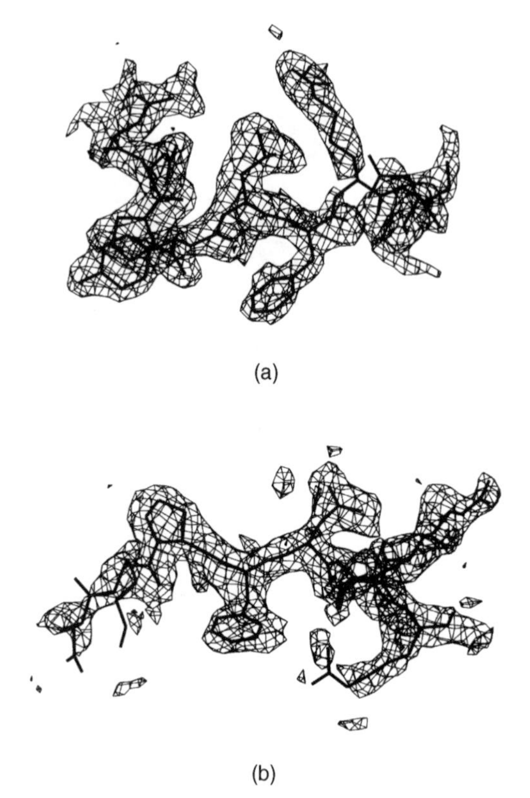

(a)

(b)

FIGURE 10.7 Computer graphics systems now allow rapid adjustment of molecular models to fit the electron density generated as three-dimensional mesh images. Two segments of polypeptide chain from a protein are seen here.

for more than 25 years (Jones, 1978), and a more recent, but increasingly popular program named COOT (Emsley, and Cowtan, 2004). Both are readily available to investigators, but it is not clear that anyone knows why they are named ***O*** and COOT. Although there are automated model-building programs that aspire to take the pain out of model building, they generally only work well with very high resolution, and near perfect data. So model building remains among the most trying and demanding tasks still left to the crystallographer. Model building requires patience and care. Diligence at this stage is rewarded later as refinement proceeds. A poor starting model may never properly refine, or it may require frequent returns to the graphics system to repair deficiencies before it does.

Experimentally determined electron density maps are never perfect because of defects in the data and phases, and as a consequence even the best of models will have errors in atomic positions, errors in dihedral angles, improper rotomers for side chains, or unacceptable contacts between atoms or chemical groups. When the quality of an electron density map is marginal or poor, serious errors in the model may occur, or the model may be fundamentally wrong. Even when the electron density map is good, and the model for most practical purposes correct, errors remain. The initial model for a protein will, in general, have rms errors in atomic positions of **1*A*** or more, and an ***R*** factor computed from the model (with the exception perhaps of models obtained from molecular replacement) on the order of **0.50**, an often discouraging result.

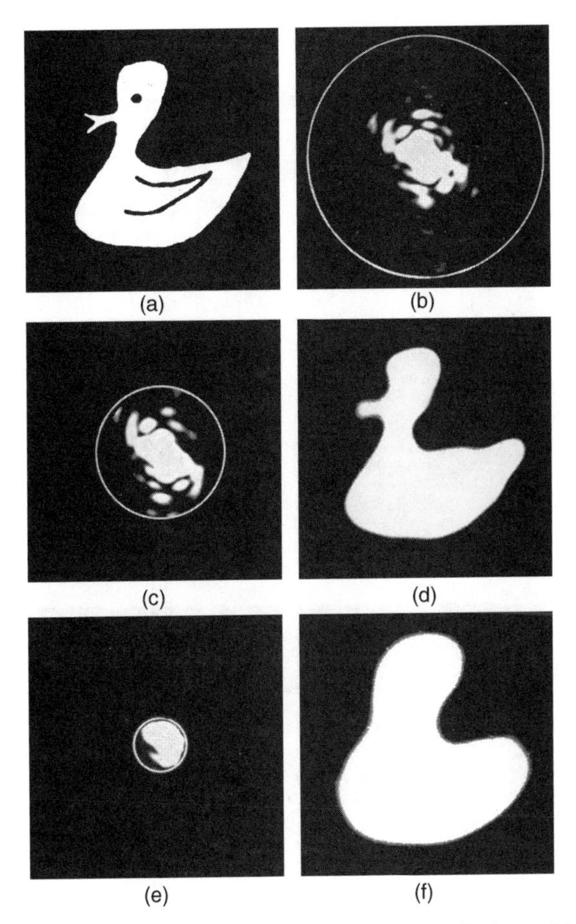

FIGURE 10.8 The now famous diffraction image of a duck from Taylor and Lipson's 1964 book, *Optical Transforms*. In (*a*) is the object of the transform, a duck, and in (*b*) its optical diffraction pattern. If that part of the transform containing only the low-resolution area in (*c*) is now Fourier transformed to recreate the object, as in (*d*), we see that only a low-resolution image of the duck appears. When virtually the entire high-resolution portion of the transform is excluded, as in (*e*), and this is backtransformed, the image can be interpreted as a duck only with application of considerable imagination. A corresponding degradation of an electron density image of a macromolecule similarly occurs when only low-resolution $\bar{F}_{hk\ell}$ (low $\sin\theta$, or structure factors with small values of $hk\ell$) are included in the Fourier synthesis $\rho(x, y, z)$.

To reduce the errors to a minimum (to zero is virtually impossible), one normally embarks on a refinement process of one sort or another. With a good starting model containing no serious errors, but only lots of small ones, and some of the powerful refinement programs available today, the entire refinement process may be completed over lunchtime. This is, however, rarely the case. Usually one refines a model, examines the resultant structure in various ways to determine if some local rebuilding is called for, and then refines that, hopefully, improved model. Thus most refinement procedures alternate computational refinement with manual rebuilding until an optimal model is achieved. An optimal model emerges when you can think of no way to improve it further.

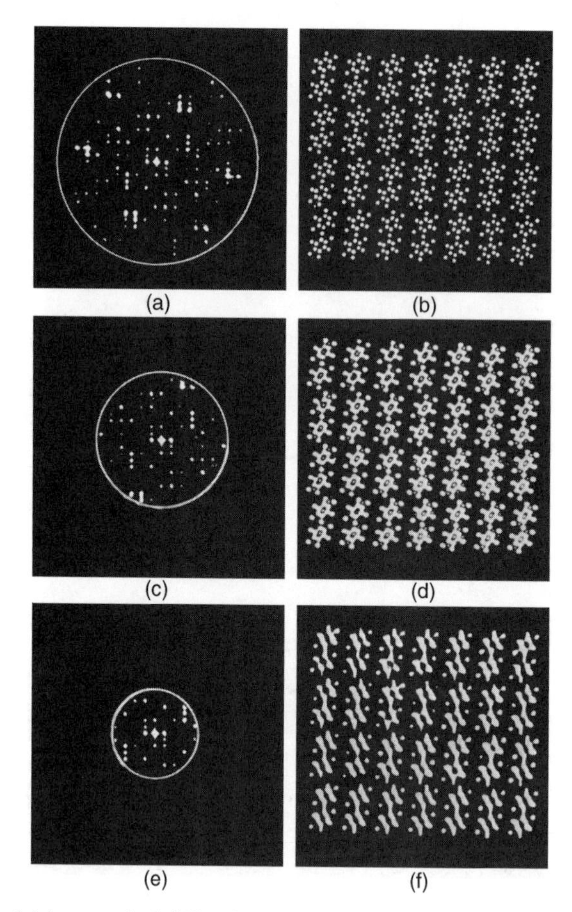

FIGURE 10.9 In (*a*) is an optical diffraction pattern of the two-dimensional molecular crystal in (*b*). In (*c*) the resolution of the terms used to produce the transform are limited, and a degraded image of the molecules results in (*d*). When only very low resolution terms are included in the backtransform as in (*e*), the periodicity of the pattern in (*f*) remains clear, but the image of the molecule making up the pattern becomes virtually uninterpretable.

MODEL REFINEMENT

There are a number of criteria used to measure the improvement of a model, such as bond lengths between atoms and bond angles that are closer to canonical values and improved stereochemistry. Indeed these are all useful and must be closely monitored during the course of refinement, but they are not the foundation for refining a structure determined by X-ray crystallography. The measured structure amplitudes F_{obs} from the crystal, and the structure amplitudes F_{calc} calculated from the positions and scattering factors of the atoms making up the molecular model, provide that foundation. Furthermore, their correlation, expressed as the conventional R factor (see Chapter 8), is, as we saw earlier, an appropriate measure of how well the model agrees with the true structure.

Notice that in the refinement process we are using the agreement of reciprocal space quantities F_{obs} and F_{calc} to tell us the extent of agreement of real space entities, the model and the real crystal structure. These two are linked to one another through the Fourier transform, as we saw in Chapter 5. A change in atomic positions x, y, z results, through the structure factor equation, in changes in the amplitudes and phases of all the structure factors in diffraction space. In general, because we cannot directly measure phases experimentally, phases of structure factors, experimental or computed, are not explicitly used in the refinement process, as are structure amplitudes. Refinement then is based on maximizing the agreement between F_{obs} and F_{calc}, or minimizing their differences $F_{obs} - F_{calc}$ overall, which is equivalent to minimizing the R value.

It is wise to remember that the term *refinement* means exactly what it says; it means the minimization of small errors. Refinement is most certainly not a stage of structure determination as is sometimes supposed. The initial model must be a good approximation to the truth. If it is distant from the actual structure, either overall or locally, refinement is unlikely to correct those errors and result in a correct structure. Thus it is important, in practice, not to build a sloppy or careless model, or a model based on a wing and a prayer, and then hope that refinement will find the right solution or fix the problems.

In practice, refinement may be carried out in either real space or reciprocal space. Any operation in one space must have some equivalent operation in the other. Often these correspondences are obscure, but they are always there. Thus minimizing the R factor in reciprocal space has a correlate in real space. That correlate amounts to minimizing the difference between the experimentally determined electron density map and the electron density computed from the model. The first approach, minimizing the R factor in reciprocal space is usually computationally intense. Minimizing differences in actual densities in maps tends to be investigator intensive; hence most investigators choose the reciprocal space route whenever possible. Using difference Fourier maps to visualize the density differences between model and experimental maps, is, however, a very useful procedure for following the real space changes wrought by reciprocal space refinement. Thus the real space evaluation of reciprocal space refinement plays a valuable role in the overall process.

Minimization of the R factor entails minimizing the sum of differences $\sum |F_{obs} - F_{calc}|$ arising from the individual structure amplitudes, and the number of these may vary from several thousand for a small protein, to many millions for a virus crystal. The F_{obs}, the observed structure amplitudes, cannot change. They are immutable. They are what you measured. They are truth. Only the F_{calc} can change. The F_{calc}, as we saw in Chapter 5, are functions of the atomic coordinates x_j, y_j, z_j, the scattering factors of the atoms at those positions f_j, and the temperature factors of the atoms B_j. A change in any of the parameters that define the model, even if it is to a single atom, will affect F_{calc}. Therefore, by carefully altering the positions of the atoms in a coordinated manner, F_{calc} can be made to more closely approach F_{obs}, with the differences reduced overall and the R factor lowered. A lower R factor implies a more accurate model.

RECIPROCAL SPACE REFINEMENT: LEAST SQUARES

The most common and most powerful methods for minimizing the R factor over an entire set of structure amplitudes $F_{hk\ell}$ are all based on the procedure of least squares. Although some significant complexities are introduced into the use of least squares when applied to crystallographic problems, they are fundamentally based on the same idea as guides us in the fitting of a straight line to a set of experimentally measured points on a graph. There are just a lot more points and a lot more variables, and the relationships between them are less readily addressed mathematically. Nonetheless, least squares procedures have proved

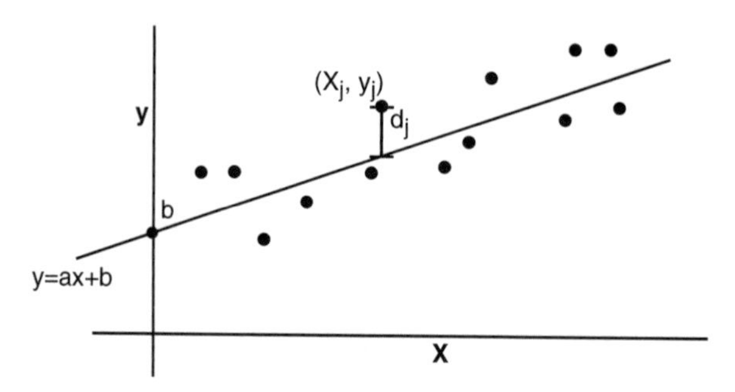

FIGURE 10.10 A familiar example of the least squares procedure for optimizing the fit of data to a mathematical model is identifying the optimal values for the slope $\boldsymbol{a}$, and the intercept $\boldsymbol{b}$, for a straight line passing through a set of measurements. The model is $\boldsymbol{y = ax + b}$, and the fit is achieved by minimizing the sum of the squares of the deviations, $\boldsymbol{d}$, from the "best line" given by $\boldsymbol{a}$ and $\boldsymbol{b}$.

to be enormously powerful in X-ray crystallography, to the point where they are sometimes dangerous when misused or overused.

The fundamental idea of the least squares procedure, as applied to X-ray structure refinement, is essentially the same as that used to fit a straight line, which can be described mathematically by the expression $\boldsymbol{y = ax + b}$ to a set of points obtained experimentally. Lagrange (I am instructed) showed that the "best" line that could be drawn through the set of points, as in Figure 10.10, was the one which minimized the sum of the squares of the distances from the points to the line. The assumption is that a straight line (the model in this case, $\boldsymbol{y = ax + b}$) in fact accurately describes the relationship, $\boldsymbol{y = f(x)}$, between $\boldsymbol{x}$ and $\boldsymbol{y}$. For data completely free of errors, the observed points $(\boldsymbol{x, y})$ would all fall exactly on some straight line, meaning the two variables $\boldsymbol{x}$ and $\boldsymbol{y}$ truly have a linear relationship.

We know, however, that in reality, for each $\boldsymbol{x_j}$ at which $\boldsymbol{y_j}$ is measured, the observed value for $\boldsymbol{y_j}$ must contain some experimental *error*, $\boldsymbol{e_j}$. According to Lagrange, the best line will be that for which $\sum_j \epsilon_j^2$ has a minimum value. For any measurement $\boldsymbol{e_j = y_j - ax_j - b}$, where $\boldsymbol{ax_j - b}$ represents the true value of $\boldsymbol{y_j}$ predicted by the model. For all measurements $\boldsymbol{y_j}$, then

$$\sum_j \epsilon_j^2 = \sum_j (y_j - ax_j - b)^2.$$

We know from differential calculus (try to remember) that all functions assume a minimum when their first derivatives are zero; hence $\sum_j \epsilon_j^2$ is minimized when the first derivatives of

$$\sum_j (y_j - ax_j - b)^2$$

in terms of $\boldsymbol{x}$ and $\boldsymbol{y}$ are zero.

We will not carry this any further, except to say that when the experimental values, the observed points, are inserted into the equation and the derivatives are set to zero, two linear equations in two unknowns are obtained. These can easily be solved to yield the slope $\boldsymbol{a}$ and the intercept $\boldsymbol{b}$ of the "best" straight line.

The same idea can equally well be applied to linear equations having more than two variables, that is, where the observed or predicted result q is described by a model of the form

$$q_j = ax_j + by_j + cz_j \ldots.$$

Again, the sum of the squares of the distances between the observed points q and the model, reflecting the experimental errors,

$$\sum_j \epsilon_j^2 = \sum_j (q_j - ax_j - by_j - cz_j \ldots)^2,$$

must be minimized. The major difference here is that partial derivatives, such as

$$\frac{\partial f(x, y, z \ldots)}{\partial x}, \quad \frac{\partial f(x, y, z \ldots)}{\partial y},$$

must be set to zero. Again, the result is a system of n equations in n unknowns that can readily be solved to provide the "best" values for a, b, c, and so on.

A serious complication arises, however, in the application of the least squares approach to structure refinement using X-ray diffraction data. (One seems to appear in almost every aspect of X-ray crystallography, doesn't it?) Here we are trying to minimize the difference between the observed and calculated structure amplitudes F_{obs} and F_{calc}; that is, $F_{obs} - F_{calc}$, or more properly, the sum of their squares taken over the entire data set.

$$\sum_{hk\ell} (F_{obs-hk\ell} - F_{calc-hk\ell})^2.$$

We know that $F_{calc-hk\ell} = \sum_j f_j \exp i2\pi(\bar{h} \cdot \bar{x}_j)$ from the structure factor expression that we derived in Chapter 5, where $\bar{h} \cdot \bar{x}_j = hx_j + ky_j + lz_j$ (we have left out the B_j factor for simplicity, but we could just as well have included it). As before, we would like to find the values for x_j, y_j, z_j, the atomic coordinates, that make F_{calc} most like F_{obs} for the entire set of data. The problem is, however, that F_{calc} is not a linear function of x_j, y_j, and z_j. The variables are related to F_{calc} by a transcendental function $\exp i2\pi(\bar{h} \cdot \bar{x}_j)$. The least squares procedure does not work in an exact sense for such functions.

The way that crystallographers, and many in other fields as well, have addressed this problem is by the Taylor expansion. The first two terms of the Taylor expansion* of any transcendental function are linear (contain no higher order terms). Thus, if we use only the first two terms of the Taylor expansion of the function $\exp i2\pi(\bar{h} \cdot \bar{x}_j)$, then we have created a linear relationship with F_{calc} that is approximately correct, the error represented by the omitted higher order terms.

We pay a steep price, however, for this bit of clever footwork. When we take the partial derivatives in terms of x_j, y_j, z_j and set them to zero to produce a system of n equations in n unknowns, and when we solve the equations for the x_j, y_j, z_j, we do not get the exact values that we might like to have. What we obtain are values of shifts in the coordinates,

* Don't worry if you don't know what this is, you can still do X-ray crystallography.

Δx_j, Δy_j, Δz_j, as well as ΔB_j. The shifts from the initial positions are toward what should be the correct values; they are not shifted all the way to the correct values because we have truncated the Taylor series in setting up the equations. Nevertheless, we can expect that application of the computed shifts will make things better.

We need not stop here, though. After carrying out the nonlinear least squares procedure above, and applying the calculated shifts, we now have a new set of presumably improved coordinates for the atoms. We can then insert these new values for x_j, y_j, z_j and repeat the entire process (which we call a cycle of least squares), and we can do it again, and again, until there is virtually no significant change in the coordinates, meaning the magnitudes of the shifts approach zero. At this point we say that the process has "converged." After each cycle of refinement we can calculate an R factor using current x_j, y_j, z_j. This provides us with a monitor of the progress of the refinement and a criterion for its success.

From this discussion of least squares we might conclude that one needs simply to plug in initial values for the positional coordinates, x_j, y_j, z_j, extracted from a model that we have built, and run the necessary number of cycles to achieve convergence, and we would then have the best approximation possible to the true structure. Let the computer do all of the heavy lifting. Ah, if only that were always true; unfortunately, it usually isn't. First of all, the totality of the shifts experienced by the variables x_j, y_j, z_j is limited (the extent of the sum of the shifts determines what we call the "radius of convergence" of the least squares procedure). If the initial values for x_j, y_j, z_j are too distant from the correct values (i.e., the model has serious errors), the least squares procedure may not have the pulling power to move them over long distances. Thus the atoms may never reach their true positions. Second, the least squares procedure is seeking to minimize the differences $|F_{obs} - F_{calc}|$, but there may be many combinations of atomic coordinates that give about the same minimum for the differences as does the true minimum. These are called false minima. It is not rare at all for the procedure to "find" one of these convenient minima and lock on to it. When that occurs, there is little probability that it can extract itself and search elsewhere for an even deeper minimum. This makes refinement a trickier business than we might expect.

In addition to the problem of false minima, the least squares procedure poses a not insignificant computational problem, one that grows by the square of the number of parameters under refinement. There have been many ingenious approaches to reducing this problem. Given the computational power available today, however, this has become less of an issue. It should be noted that the number of terms summed is proportional to the number of F_{obs} available. Thus the higher the resolution of the data, the more F_{obs} there are, and the more powerful is the method. In addition, as the resolution of the data increases, the differences between the F_{obs} and the F_{calc} become increasingly sensitive to small shifts in the positional coordinates; that is, high-resolution data leads to more precise positioning of the atoms.

REAL SPACE REFINEMENT: DIFFERENCE FOURIER SYNTHESES

We think of least squares refinement as occurring in reciprocal space. Although improvement of real space quantities x_j, y_j, z_j are the object of the endeavor, we achieve this by minimizing reciprocal space quantities, namely $|F_{obs} - F_{calc}|$ over all $hk\ell$. Real space refinement focuses on altering the same real space quantities, x_j, y_j, z_j, but does so by trying to minimize real space quantities, the differences between the electron density calculated using the F_{obs}, and the electron density calculated using the F_{calc}, that is, $\rho(x, y, z)$

experimental minus $\rho(x, y, z)$ model. We do not know the correct phases for the F_{obs}, so we cannot calculate $\rho(x, y, z)$ experimental exactly. A good approximation to the difference between the two electron densities can nonetheless be obtained from

$$\rho(x, y, z)_{expt} - \rho(x, y, z)_{model} = \sum_{hk\ell} (F_{obs} - F_{calc})_{hk\ell} \cos 2\pi(hx + ky + lz + \phi_{calc-hkl}),$$

where $\phi_{calc-hkl}$ of each difference is calculated from the current model coordinates.

The difference Fourier map calculated from this equation should be near zero in value and featureless where the model corresponds to the correct structure, as the two electron densities would essentially subtract to zero (assuming F_{obs} and F_{calc} are scaled properly to one another). Where they differ, however, is where we would expect features to appear. At those places where the model contains atoms, and therefore electron density, but the true structure does not, then $\rho(x, y, z)_{expt} - \rho(x, y, z)_{model}$ will be negative. Negative density will appear at that location in the difference Fourier map. If the model lacks atoms, hence electron density, at places occupied by atoms in the true structure, the $\rho(x, y, z)_{expt} - \rho(x, y, z,)_{model}$ will produce positive density.

Consider then what we would expect to see if, in interpreting the native electron density map and building the molecular model, we had misplaced an arginine side chain, as in Figure 10.11. What we would see in the difference Fourier map, probably displayed on a computer graphics screen, is a mass of negative density superimposed on the current arginine side chain position ($\rho(x, y, z)_{expt} - \rho(x, y, z)_{model}$ is negative) and a positive peak having roughly the shape of the side chain in the correct and likely nearby location ($\rho(x, y, z)_{expt} - \rho(x, y, z)_{model}$ is positive). The proper response of the investigator upon

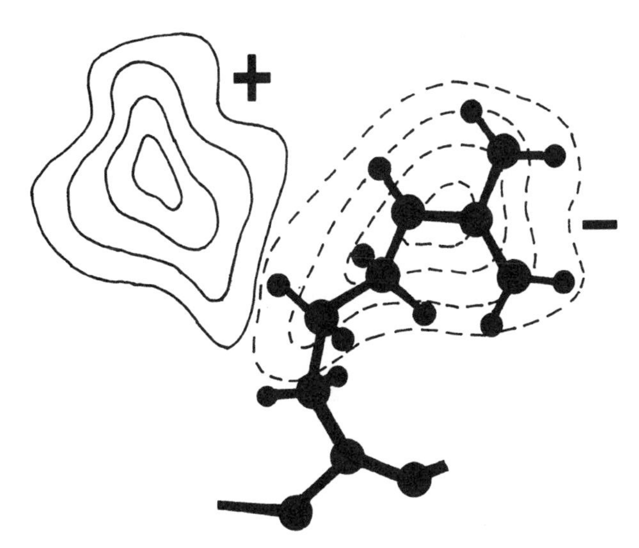

FIGURE 10.11 Here is a two-dimensional representation of what one would likely see in an $Fo - Fc$ difference Fourier map, where the Fc were calculated from a model that includes an arginine side chain, shown here, which was misplaced. A region of negative difference electron density (dashed contour lines) would superimpose upon the position incorrectly occupied by the side chain, while the positive density (solid contour lines) would indicate the location to which it should be shifted.

seeing this would be to manually reposition the side chain out of the negative density and into the positive density. This is what we commonly call a "manual rebuild."

One might ask whether application of least squares would not have led to the same result. That is, at convergence, wouldn't the atoms of the arginine side chain have moved, as a group, to the correct position? The answer is, probably not. Changes of this magnitude generally exceed the capacity, the "radius of convergence" of the procedure, to move groups of atoms long distances. By least squares standards, this is a pretty long distance. Again, *refinement* is just what it says; it is not a model-building program!

As noted already, most X-ray crystal structure refinement incorporates both real and reciprocal space refinement approaches, generally alternating one with the other. At the end of the refinement process, one should have both a low $\boldsymbol{R}$ factor and a featureless difference Fourier map.

Difference Fourier methods are important not only in detecting minor errors but in making major corrections to a model, or even finding missing parts of a structure. Frequently in the later stages of structure determination some segment of polypeptide will be ambiguous, or even invisible in the electron density maps. In these cases a common approach is to drop the atoms of this segment from the calculation of $\boldsymbol{F_{calc}}$. A difference Fourier map using these abbreviated $\boldsymbol{F_{calc}}$, commonly called an omit map, may reveal the disposition of the ambiguous peptide segment. The same principle may be applied to individual side chains or other parts of the model.

One of the most commonly employed types of difference Fourier synthesis uses as coefficients $\boldsymbol{2F_{obs} - F_{calc}}$, and phases calculated from the current model. This shows the investigator positive density superimposed on the model where the model is correct, with no density appearing where the model contains incorrectly placed residues or atoms and positive density appears where the model should be. There are also other higher order difference syntheses, but they are less frequently used.

In any difference Fourier synthesis using structure amplitudes, $\boldsymbol{F_{hk\ell}}$, be they observed or calculated from a model, and phases based on a current model, there will be some degree of model bias. This is inevitable because the phases strongly dominate the synthesis, and the contribution of the entire model is properly accounted for in the calculated phases. A model bias will exist whether the difference Fourier synthesis is an $\boldsymbol{F_{obs} - F_{calc}}$ synthesis or some variation such as $\boldsymbol{2F_{obs} - F_{calc}}$. As a consequence one must view with some caution any map or features of a map that support the existing model but suggest no alterations. A more conservative approach is to address those regions of the map where there is clearly a conflict or some disagreement is evident. In the early stages of refinement, when the model is rough and does not provide phases that are good approximations to the true phases, this problem is particularly acute, and often it is difficult to break through this phase bias and advance the model. Phase bias appears to be less of a problem in the later stages where the phases have become more reliable and $\boldsymbol{\phi_{calc}}$ more closely approximate the true $\boldsymbol{\phi}$.

ADDITIONAL CONSIDERATIONS IN REFINEMENT

As noted already, most refinement processes use both difference Fourier and least squares approaches, alternating between the two. We can think of the former as pointing up large changes that need be made, and the latter as polishing the result to a high finish. Obviously, if the initial model is very good, nothing more need be done than polishing, but the difference Fourier is always a good final check on a model.

Because least squares procedures are so powerful, so commonly used, and so versatile, and because their power continues to increase as new algorithms are developed and computing power soars, it is worth examining some of the ways that the simple least squares idea has been sharpened for X-ray structure analysis. First of all, there is a problem with least squares refinement in X-ray crystallography. Although we might have a very large set of observed structure amplitudes, $F_{hk\ell}$, we also have an awful lot of atoms. Each atom has three refineable positional parameters x_j, y_j, z_j, and usually a temperature factor B_j as well. Mathematicians would tell us that for the least squares procedure to work, we should have at least three (preferably many more) observations, F_{obs}, for each parameter we refine. Hence we should have, at minimum, four times the number of $F_{hk\ell}$ as we do atoms, and if possible, many more. In general, we do not. Therefore we must resort to ingenuity, and this must be incorporated into the mathematics of the least squares procedures.

We get around the observations to parameters ratio by noting that we have other kinds of observations, in addition to the F_{obs}, that constrain the parameters that define the model under refinement. We know, for example, that all of the atoms comprising an aromatic ring must lie in a plane, that all amino acids must have the l configuration, that peptide bonds must be essentially planar, that bond lengths and angles must approach canonical values, that the atomic radii of nonbonded atoms cannot overlap, and that there are many others of a chemical and stereochemical nature. These constraints can in fact be parameterized and incorporated as observational equations in the least squares procedure. This way the number of observations can be brought into balance with the number of parameters. Another way to increase observations is to collect more data, particularly data to higher resolution. Intuitively we would expect that higher resolution X-ray data would also increase the precision of a structure refinement, as higher resolution terms with larger values of $hk\ell$ arise from Bragg planes of smaller spacing and therefore carry information of greater structural detail.

Another approach like stereochemical constraints that can also be incorporated into least squares programs is the inclusion of energy constraints. We know that all chemical systems tend to assume a state of minimum free energy, and this is true of the conformation of a macromolecule as well. Thus a refinement procedure may simultaneously press the model to better predict the observed structure amplitudes F_{obs} by reducing the $|F_{obs} - F_{calc}|$, conform better to ideal geometry, and assume an overall state of lower free energy. A subject of some controversy is, as one might expect, how to weight the importance of each of these targets.

One interesting approach that has been introduced to overcome the shortcomings of least squares methods and correct larger errors in a model (increase radius of convergence) is called simulated annealing (Brunger, 1991). In this procedure the model is mathematically "heated up" to a high temperature. That is, the model is effectively jostled and slightly disordered from its initial state. The limited disorder is introduced into the coordinates by either what is called "metropolis Monte Carlo," where coordinates are slightly randomized, or by molecular dynamics where random trajectories are assigned to atoms at an initial time. This has the possible effect of bumping the model out of a stable but incorrect conformation, globally, or locally. The tip of a beta loop that the normal least squares procedure is unable to move may be freed to move in the direction of the correct conformation. This destabilization by heating is then followed by "annealing," consisting of cycles of standard least squares minimization. Simulated annealing has become almost universally popular, but again, caution should be exercised. Even with this approach, large errors in models are likely to remain uncorrected. It is not a substitute for careful inspection and rebuilding.

Although the objective of most refinement procedures is to improve the precision of the x, y, z coordinates of the atoms, a model can also be described not as point atoms but as a series of atomic centers or groups related by dihedral angles. An alternative approach then becomes the adjustment of dihedral angles rather than coordinates themselves. This, in most cases, reduces the number of parameters and may accelerate the process toward convergence. In addition to the positional coordinates, usually an isotropic temperature factor is also associated with each atom, although in some cases it may be best to use a single such factor for an entire side chain. That is, the temperature factors of bonded atoms are correlated. For very high resolution structure refinement, say beyond 1.5 Å resolution, the observation to parameter ratio may be sufficiently favorable that anisotropic temperature factors can be incorporated. It is important to note that not only positions and temperature factors can be refined but other parameters that may impact the agreement between F_{obs} and F_{calc}. These include for example, occupancies of atoms (as in bound ligands or disordered regions of the protein), scale factors between the F_{obs} and F_{calc}, and measures of data decay.

Macromolecular crystals contain anywhere from 30% to 90% solvent, and many of the water molecules, and often ions as well, are observed in difference Fourier maps to be ordered in the lattice. These are commonly termed "structural waters". Structural waters are refined in the same way as atoms comprising the macromolecules, as they also contribute to scattering material in the unit cell and hence, to the F_{calc}. Hydrogen atoms possess only a single electron, and often this is unfavorably shared with an electronegative atom. Because it scatters X rays so poorly, the hydrogen atom is usually ignored in refinement. The problem with this is that there are a lot of hydrogen atoms; hence, cumulatively, they may have a significant effect on the structure amplitudes. In addition contacts between groups in a macromolecule will be different when the hydrogens are considered. There is a growing tendency now not to ignore hydrogen atoms in refinement and analysis.

THE FREE *R* FACTOR

Whether for good, or otherwise, algorithms have been devised and programs written that have now become so powerful that they can drive the R factor to acceptably low values even when the model contains unacceptably large errors. For example, a protein of 30 kD, and only 65% identical in sequence to a homologous protein from which crystallographic data were collected, was used as a starting model in least squares refinement. With no change in sequence, an R value of 0.19, which is a very respectable result, was achieved, yet 35% of the side chains were incorrect.

Blind trust in the power of least squares programs has often been betrayed, and it has been responsible for many, if not most of the errors that have crept into protein structure determination. This problem was largely remedied, however, in the early 1990s by the introduction by Brunger (1992) of the R_{free}. The R_{free} is basically free of the tendency of programs to "over-refine" structures, to produce low R factors (now commonly called the "working" R factor) and incorrect structures.

The idea with the R_{free} is to omit from the least squares calculations, from the observational equations, a subset of the observed structure amplitudes. This set usually constitutes about 10% of the total data and is chosen to be representative of all resolution and magnitude ranges of the reflections. The least squares procedure is then executed as before, but this time only the reflections in the omitted subset are used in the calculation of R_{free}. Two things are true. If the refinement indeed led to an improved model that is closer to the truth, then R_{free} should decline in parallel with the "working" R, or conventional R factor based on the **90%** of the data actually used in the refinement. Second, the omitted reflections, and hence the R_{free}, must be uncontaminated by inclusion in the refinement process at any time, that is, they should provide an independent and unbiased measure of the course of refinement.

The use of the R_{free} has had a near revolutionary effect on protein structure refinement. It is now *de rigueur* to report R_{free} values for any protein structure refinement. The only point that remains in contention is what value constitutes an appropriate R_{free}, given a particular resolution, data quality, molecular weight, and so on.

SOME GENERAL OBSERVATIONS

The refinement of protein models is an effort to reconcile two frequently competing objectives: the minimization of the R factor, or some equivalent function such as a correlation coefficient, and the optimization of stereochemical properties of the model. The latter, optimized in terms of bond lengths, bond angles, and dihedral angles is essentially equivalent to the minimization of the energy of the system, the model. Geometrical requirements can be expressed, or parameterized, in terms of energy gains as a bond length or angle deviates from its ideal value. If one is intent on driving the R factor to a low number, this can often be done by neglecting the geometric quality of the model. Too stringent demands for geometric ideality, however, may prevent the R factor from finding an acceptable minimum. The model may become effectively "locked up." In the mean time difference Fourier maps provide effective reality checks. Balancing the two is part of the art of structure refinement.

At the end of refinement we often engage in a final process that is termed "validation" (Eisenberg et al., 1997). The idea here is to examine the final model quantitatively to evaluate how well it conforms to stereochemical reality. The Ramachandran plot, for example, or some improved version, is calculated to determine how well the Φ and Ψ angles (the dihedral angles of the polypeptide backbone) agree with allowed values. Side chain rotomers are examined against libraries of favored forms, average bond lengths and angles computed, and outliers flagged. There are now programs that patiently search through the entire structure, hydrogens in place, looking for any improper contacts. These procedures are well advised, as they serve as sentinels against errors and occasions of negligence.

CRITERIA FOR JUDGING A STRUCTURE DETERMINATION

In evaluating a structure determination for accuracy and precision, it is usually prudent to inquire as to the quality of the electron density map according to which the final model was built, and to the properties and quality of the final model. The former question can be addressed in real space, that is, how good is the electron density map, and/or in reciprocal space, that is, how well determined were the phases, and how well measured the structure amplitudes that contributed to the map. The model, of course, can be judged by how well it predicts the diffraction intensities (the R factor), by how it explains chemical and biochemical questions, and how well it agrees with canonical stereochemical properties, such as bond lengths and angles.

If the electron density map was originally phased using isomorphous replacement or anomalous dispersion methods, then a number of statistical criteria are usually available, even before the map itself is calculated. For example, if MIR was employed, then the cullis ***R*** factor and the centric ***R*** factor for each derivative give measures of agreement between the description of the heavy atom contribution for that derivative, and the observed differences between derivative and native structure amplitudes (Blundel and Johnson, 1976; Dickerson et al., 1968). The phasing power $f_{hk\ell}/\Delta F_{hk\ell}$ provides a measure of how much each derivative contributes to phase determination. Generally, all these values are calculated as a function of resolution, $\sin^2\theta/\lambda^2$ to illustrate how the quality of phasing changes at higher Bragg angles.

The figure of merit, ***figm***, which is commonly cited in MIR analyses, is a measure of the precision, or sharpness with which the phases have been determined. Although it does not directly correlate with the accuracy of the phases, it suggests to us a level of confidence or precision that we might reasonably assume. At the stage of an electron density map we cannot know where errors in the phases lie, but knowing the figure of merit does allow us to at least calculate the estimated error of the electron density map.

In real space, by critically examining the electron density map, we can also evaluate quality. The questions we ask should test the physical reality of the map. For example, are rotational symmetry axes free of density? Are the boundaries of individual molecules clearly marked by low-density solvent regions? Can we follow segments of the polypeptide chain? Is the direction of chain evident from visible carbonyl oxygen positions? Can we identify characteristic secondary structural features such as alpha helices and beta sheets like those in Figures 10.12 and 10.13? Can individual amino acid side chains like those in Figure 10.14 be recognized?

Ultimately the question of electron density map quality is answered by whether we can trace a single polypeptide or polynucleotide chain through the density in a manner consistent with the known amino acid or nucleotide sequence. In doing so, we consider the agreement between amino acid side chains and the density assigned to them, whether selenium atoms in a map experimentally determined by single (SAD) or multiple anomalous dispersion

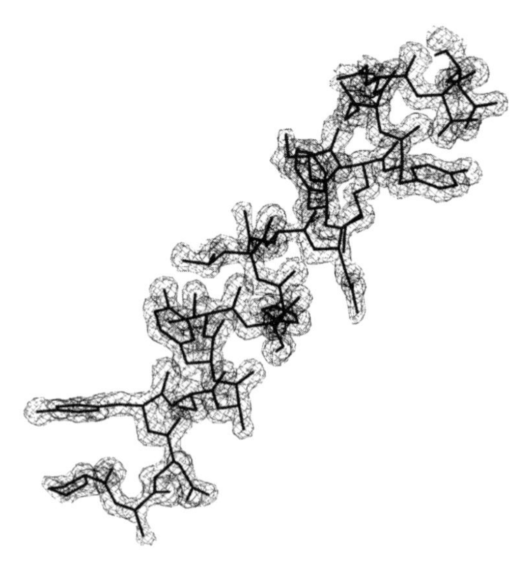

FIGURE 10.12 Computer graphics representation of the electron density, calculated at **1.4 Å** resolution, of an α helix from a bacterial xylanase.

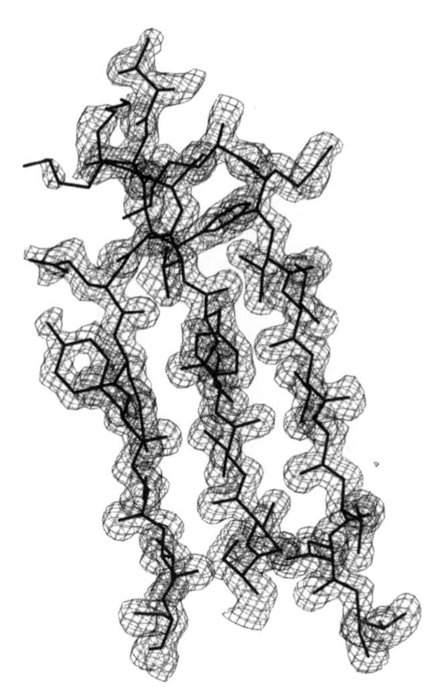

FIGURE 10.13 Computer graphics representation of the electron density corresponding to several strands of a large parallel ***β*** barrel from bacterial xylanase.

(MAD) methods fall at predicted positions. In general, we ask whether the structure makes chemical, biochemical, and physical sense. The questions under consideration must be posed in the context of map resolution. As resolution increases, we would expect the answers to emerge with increasing clarity.

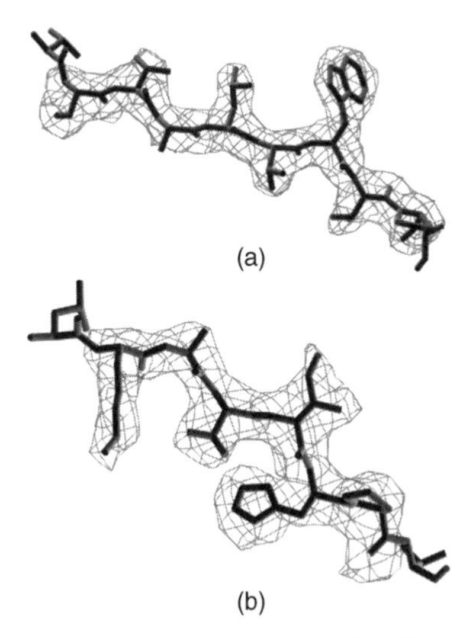

FIGURE 10.14 In (*a*) and (*b*) are two segments of the protein xylanase from bacteria superimposed on computer graphic representations of their corresponding electron density. The tryptophan in (*a*) and the histidine sidechain in (*b*) are particularly characteristic markers along the chain.

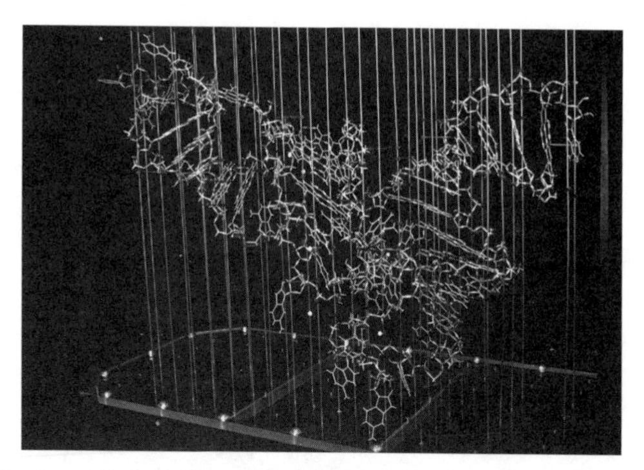

FIGURE 10.15 A model of yeast phenylalanine tRNA built according to an electron density map computed using **2.8 Å** resolution X-ray data. The model is constructed of Kendrew parts connected together by tiny screws. It required several months to build and is roughly 6 feet by 4 feet by 3 feet in size. It was constructed at MIT in 1972 by the author.

In summary, a model of the protein, nucleic acid, or virus is eventually constructed according to an electron density map; an ancient example is seen in Figure 10.15. The model is refined by one or more procedures. In general, these are based on some form of least squares minimization of the differences between the predicted, or calculated structure amplitudes, $\boldsymbol{F}_{calc}$, and those actually measured experimentally, $\boldsymbol{F}_{obs}$. Stereochemical constraints reflecting ideal geometry are also incorporated into the minimization procedures in the form of additional observational equations. The target function has traditionally been the conventional, or in some cases weighted $\boldsymbol{R}$ factor, but others such as a correlation function have also been used. Increasingly sophisticated and complex functions are now being introduced, based on what is known as "maximum likelihood," that may significantly improve the refinement procedure (see Lamson and Wilson, 1997).

BIOCHEMICAL EXPERIMENTS USING X-RAY CRYSTALLOGRAPHY

Structural models of protein and nucleic acid molecules derived by X-ray crystallography are extremely interesting in themselves, each being a representative member of some architectural class of macromolecule shaped by evolutionary time and process toward the optimal completion of a specific cellular or metabolic task. They are nevertheless static objects. Because the catalytic functions they perform depend on dynamic events involving the interaction of the macromolecules with substrates, effectors, inhibitors, and other cellular components, we are constantly searching for techniques that will allow us to visualize the macromolecules in some intermediate stages of a biochemical or physiological activity.

By comparing multiple static images representative of different points in the course of catalysis or other functional role, we stand a better chance of discerning the sequence of micro events reflected in the molecule's unique structure and chemical capability. In addition we gain some appreciation of the significance of molecular substructures, their features, and how they interact. In general, the most important images, other than that of the native macromolecule itself, are those of the macromolecule associated with its substrate, coenzyme, ligands, or with another macromolecule. By observing directly the distribution of bonds and the juxtaposition of chemical groups, the structural and steric complementarity, and the alterations from the native structure required by the event, a great deal can be learned of the purposes of the individual amino acid residues in the structure and what catalytic or binding responsibility governs their placement. From such images the molecular scheme of things, in precise chemical terms, begins to emerge and the intentions and principles underlying macromolecular design and catalytic function are revealed.

THE DIFFERENCE FOURIER METHOD

The principal crystallographic tool for directly visualizing complexes between macromolecules and ligands is the difference Fourier synthesis (Henderson and Moffat, 1971; McPherson, 1987). This has been discussed above in conjunction with refinement in real space. The attraction of difference Fourier experiments is that more than half of the necessary data is already in hand when the experiment begins. This is in the form of a native crystal X-ray diffraction data set obtained in the course of the structure solution, and an accurate set of corresponding phases. These latter components, the phases, are usually obtained by Fourier transform of the refined protein structural model, that is, from the calculated $\bar{F}_{hk\ell}$. Such phases, ϕ_{calc}, are by most measures, far more accurate than experimental phases estimated from heavy atom derivatives or other sources. The missing component is a set of diffraction intensities from an isomorphous crystal of the protein when complexed with the substrate or ligand under investigation. If such a set of intensities can be obtained, then it is a simple matter to subtract the structure amplitudes of the native crystals, essentially its diffraction pattern, from the structure amplitudes recorded from the complex crystals, and incorporate the resultant differences, $\Delta F's$, as the coefficients in a Fourier synthesis. Because the phases of these differences are reasonably well approximated by the computed phases from the native structure, the calculation is straightforward. No MIR, SIR, or MAD techniques need to be applied to find new phases, and the entire experiment is, by crystallographic standards, relatively simple. The only problem in conducting such an experiment is obtaining isomorphous crystals of the macromolecule complexed to the ligand or substrate of interest. It is at this point that the otherwise perverse and difficult character of protein crystals becomes an advantage.

Protein crystals (for methods utilized to grow protein crystals; see Chapter 2), and indeed all macromolecule crystals, are composed of approximately 50% solvent, though this amount may vary anywhere from 30% to 90% depending on the particular macromolecule. The protein or nucleic acid occupies the remaining volume so that the entire crystal is in many ways an ordered gel with vast interstitial spaces through which solvent and other small molecules may freely diffuse. Macromolecular crystals are always grown from solution, since complete hydration is essential for the maintenance of structure, and therefore they are always, even during X-ray data collection, bathed in a liquid medium that we refer to as the mother liquor. The open channels and large solvent cavities that characterize macromolecular crystals, because they allow uninhibited diffusion, prove invaluable for the formation of crystalline complexes between macromolecules and their relevant ligands.

There are basically two approaches for forming crystals of a protein–ligand complex. One technique is to simply mix the protein with the ligand in solution at a ratio that ensures virtually all protein molecules will have ligand bound. This will be a function of the association constant, or binding affinity, something usually obtainable from various kinds of solution experiments. In the presence of the ligand the crystallization conditions are re-established by conventional techniques of protein crystallization (McPherson, 1999, and Chapter 2). If good fortune prevails, the protein–ligand complex will crystallize isomorphous to the native protein alone. The crystal will have the identical unit cell dimensions and symmetry properties. If this occurs, the X-ray data can be collected in a straightforward manner from the complex crystal, differences formed, a difference Fourier image calculated, and the ligand can be visualized directly as it is bound to the protein. In addition, any parts of the protein, or individual amino acids that have altered position during the binding process will ideally appear as pairs of closely associated negative and positive density as was seen in Figure 10.11. The negative density represents the position occupied by the group in the original model, and the positive density that position to which it has moved in the complex. Thus, in carrying out a difference Fourier experiment, one sees not only how the ligand fits to the surface of the protein but how the protein alters conformation to accommodate and optimize its interactions with the ligand.

Unfortunately, difference Fourier studies predicated on successful "cocrystallization" as described above are only occasionally successful. This is because the binding of the ligand frequently perturbs either the conformation of the protein or the chemical groups on the protein's surface that are essential in forming lattice contacts required for isomorphous crystallization. Even very minor alterations can result in the macromolecule failing to crystallize at all, or as is more often the case, crystallizing in a unit cell different from that of the native protein. The new cell may be entirely different in terms of dimensions and symmetry, or only in the lengths of cell edges. Because the crystal lattices are different, however, the diffraction patterns from the two crystals are not comparable in terms of either the amplitudes or the phases.

A more common experimental procedure utilizes the porous character of macromolecular crystals and their internal channels, which are occupied by aqueous mother liquor. In these "soaking" experiments the ligand of interest is added directly to the mother liquor of the protein crystal, generally in incremental amounts over an extended period of time, until the concentration of ligand in the solvent reaches a concentration sufficient to saturate the binding sites. Exposures may vary from no more than a day's time to several months in unusual cases. Longer times may be essential when ligand binding sites are obstructed by neighboring molecules in the crystal lattice, or when the solvent content of the crystal is particularly low. For some crystalline enzymes whose structures have been solved by X-ray diffraction, obstruction has been so severe that the ligand simply cannot be induced to bind even when its concentration is far in excess of saturation and very long time periods are employed. In such instances the only recourse is to solve the structure of the macromolecule in some other crystalline unit cell that is more accommodative toward ligand binding, often obtained by "co-crystallization. Fortunately, once the structure of a protein is known in one crystalline unit cell, its structure determination in a second crystal form is appreciably simplified through the use of molecular replacement methods (see Chapter 8).

The nature of the mother liquor from which the crystals have been grown may be of particular importance in difference Fourier experiments. Many protein crystals are produced from solutions of very high salt content, such as ammonium sulfate, that may range up to three molar or more. The high concentration of charged ions may weaken the affinity of the protein for its ligand, it may induce minor conformation changes in the protein that perturb normal binding, or, when the binding of ligand to protein involves primarily electrostatic interactions, the salt ions may compete for the binding site. The net effect, in any case, is a reduction, possibly to zero, of the number of sites in the crystal occupied by ligand, and a concomitant difficulty in visualizing the ligand by X-ray diffraction techniques. Difference Fourier experiments are, therefore, most successful when they can be carried out in low ionic strength mother liquor at a permissible pH, and one that does not contain other molecules that would interfere with the protein–ligand association.

Once the crystal has been saturated with ligand, then X-ray diffraction data collection can commence. If the ligand under study is colored, then this condition can often be recognized by the degree of color exhibited by the crystal. Even when the ligand is not colored, the change in birefringence of the crystal, observable under a low-power microscope with polarized light, may serve as a useful indicator of complexation.

If possible, the investigator should test for ligand substitution by collecting some small subset of the diffraction data from the derivatized crystal and comparing it with the native diffraction pattern. From this it is usually possible to determine if the unit cell dimensions are the same as for native crystals, and if other criteria for isomorphism have been met. Calculation of an R factor between the two data sets based only on a subset of reflections can furthermore give some quantitative indication of the degree of ligand substitution.

Data are collected from the macromolecule–ligand complex crystals as for any other crystal and processed in the usual ways. The complex crystal data set, however, must be placed on a common numerical scale with the native crystal data so that relative differences between individual reflections can be formed by subtraction of the two. This is relatively straightforward since the differences introduced by the ligand are generally randomly distributed, and for the purpose of scaling, the two data sets can be assumed to have been produced from the same crystal. The residual after scaling and difference formation is

$$R = \frac{\sum_{hk\ell} ||\bar{F}_{hk\ell}|_{nat} - K|\bar{F}_{hk\ell}|_{cmplx}|}{\sum_{hk\ell} |\bar{F}_{hk\ell}|_{nat}},$$

where K is a scale factor.

The value of R serves as a measure of the diffraction differences produced by the atoms of the ligand in the crystal, but also contains the observational errors inherent in the experiment. If it is assumed that the random observational error between two data sets from identically the same crystal is about 3% to 5% of the mean value of $F_{hk\ell}$, then the residual produced by a successful ligand diffusion experiment may be of the order of 8% to 15% depending on the relative electron complement of the macromolecule and the ligand, and on the extent of substitution. If an unusually high residual of the order of 20% or more is observed, particularly when a relatively small ligand is diffused into a large protein, then one must assume that a substantial conformational change occurred in the protein as a consequence of binding. As noted above, however, careful analysis of the resultant difference electron density map can allow delineation, in some cases very precisely, of the conformation changes that occurred.

Once differences in structure amplitudes, ΔF, have been obtained by comparison of the diffraction data from ligand-saturated and native crystals, a Fourier synthesis can be computed in the normal way with phase angles calculated from the protein structure alone. These phases are not exactly the phases that should be attached to the observed differences forming the Fourier coefficients. The correct phases would be those calculated from the ligand correctly disposed in the crystal unit cell, which is, of course, what is being sought, what is not known. Using phases calculated from the native structure in conjunction with

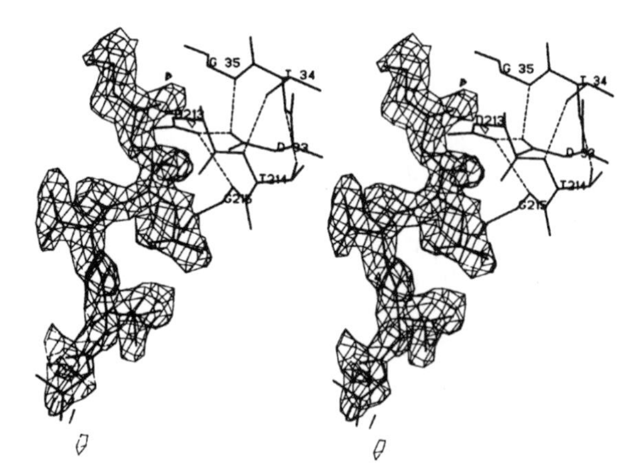

FIGURE 10.16 Stereodiagram of a portion of a difference electron density map of a polypeptide inhibitor bound at the active site of penicillopepsin. A portion of the active site derived from the native structure is also shown (from James and Sielecki, 1987).

observed differences $\mathbf{\Delta F}$ corresponds to a Fourier synthesis whose coefficients are the components of the true Fourier coefficients having the same phase as the native $\bar{F}_{hk\ell}$. Since, on average, this is $\mathbf{1/\sqrt{2}}$ that of the correct terms, calculated phases are generally adequate (Luzzati, 1953). Although reduced in magnitude by $\mathbf{1/\sqrt{2}}$, Fourier maps containing difference electron density peaks are particularly error free (Henderson and Moffat, 1971). Thus they are usually interpretable. Furthermore, since one generally knows the structure of the ligand that has been diffused into the crystals, the problem is really reduced to one of fitting a small familiar structure into an irregular, but hopefully suggestive volume of difference electron density.

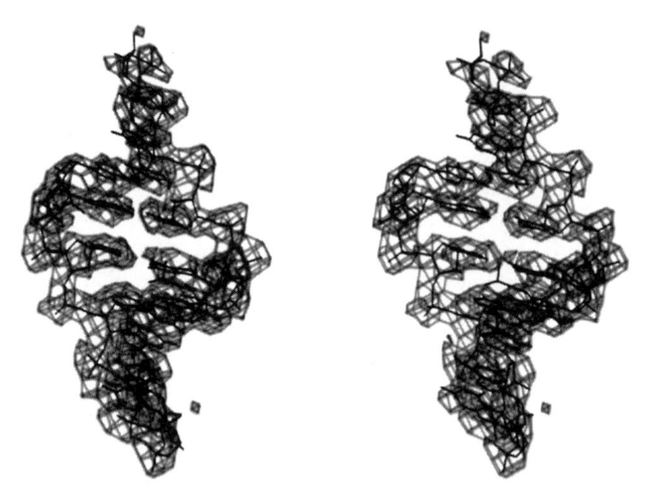

FIGURE 10.17 Stereodiagram of a segment of helical double-stranded RNA, seven base pairs in length, is shown superimposed on its corresponding difference electron density derived from crystalline satellite tobacco mosaic virus.

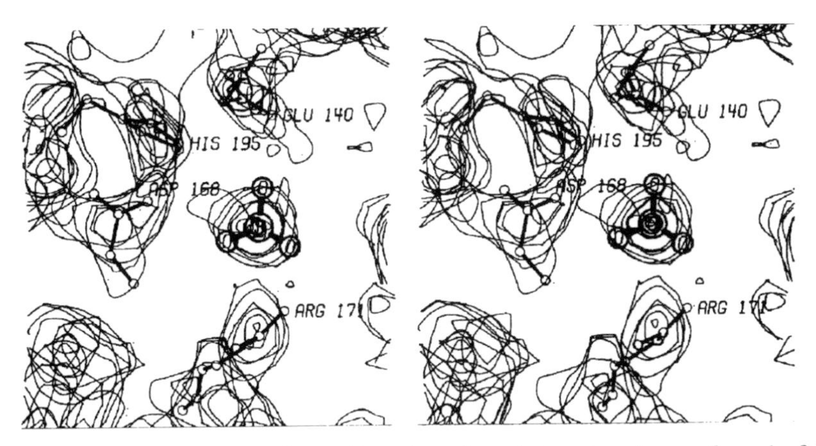

FIGURE 10.18 A stereo image of a small portion of an electron density map from dogfish M4 lactate dehydrogenase showing the position of a bound sulfate ion, circa 1969.

The range of possible inhibitors, substrates, substrate analogues, coenzymes, ions, and other effectors that can be studied in the crystalline state using the difference Fourier approach is almost limitless. Figures 10.16, 10.17, and 10.18 present a few examples. The method is most ideal for investigating ligands having a large complement of electrons concentrated in a small volume, such as metal ions, large anions, or molecules containing ring systems. Because the magnitude of difference electron density peaks will be proportional to the electron density of the ligand, such ligands can often be easily located even in relatively low resolution difference Fourier maps, and even when there is a substantial degree of nonisomorphism.

BIBLIOGRAPHY

Arndt, W. W., and Willis, B. T. M. 1966. *Single Crystal Diffractometry.* Cambridge University Press, Cambridge.

Bard, J., Ercolani, K., Svenson, K., Olland, A., and Somers, W. 2004. Automated systems for protein crystallization. In *Methods: A Companion to Methods in Enzymology; Macromolecular Crystallization* (ed. A. McPherson) 34(3): 329–347.

Bergfors, T. M. 1999. *Protein Crystallization: Techniques, Strategies and Tips.* International University Line, La Jolla, CA.

Bernal, J. D., and Crowfoot, D. 1934. X-ray photographs of crystalline pepsin. *Nature (London)* 133: 794.

Blow, D. M., and Crick, F. H. C. 1959. The treatment of errors in the isomorphous replacement method. *Acta Cryst.* 12: 794.

Blow, D. M., and Rossmann, M. G. 1961. The single isomorphous replacement method. *Acta Cryst.* 14: 1195.

Blundell, T. L., and Johnson, L. N. 1976. *Protein Crystallography.* Academic Press, New York.

Bourne, P. E., Berman, H. M., McMahon, B., Watenpaugh, K. D., Westbrook, J. D., and Fitzgerald, P. M. D. 1997. Macromolecular crystallographic information file. In *Meth. Enz. Macromolecular Crystallography* (eds. C. W. Carter and R. W. Sweet), vol. 277, pp. 571–589.

Boyes-Watson, J., Davidon, E., and Perutz, M. F. 1974. An X-ray study of horse methemoglobin. *Proc. Roy Soc. (London)* A191: 83.

Bragg, W. L., and Pertuz, M. F. 1954. The structure of hemoglobin: VI. Fourier projections on the 010 plane. *Proc. Roy. Soc. (London)* A225: 315.

Bricogne, G. 1976. Methods and programs for direct-space exploitation of geometric redundancies. *Acta Cryst.* A32: 832–847.

Brunger, A. T. 1992. The free R value: A novel statistical quantity for assessing the accuracy of crystal structures. *Nature (London)* 355: 527–474.

Brünger, A. T., and Kuriyan, J. 1987. Crystallographic R factor refinement by molecular dynamics. *Science* 235: 458–460.

Buerger. 1960. *Crystal Structure Analysis.* Wiley, New York.

Buerger, M. J. 1942. *X-ray Crystallography.* Wiley, New York.

Buerger, M. J. 1944. *The Photography of the Reciprocal Lattice.* American Society for X-ray and Electron Diffraction, Washington, DC.

Buerger, M. J. 1960. *Crystal Structure Analysis*, pp. 198–200. Wiley, New York.

Cohn, E. J., and Ferry, J. D. 1950. The interactions of proteins with ions and dipolar ions. In *Proteins* (ed. J. T. Edsall). p. 586. Reinhold, New York.

Crick, F. H. C., and Magdoff, B. S. 1956. The theory of the method of isomorphous replacement for protein crystals. *Acta Cryst.* 9: 901.

Czok, R., and Buecher, T. 1960. Crystallized enzymes from the myogen of rabbit skeletal muscle. *Adv. Protein Chem.* 15: 315–415.

Dale, G. E., Oefner, C., and D'Arcy, A. 2003. The protein as a variable in protein crystallization. *J Struct. biol.* 142: 88–97.

de la Fortelle, E., and Bricogne, G. 1997. Maximum likelihood heavy-atom parameter refinement for MIR and MAD methods. In *Methods Enzymology; Macromolecular Crystallography* (eds. C. W. Carter and R. W. Sweet), vol. 276, 472–493, Academic Press, NY.

deJong, W. F., and Bouman, J. 1938. Das Photographieren von Resiproken Netzenbenen eines Kristalles mittles Rontgenstrahlen. *Physica* 5: 220.

DeLucas, L. J., Bray, T. L., Nagy, L., McCombs, D., Chernov, N., Hamrick, D., Cosenza, L., Belgovskiy, A., Stoops, B., and Chait, A. 2003. Efficient protein crystallization. *J. Struct. Biol.* 142: 188–206.

Derewenda, Z. S. 2004. The use of recombinant methods and molecular engineering in protein crystallization. *Methods: A Companion to Methods in Enzymology; Macromolecular Crystallization* (ed. A. McPherson) 34(3) 354–363.

Dickerson, R. E. 1992. A little ancient history. *Protein Sci.* 1: 182–186.

Dickerson, R. E., Kendrew, J. C., and Strandberg, B. E. 1961. The crystal structure of myoglobin: Phase determination to a resolution of 2 A by the method of isomorphous replacement. *Acta Cryst.* 14: 1188.

Dickerson, R. E., Weinzierl, J. E., and Palmer, R. A. 1968. A least-squares refinement method for isomorphous replacement. *Acta Cryst.* B24: 997.

Drenth, J. 1999. *Principles of Protein X-ray Crystallography*. Springer-Verlag, NY.

Ducruix, A., and Giége, R. 1992. *Crystallization of Nucleic Acids and Proteins: A Practical Approach.* IRL Press, Oxford.

Eisenberg, D., Luthy, R., and Bowie, J. U. 1997. VERIFY3D: Assessment of protein models with three-dimensional profiles. In *Methods in Enzymology, Macromolecular Crystallography* (eds. C.W. Carter, and R.W. Sweet), vol. 277, pp. 396–406.

Emsley, P., and Cowtan, K. 2004. COOT model-building tools for molecular graphics. *Acta Cryst.* D60: 2126–2132.

Ewald, P. P. 1921. Das "Resiproke Gitter" in der Strukturtheorie. *Z. Krist.* 56: 129.

Fitzgerald, P. M. D., and Madsen, N. B. J. 1986. Improvement of limit of diffraction and useful X-ray lifetime of crystals of glycogen debranching enzyme. *J. Cryst. Growth* 76: 600–606.

Fury, W., and Swaminathan, S. 1997. Phases-95: A program package for processing and analyzing diffraction data from macromolecules. In *Methods in Enzymology, Macromolecular Crystallography* (eds. C. W. Carter and R. W. Sweet), vol. 277, pp. 590–619.

Garmen, E. F., and Schneider, T. R. 1997. Macromolecular cryocrystallography. *J. Appl. Cryst.* 30: 211–237.

George, A., and Wilson, W. W. 1994. Predicting protein crystallization from a dilute solution property. *Acta Cryst.* D50: 361.

Glusker, J. P., Patterson, B. K., and Rossi, M. 1987. Patterson and Pattersons: Fifty years of the Patterson function. Chapter 1 In *IUCr Crstallographic Symposia, Int. Union of Cryst.* Oxford Science, Oxford, UK.

Granick, S. 1941. Physical and chemical properties of horse spleen ferritin. *J. Biol. Chem.* 146: 451.

Hampel, A., Labanauskas, M., Connors, P. G., Kirkegard, L., RajBhandary, U. L., Sigler, P. B., and Bock, R. M. 1968. Single crystals of transfer RNA from formylmethionine and phenylalanine transfer RNA's. *Science* 162: 1384–1387.

Harker, D. 1956. Determination of the phases of the structure factors of noncentro-symmetric crystals by the method of double isomorphous replacement. *Acta Cryst.* 9: 1.

Harker, D., and Kasper, J.S. 1947. Phases of Fourier coefficients directly from crystal structure data. *J. Chem. Phys.* 15: 882–884.

Hatefi, Y., and Hanstein, W. G. 1969. Solubilization of particulate proteins and nonelectrolytes by chaotropic agents. *Proc. Nat. Acad. Sci. USA* 62: 1129–1136.

Hauptman, H., and Karle, J. 1952. Crystal structure determination by means of a statistical distribution of interatomic vectors. *Acta Cryst.* 5: 48.

Helliwell, J. R. 1992. *Macromolecular Crystallography with Synchrotron Radiation.* Cambridge University Press, Cambridge, UK.

Henderson, R., and Moffat, J. K. 1971. The difference Fourier technique in protein crystallography; errors and their treatment. *Acta Cryst.* B27: 1414.

Hendrickson, W. A., and Konnert, J. H. 1981. Stereochemically restrained crystallographic least squares refinement of macromolecule structures. In *Biomolecular Structure, Conformation, Function and Evolution* (ed. R. Srinivasan). Pergamon Press, New York.

Hendrickson, W. A., Smith, J. L., and Sheriff, S. 1985. Diffraction Methods for Biological Macromolecules. In *Methods in Enzymology* (eds. S. P. Colowick and N. O. Kaplan), vol. 115, pp. 41–54. Academic Press, New York.

Hofmeister, T. 1888. Zur Lehre von der Wirkung der Saltz. Naunyn-Schmiedebergs. *Arch. Exp. Pathol. Pharmakol.* 24: 247–260.

Hosfield, D., Palan, J., Hilgers, M., Scheibe, D., McRee, D. E., and Stevens, R. C. 2003. A fully integrated protein crystallization platform for small-molecule drug discovery. *J. Struct. Biol.* 142: 207–217.

Jacoby, W. B. 1968. A technique for the crystallization of proteins. *Anal. Biochem.* 26: 295.

Jones, A. T. 1978. A graphics model building and refinement system for macromolecules. *J. Appl. Crystallog.* 11: 268–272.

King, M. V. 1954. An efficient method for mounting wet protein crystals for X-ray studies. *Acta Cryst.* 7: 601.

Lamsin, V. S., and Wilson, K. S. 1997. Automated refinement for protein crystallography. In *Methods in Enzymology Macromolecular Crystallography* (eds. C. W. Carter and R. W. Sweet), vol. 277, pp. 269–305. Academic Press, NY.

Larson, S. B., Day, J. S., Glaser, S., Braslawsky, G., and McPherson, A. 2005. The structure of an antitumor CH-2-domain-deleted humanized antibody. *J. Mol. Biol.* 348: 1177–1190.

Laufberger, V. 1937. Crystallization of ferritin. *Bull. Soc. Chim. Biol.* 19: 1575.

Luft, J. R., Collins, R. J., Fehrman, N. A., Lauricella, A. M., Veatch, C. K., and DeTitta, G. T. 2003. A deliberate approach to screening for initial crystallization conditions of biological macromolecules. *J. Struct. Biol.* 142: 170–179.

Luzzati, V. 1953. Resolution d'une structure cristalline lorsque les positions d'une partie des atomes sont connues: traitement statistique. *Acta Cryst.* 6: 142.

Matthews, B. W. 1968. Solvent content of protein crystals. *J. Mol. Biol.* 33: 491–497.

McPherson, A. 1976. Crystallization of proteins from polyethylene glycol. *J. Biol. Chem.* 251: 3600–3603.

McPherson, A. 1976. The growth and preliminary investigation of protein and nucleic acid crystals for X-ray diffraction analysis. *Meth. Biochem. Anal.* 23: 249–345.

McPherson, A. 1982. *The Preparation and Analysis of Protein Crystals.* Wiley, New York.

McPherson, A. 1985. Crystallization of macromolecules: general principles. *Meth. Enzymol.* 114: 112–120.

McPherson, A. 1987. Interactions of biological macromolecules visualized by X-ray crystallography. In *Crystallography Reviews.* (ed. M. Moore), vol. 1, pp. 191–250. Gordon and Breach, London.

McPherson, A. 1999. *Crystallization of Biological Macromolecules.* Cold Spring Harbor Laboratory Press, Cold Spring Harbor, NY.

McPherson, A. 2000. In situ X-ray crystallography. *J. Appl. Crystallog.* 33: 397–400.

McPherson, A. 2004. Introduction to protein crystallization. *Methods: A Companion to Methods in Enzymology, Macromolecular Crystallization* (ed. A. McPherson) 34(3), 254–265.

McPherson, A., and Cudney, B. 2006. Searching for silver bullets: an alternative strategy for crystallizing macromolecules. *J. Struct. Biol.* 156: 387–406.

McPherson, A., Koszelak, S., Axelrod, H., Day, J., Williams, R., Robinson, L., McGrath, M., and Cascio, D. 1986. An experiment regarding crystallization of soluble proteins in the presence of beta-octyl glucoside. *J. Biol. Chem.* 261: 1969–1975.

McPherson, A., and Shlichta, P. 1988. Heterogeneous and epitaxial nucleation of protein crystals on mineral surfaces. *Science* 239: 385–387.

McRee, D. E. 1999. *Practical Protein Crstallography.* Academic Press, New York.

Michel, H. 1990. General and practical aspects of membrane protein crystallization. In *Crystallization of Membrane Proteins.* (ed. H. Michel), pp. 73. CRC Press, Boca Raton, FL.

Patterson, A. L. 1935. A direct method for the determination of components of interatomic distances in crystals. *Z. Krist.* 90: 517.

Pflugrath, J. W. 2004. Macromolecular cryocrystallography—Methods for cooling and mounting protein crystals at cryogenic temperatures. *Methods: A Companion to Methods in Enzymology, Macromolecular Crystallization* (ed. A. McPherson) 34(3), 415–423.

Ray, W., and Bracker, C. J. 1986. Polyethylene glycol; catalytic effect on the crystallization of phosphoglucomutase at high salt concentration. *J. Cryst. Growth* 76: 562–576.

Read, R. J. 1996. As MAD as can be. *Structure* 4: 11–14.

Read, R. J. 1997. Model phases: Probability and bias. In *Methods in Enzymology, Macromolecular Crystallography* (eds. C. W. Carter and R. W. Sweet), vol. 276, pp. 110–130.

Rossmann, M. G. 1960. The accurate determination of the position and shape of heavy atom replacement groups in proteins. *Acta Cryst.* 13: 221.

Rossmann, M. G. 1972. *The Molecular Replacement Method.* Gordon and Breach, London.

Rossmann, M. G., and Blow, D. M. 1962. The detection of subunits within the crystallographic asymmetric unit. *Acta Cryst.* 15: 24.

Rossmann, M. G., and Blow, D. M. 1963. Determination of phases by the conditions of non-crystallographic symmetry. *Acta Cryst.* 16: 39.

Salemme, F. R. 1972. A free interface diffusion technique for the crystallization of proteins for X-ray crystallography. *Arch. Biochem. Biophys.* 151: 533–539.

Sayre, D. 1952. The squaring method: a new method for phase determination. *Acta Cryst.* 5: 60–65.

Smith, J. L., Hendrickson, W. A., Terwilliger, T. C., and Berendzen, J. 2001. MAD and MIR. In *International Tables for Crystallography*, vol. F. (eds. M. G. Rossmann and E. Arnold), pp. 299–310. Kluwer, London.

Sussmann, J. L., Holbrook, S. R., Church, G. M., and Kim, S. H. 1977. A structure-factor least-squares refinement procedure for macromolecular structures using constrained and restrained parameters. *Acta Cryst.* A33: 800.

Taylor, C. A., and Lipson, H. 1964. *Optical Transforms: Their Preparation and Application to X-ray Diffraction Problems.* Cornell University Press, Ithaca, NY.

Thaller, C., Eichele, G., Weaver, L. H., Wilson, E., Karlsson, R., and Jansonius, J. N. 1985. Seed enlargement and repeated seeding. *Methods for Biological Macromolecules: Diffraction Methods in Enzymology* (eds. S. P. Colowick and N. O. Kaplan), 114: 132–135.

Thaller, C., Weaver, L. H., Eichele, G., Wilson, E., Karlsson, R., and Jansonius, J. N. 1981. Repeated seeding technique for growing large single crystals of proteins. *J. Mol. Biol.* 147: 465–469.

Wang, B. C. 1985. Resolution of Phase ambiguity in macromolecular crystallography. In *Meth. Enz. Diffraction Methods for Biological Macromolecules* (eds. S. P. Colowick and N. O. Kaplan), pp. 90–111. Academic Press, New York.

Weber, B. H., and Goodkin, P. E. 1970. A modified microdiffusion procedure for the growth of single protein crystals by concentration-gradient equilibrium dialysis. *Arch. Biochem. Biophys* 141: 489–498.

Wiener, M. C. 2004. A pedestrian guide to membrane protein crystallization. *Methods: A Companion to Methods in Enzymology, Macromolecular Crystallization* (ed. A. McPherson) 34(3), 364–372.

Wilson, W. W. 2003. Light scattering as a diagnostic for protein crystal growth—A practical approach. *J. Struct. Biol* 142: 56–65.

Wood, E. A. 1977. *Crystals and Light: An Introduction to Optical Crystallography.*, Dover, New York.

Zeppenzauer, M. 1971. Formation of large crystals. *Meth. Enzymol.* 22: 253.

Zeppezauer, M., Eklund, H., and Zeppezauer, E. S. 1968. Micro diffusion cells for the growth of single protein crystals by means of equilibrium dialysis. *Arch. Biochem. Biophys.* 126: 564–573.

INDEX

Introduction to Macromolecular Crystallography, Second Edition By Alexander McPherson